Heat and Power
Thermodynamics

Online Services

Delmar Online
To access a wide variety of Delmar products and services on the World Wide Web,
point your browser to:
     http://www.delmar.com/delmar.html
     or email: info@delmar.com

thomson.com
To access International Thomson Publishing's
home site for information on more than 34 publishers
and 20,000 products, point your browser to:
     http://www.thomson.com
     or email: findit@kiosk.thomson.com

A service of I⊕P®

Heat and Power Thermodynamics

James Kamm

Department of Engineering Technology
University of Toledo
Toledo, Ohio, USA

Delmar Publishers

I(T)P   An International Thomson Publishing Company

Albany • Bonn • Boston • Cincinnati • Detroit • London • Madrid • Melbourne
Mexico City • New York • Pacific Grove • Paris • San Francisco • Singapore • Tokyo
Toronto • Washington

NOTICE TO THE READER

Cover photo courtesy of Westinghouse Electric Corporation
Cover Design: Brucie Rosch

COPYRIGHT © 1997
By Delmar Publishers
a division of International Thomson Publishing Inc.

The ITP logo is a trademark under license.

Printed in the United States of America

For more information, contact:

Delmar Publishers
3 Columbia Circle, Box 15015
Albany, New York 12212-5015

International Thomson Publishing Europe
Berkshire House 168-173
High Holborn
London, WC1V 7AA
England

Thomas Nelson Australia
102 Dodds Street
South Melbourne, 3205
Victoria, Australia

Nelson Canada
1120 Birchmont Road
Scarborough, Ontario
Canada, M1K 5G4

Delmar Staff
Publisher: Robert Lynch
Senior Administrative Editor: John Anderson
Developmental Editor: Kathleen Tatterson
Senior Project Editor: Christopher Chien
Production Manager: Larry Main
Art and Design Coordinator: Nicole Reamer
Editorial Assistant: John Fisher

International Thomson Editores
Campos Eliseos 385, Piso 7
Col Polanco
11560 Mexico D F Mexico

International Thomson Publishing GmbH
Königswinterer Strasse 418
53227 Bonn
Germany

International Thomson Publishing Asia
221 Henderson Road
#05-10 Henderson Building
Singapore 0315

International Thomson Publishing—Japan
Hirakawacho Kyowa Building, 3F
2-2-1 Hirakawacho
Chiyoda-ku, Tokyo 102
Japan

1 2 3 4 5 6 7 8 9 10 XXX 01 00 99 98 97 96

Library of Congress Cataloging-in-Publication Data
Kamm, James.
    Heat and power thermodynamics / James Kamm.
        p.   cm.
    ISBN 0-8273-7257-4
    1. Thermodynamics—Mathematics.   2. Thermodynamics—Industrial
applications.   3. Heat engineering—Mathmatics.   I. Title.
QC311.K27   1996
536.7—dc20
                                                                                            96-15429
                                                                                                  CIP

Table of Contents

Preface

To the Student: If this is your first experience with thermodynamics, it won't be your last. You may not take another class in thermodynamics, and if you don't, this class will prepare you well for most of what you need in an applications-oriented engineering or technology career. But no matter what field of technology you have chosen—mechanical, industrial, chemical, civil, or even electronic—you will experience the principles and properties of thermodynamics over and over in your daily work.

Some of you will find jobs that deal entirely with thermodynamics. If you work in, design for, or make equipment required by electrical power generating plants, your job will be almost completely in thermodynamics. If you design mechanical systems for buildings such as systems of heating, ventilating, and air-conditioning, thermodynamics will be your tool. If you work on a team that designs power production devices such as automobile engines, jet engines, and gas turbines, the mathematics of thermodynamics will become second nature.

Even if you don't deal with thermodynamics every day, if you are in manufacturing, or electronic prototyping, or structural design, the principles of thermodynamics will crop up periodically in your deliberations and you will have to deal with them. You will see thermodynamics again.

This subject is not a simple one. It deals with quantities that you are familiar with: volume, temperature, energy. But the importance of thermodynamics is power, and power comes from molecules and how they behave. Power cannot be seen, smelled, or heard; it is different from anything you have experienced. It will be difficult at first to think in brand-new terms.

Reading this text is important, but far more important is working the problems. The calculations are what will carry you into your careers and, in the short term, into your testing for this class. Study the sample problems carefully and understand their solutions, line by line. There are many of them because they will provide the key to solving the chapter end problems and help with any design calculations you will be called on to make. This text uses few fundamental equations, just the perfect gas law and the energy equation, so there won't be many equations to "memorize." In the last six chapters, you will eliminate the lengthy calculations and reduce all thermodynamics to simple addition and subtraction with the help of tables and graphs.

A computer disk is provided that presents full solutions to odd-numbered problems at the end of each chapter (Microsoft Excel is required) and eight exercises that demonstrate the spreadsheet as a useful tool for analyzing thermodynamic problems.

Keep up with the work, and enjoy yourself.

To the Instructor: I hope that the most critical statement about this textbook is that it is too simple. Students deserve thermodynamics served simply. Save your judgment until Chapter 20 or later, then ask whether the material has been covered. If this text covers the important material and is simple, then my objective, as well as yours, has been met.

This text takes a new path in looking at the subject, a direction that will allow the student to grasp the material more easily. This approach is not new, but it is new to thermodynamics. It is called the "building block approach," where no concept is developed before its time. No concept is developed before it is a natural consequence and before it is to be used for applications. Taking this approach means meeting the students at the beginning of each chapter at their level of understanding and expanding their understanding from that point. The best illustration of this can be found in the first several chapters. They meet the students at their level of understanding as they walk in the class for the first time. Their life experience with thermodynamics has been heating water on the stove and turning up the household thermostat. *Heat* is a word that is part of their vocabulary, but their notion of heat is *temperature*. This text attacks this concept in the chapter that develops the first law of thermodynamics by naming the chapter "Heat Is Not Temperature."

Further illustration of the building block approach is that the second law of thermodynamics is introduced not with the Carnot cycle, but rather with the Otto cycle, a device with which the student is much more familiar. The Carnot cycle gets its play, but after Otto. The text examines the Otto cycle in depth, in an attempt to show the student the power and excitement of thermodynamics, in a chapter that relates the thermodynamic history of the Grand Prix racing engine, including the introduction of supercharging and turbocharging, high RPM engines, nitrous fuels, the turbine car, and much more.

Before you get the impression that this text is written more for nonscience majors than for engineers and technicians, notice that the emphasis is on problem solving. Each chapter contains an average of ten sample problems. The objective of each chapter is to expand the student's ability to design and calculate in thermodynamics. In fact, the first half of the book dwells on problem solving only with the first law. The thermodynamics that most engineers and technicians will be using in their careers is first law thermodynamics. In these chapters, the topics include a three-chapter sequence in the perfect gas law (PGL) involving a fundamental understanding of the kinetic theory of gases, an understanding of consistent units and unit conversion, and some difficult problems with the PGL. There is also a sequence of five chapters covering the fundamental thermodynamic processes that have ready solutions within the perfect gas model, from constant volume to polytropic. These chapters apply the results to realistic problems such as quenching processes, pneumatics, compressor design, compressed gases, industrial furnaces, hydronic heat, and many more.

Irreversible thermodynamics plays a role in this text and is made understandable by a technique that is not new to this text, but resurrected from older works: free expansions. By presenting the governing relations for free expansions ("totally" irreversible processes), the text establishes an organized manner of dealing with partially irreversible processes (the general case). The method works so well that by using it, the author has found errors in problem solving in many, if not most, of the textbooks on the market today and in scholarly articles written by thermodynamicists. Take note of Section 11–7. These solutions are totally new with the publishing of this book.

On the topic of open processes, the author departs from the conventional approach by deemphasizing the terms *open system WORK* versus *closed system WORK* in relation to *boundary WORK*. In the end, these concepts leave the impression that WORK is defined differently for open and closed systems. In fact, there is boundary WORK in closed as well as open systems: WORK done against whatever is pushing the piston from the opposite side (the surroundings). This text takes the more understandable approach that there are two viewpoints on WORK: *WORK done* by the gas and *WORK delivered* to the WORK-generating device. Both of these forms of WORK exist in both closed and open systems. We usually use the former in discussions of closed systems (probably erroneously, since in problem 9 at the end of Chapter 20, we leave the student with the idea that raising a 10-lb block twenty feet requires 200 ft-lb of WORK done by the gas, an answer that is off by a multiple of 100), and we use the latter in discussions of open systems (WORK done by the gas can be calculated but has no relevance for open systems).

Because the chapters on the first law and the introductory chapters on the second law are written exclusively using the perfect gas model, real gases and, therefore, the need for enthalpy and entropy are not introduced until Chapter 16. There will be some disagreement on delaying these topics so long, but this is the impact of the building block approach: the student is now ready to accept these concepts and is eager for them because with enthalpy and entropy, the student is shown how to reduce computation time to virtually nothing, i.e., with graphic/tabular techniques. There is very little said about the nature of enthalpy and entropy. There are no calculations of either quantity for a given state nor closed form calculations of the change in either for simple cases. Rather, the point is made that the beauty of these two quantities is that they have been previously determined and tabularized and need no calculation (state variables). Time is spent instead in the understanding of the tables and the Mollier diagrams of P-h and T-s and their application to systems of gas-only and liquid-vapors.

The mathematics of this text is algebra, but there are stand-alone sections at the end of four chapters that demonstrate the application of calculus to thermodynamics. The topics covered are: the differential form of the first law (including the definition of WORK), the energy equation for variable specific heat, integrating the WORK formula for constant pressure process, integrating the WORK formula for constant temperature processes, deriving the state relations for isentropic processes, deriving the state equations for a specific irreversible process, and deriving cycle relationships using closed integrals. These sections are in no way a requirement for understanding the flow of the text.

The text is intended as a complete course in thermodynamics (but not heat transfer) for engineering technology, in both two-year and four-year programs, and as an introductory text for application-oriented engineering colleges. All twenty-two chapters can be covered comfortably in a semester. For classes that are limited on time, Chapters 10, 11, 15, 21, and 22 can be eliminated without any disturbance in continuity.

I would like to thank and acknowledge the many professionals who reviewed the manuscript to help me publish this text. A special acknowledgment is due the following instructors who reviewed the chapters in detail:

Tsu-Chien Cheu
Gateway Community-Technical College

Mike DeVore
Cincinnati State Technical College

John R. McCravy, Jr.
Piedmont Technical College

Bonnie Mills
Augusta Tech

Frank Rubino
Middlesex County College

Cyprian Ukah
Gateway Community-Technical College

Arthur Neal Willoughby
Morgan State University

James Kamm
Toledo, OH

Dedication

This book is dedicated to the mechanical engineering technology students at the University of Toledo who for twenty years used this text in ever-expanding copied form and were patient with it by pointing out the calculation errors, typographical errors, and worse. Also it is dedicated to those who weren't so patient.

"To engage in experiments on Heat was always one of my most agreeable occupations . . . I was often prevented by other matters from devoting my full attention to it, but whenever I could snatch a moment I returned to it anew, and always with increased interest."

—Benjamin Thompson, the Count von Rumford

CHAPTER

1

Mathematical Refresher for Thermodynamics

PREVIEW:
This textbook explains a difficult subject using relatively simple mathematics. The text is extremely oriented toward problem solving, but most of the calculations can be accomplished with algebraic equations. Calculus is not required. The first chapter reviews the pertinent topics in algebra that will be utilized in the text. The student should use this chapter as a review.

OBJECTIVES:
❑   Review, section by section, selected topics of algebra.

1.1 INTRODUCTION

This text investigates the effect of thermal changes on special systems: the heating of a steam boiler, the ignition of fuel in an automobile engine, and others. The approach taken is to predict what will happen without actually doing it. This is the requirement of a designer or inventor who must make predictions on paper before an actual system is built. It is also the requirement of the technician who must troubleshoot or specify such systems.

The tools they use to make such an analysis are mathematical equations. Predictions will come from the solutions to these equations; therefore, algebra is such an important prerequisite that we begin by reviewing the specific topics of algebra used in the sample problems and chapter problems of this text.

1.2 QUADRATIC EQUATIONS

All equations of algebra include at least one unknown quantity, called a *variable*, for which the value is to be solved. Some equations are very easy to solve. Others, like those that contain the square of the variable (x^2) but no higher power of the variable, do not offer a straightforward solution except by applying a universal formula.

These equations are called *quadratic equations*. Examples are:

$$x^2 = 36$$

$$3x - x^2 = 7$$

$$5x^2 + x - 6 = 20$$

The general form of all quadratic equations is:

$$ax^2 + bx + c = 0$$

where a, b, and c represent any positive or negative numbers, or zero.

The solutions of these quadratic equations can be written as follows:

$$x = \frac{-b \pm \sqrt{b^2 - 4ac}}{2a}$$

which is referred to as the *quadratic formula*. The sign $\pm$ indicates that one solution is found using the addition sign only, while the second solution is found using the subtraction sign.

SAMPLE PROBLEM 1-1

Find the value(s) of x that satisfy

$$x^2 - 3.5x + 1.8 = 0$$

SOLUTION:

Using the quadratic formula with $a = 1$, $b = 3.5$, and $c = 1.8$

$$x = \frac{-(3.5) \pm \sqrt{(-3.5)^2 - 4 \times 1 \times 1.8}}{2} = 2.87, .625$$

1.3 SYSTEMS OF EQUATIONS OF TWO UNKNOWNS

A different situation exists if there are two equations, each with two unknowns. Consider:

$$x + y = 12$$

$$2x - y = 0$$

Each of the equations has many possible solutions, but the solution that satisfies both equations is unique. Throughout this text, the method that will be used for this situation involves solving one of the equations for one of the solutions in terms of the other, then substituting this relation into the other equation to eliminate that variable. This is the method of *substitution*.

SAMPLE PROBLEM 1–2

Find x and y if

$$x + 4y = 8 \quad \text{(a)}$$

$$\frac{1}{4}x + 4y = 1 \quad \text{(b)}$$

SOLUTION:

From (a) $x = 8 - 4y$

Substitute into (b)

$$\frac{1}{4}(8 - 4y) + 4y = 1$$

Simplifying

$$2 - y + 4y = 1$$

and collecting like terms

$$2 + 3y = 1$$
$$3y = -1$$

Then

$$y = -\frac{1}{3}$$

Substituting this result into one of the original equations

$$x = 8 + 4\left(-\frac{1}{3}\right) = \frac{20}{3}$$

or

$$x = 8 - 4\left(-\frac{1}{3}\right) = \frac{28}{3}$$

1.4 EQUATIONS OF DIRECT PROPORTION

Equations are used to show a relationship between two variables. For instance, suppose that some chairs are being stored in a room. Each chair has four legs, so when there are two chairs in the room, there are eight legs; when there are five chairs, there are twenty legs, and so on. There are always four times as many legs as there are chairs. In an equation, this can be stated

$$y = 4x$$

where x is the number of chairs and y is the number of legs. The above equation is an example of a direct proportion; as x increases, so does y.

SAMPLE PROBLEM 1–3

The approximate cost of building a commercial building is $120 per square foot. What is the equation that represents the total cost, y, of a building that will have x square feet?

SOLUTION:

$$y = 120x$$

1.5 EQUATIONS OF INVERSE PROPORTION

A completely different relationship between variables is found in the situation where x people will be sharing equally a total of 20 ounces of mixed nuts. If each person's share is y ounces, then the following relationship between x and y exists:

$$xy = 20$$

or

$$x = \frac{20}{y}$$

In this situation, as the number of people (x) increases, the amount of their share (y) decreases. In this example, x and y vary inversely. The general equation for inverse variation is:

$$x = \frac{b}{y}$$

or

$$xy = b$$

SAMPLE PROBLEM 1–4

If there are 20 ounces of nuts to be split among x people, plot the amount of nuts that each person gets, y, if the number of sharing people increases from 1 to 100.

SOLUTION:

Using the equation $y = 20/x$, we can substitute for various numbers of participants and create a table of corresponding shares.

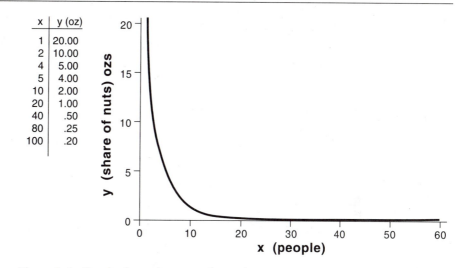

x	y (oz)
1	20.00
2	10.00
4	5.00
5	4.00
10	2.00
20	1.00
40	.50
80	.25
100	.20

Figure 1–1. Graph of people versus share of nuts.

The graph in Figure 1–1 illustrates that no matter how many people share in the nuts, there will always be a (diminishingly small) share for each. Mathematically, the curve that is represented is called a *hyperbola*, from the equation xy = constant.

1.6 EQUATIONS OF JOINT PROPORTION

Sometimes three variables x, y, z may be related to each other in a complex way. For instance, x may be directly proportional to y and inversely proportional to z:

$$x = ay$$

$$x = \frac{b}{z}$$

These two relationships may be written as one equation of joint proportionality.

$$x = \frac{cy}{z}$$

By inspection of this equation, x is directly proportional to y and inversely proportional to z. The constant c is neither a nor b nor a/b.

SAMPLE PROBLEM 1-5

A proposed office building will cost $120 per ft^2 to build and that cost is to be shared by a group of investors. If the building will eventually be x ft^2 and the number of investors is y, then the equation that describes each investor's contribution, z, is

$$z = 120\frac{x}{y}$$

Find the number of investors necessary to keep each share below $50,000 if the building is going to be 2000 ft^2.

SOLUTION:

Solving the equation for y:

$$x = 120 \times \frac{2000}{50000} = 4.8 \text{ investors}$$

Realistically, this means five investors.

1.7 ADVANCED MANIPULATION OF ALGEBRAIC EQUATIONS

Solving problems in this text often involves manipulating equations, that is, taking an equation and changing its form but not its solution. Manipulating equations is not the same as finding solutions; it is the art of working with equations to mold them into any form desired.

Start with the sample equation

$$x = \frac{cy}{z} \quad \text{(a)}$$

It has all these different forms:

(b) $z = \dfrac{cy}{z}$;   (c) $y = \dfrac{xz}{c}$;   (d) $1 = \dfrac{cy}{zx}$;

(e) $xz = cy$;   (f) $\dfrac{1}{z} = \dfrac{x}{cy}$;   (g) $\dfrac{1}{xz} = \dfrac{1}{cy}$;

(h) $\dfrac{z}{y} = \dfrac{c}{x}$;   (i) $1 - \dfrac{cy}{zx} = 0$

Those aren't all; there are many more. The original equation has been *manipulated.*

There is one rule and one rule only for rearranging equations. Any mathematical step that you perform on the left side must be duplicated on the right side. If this rule is followed faithfully, any new equation that is the result will have the same solution as the original one. Therefore, if you start with equation (a) above and divide both sides by x, the result is equation (d). Multiplying both sides by z results in equation (e). You might inspect each of the results above, equations (b) through (i), to see how they were obtained from the original equation (a).

SAMPLE PROBLEM 1-6

Show by simple manipulation that if you invert both sides of an equation, the equation has the same solutions as the original—that is, if $x = cy/z$, then it also must be true that $1/x = z/cy$.

SOLUTION:

Start with equation

$$x = \frac{cy}{z}$$

Divide both sides by x

$$1 = \frac{cy}{zx};$$

Multiply both sides by z/cy

$$\frac{z}{cy} \times 1 = \frac{z}{cy} \times \frac{cy}{zx}$$

Cancel common terms in numerator and denominator

$$\frac{z}{cy} = \frac{1}{x}$$

Both sides are now inverted from the original.

SAMPLE PROBLEM 1-7

If $x/c = y/z$
Find $c/x = ?$

SOLUTION (1):

Invert both sides of the original equation (we proved this is acceptable in Sample Problem 1–6)

$$\frac{c}{x} = \frac{z}{y}$$

SOLUTION (2):

Multiply both sides by c

$$x = \frac{cy}{z}$$

Divide both sides by y and cancel like terms in the numerator and denominator

$$\frac{x}{y} = \frac{c}{z}$$

SAMPLE PROBLEM 1-8

Solve this equation for y

$$x = 3y^2 + 4$$

SOLUTION:

$$x - 4 = 3y^2$$

$$\frac{(x-4)}{3} = y^2$$

$$\sqrt{\frac{(x-4)}{3}} = y$$

Why wouldn't the quadratic formula work for this case?

1.8 WORKING WITH EXPONENTS

In this text we will use equations with exponents. Manipulating equations with exponents will be a skill that the student will have to use often.

Using exponents is a shortcut method of writing repeated multiplication. For example:

$$a \times a = a^2$$

It is written that way purely for simplicity. Some consequences of using this notation are easily found. Since

$$a^2 \times a^3 = (a \times a) \times (a \times a \times a) = a^5 \qquad \text{(a)}$$

then it follows that in general

$$a^n \times a^m = a^{n+m} \qquad \text{(b)}$$

But equally true is the result

$$\frac{a^n}{a^m} = a^{n-m}$$

which suggests that

$$\frac{2^5}{2^3} = 2^2$$

One of the aspects of exponents that is intriguing is the definition of a zero exponent:

$$a^0 = 1 \quad (c)$$

This implies that $2^0 = 1$ and $5^0 = 1$ also!

Multiple exponents can be understood by referring to the basic definition:

$$(a^3)^2 = a^3 \times a^3 = a^{3+3} = a^6$$

In general then

$$(a^n)^m = a^{n \times m} \quad (d)$$

SAMPLE PROBLEM 1-9

Evaluate $(4^2)^3$

SOLUTION:

$$(4^2)^3 = 4^6$$

Since

$$\frac{a^n}{a^m} = a^{n-m}$$

Consider what happens if $n = 0$

$$\frac{a^0}{a^m} = a^{0-m} = a^{-m}$$

But also

$$\frac{a^0}{a^m} = \frac{1}{a^m}$$

So

$$a^{-m} = \frac{1}{a^m}$$

SAMPLE PROBLEM 1-10

What is the value of 2^{-2}?

SOLUTION:

$$2^{-2} = \frac{1}{2^2}$$
$$= \frac{1}{4}$$

Another manipulation that the student will often use involves fractional exponents—for example, $a^{1/2}$. This is a shorthand way to refer to the "root of a." $a^{1/2}$ is the square root of a and $a^{1/3}$ is the cube root of a. Fractional exponents follow the same rules as integer exponents.

$$\frac{a^{\frac{1}{n}}}{a^{\frac{1}{m}}} = a^{\left(\frac{1}{n}-\frac{1}{m}\right)}$$

The last topic on exponents deals with solving equations in which exponents are a part. As example, consider $a^3 = y$. To find a in terms of y, the usual manipulation of doing the same steps to each side of the equation still applies. To rid the left side of the exponent, take the cube root of both sides.

$$a^3 = y$$

$$\left(a^3\right)^{\frac{1}{3}} = y^{\frac{1}{3}}$$

Employing equation (d) to simplify the left side

$$\left(a^3\right)^{\frac{1}{3}} = a^{3\times\frac{1}{3}} = a$$

and

$$a = y^{\frac{1}{3}}$$

In general then

$$y = a^n \text{ implies that } a = y^{\frac{1}{n}} \quad \text{(e)}$$

SAMPLE PROBLEM 1-11

If $a^n/a^m = y$
What is a in terms of y?

SOLUTION:

$$\frac{a^n}{a^m} = a^{n-m} = y$$

Then

$$a = y^{\frac{1}{(n-m)}}$$

1.9
LOGARITHMS

One special function used in this text is the logarithm, ln. "ln" is termed the logarithm to the base "e." It is used to describe the way that one variable changes relative to another. The logarithm to the base "e" can be found by computing it on a calculator as shown in Sample Problem 1–12. The number that is to have its logarithm computed must not have any units, and the resulting value also will have no units. The number whose logarithm is to be found must not be negative, although the result may turn out negative.

Review the mathematical meaning ln in your mathematics book but, for the material in this text, be prepared to calculate the logarithm from your calculator. Be sure to use the "ln" function on the calculator, and not the "log" function.

SAMPLE
PROBLEM 1–12

Compute the ln of the following numbers:
a.   6.00
b.   1.00
c.   .25

SOLUTION:
a.   1.79
b.   0.00
c.   −1.38

Logarithms are handy for manipulating equations with variables in the exponent of a function. For example, the equation

$$6 = 2^x$$

has the unknown or variable as an exponent. It can be solved by taking the logarithm of both sides.

$$\ln 6 = \ln(2^x)$$

and then using the rule of logarithms that states that

$$\ln(a^x) = x \ln(a)$$

Therefore,

$$\ln 6 = \ln(2^x) = x \ln(2)$$

or

$$x = \frac{\ln 6}{\ln 2}$$

1.10 THE RELATIONSHIP BETWEEN EQUATIONS AND UNITS

An equation is a relationship between numbers. If $z = xy$, then when $y = 1$ and $x = 2$, z must equal 2. But in engineering, this equation means more because the variable y represents some measurable quantity: distance in feet or meters, time in seconds or days, weight in pounds or newtons.

When a variable has units, any equation that it is used in has two meanings. The first is that the numbers must balance, the second is that the units must balance. A perfect example of this is the equation that relates the distance traveled by a vehicle moving at constant speed; $d = vt$. At a velocity of 3 feet per second and a time of 2 seconds, the equation says that the distance traveled is 6 and that the units of distance must be found from a unit balance equation.

$$\text{Units: } d = \text{Units: } vt$$

To find the units of d, the units of v must be multiplied by the units of t. How can you multiply units? Just as if they were numbers:

$$\text{Units: } vt = \frac{ft}{sec} \times sec = \frac{ft - sec}{sec} = ft$$

The units of vt are feet, and this must be the unit for d. So the solution to the original problem is a distance of 6 feet.

What would happen if the velocity were 3 feet per second and we wanted to know the distance traveled after 2 hours? The equation says that $d = 6$ again, but what are the units of d? The unit equation says:

$$\text{Units: } d = vt = \frac{ft}{sec} \times hr = \frac{ft - hr}{sec}$$

The unit ft-hr/sec is a unit of distance. We do not usually use these units of distance, and no one really knows how far a ft-hr/sec is. Therefore, to say that the vehicle travels 6 ft-hr/sec is technically correct, but practically it is useless.

This example illustrates that engineering equations give numerical results that are understandable only if the units of the result are recognizable. One technique to get recognizable results is to be sure that all values used are specified in the same fundamental units. For example, if velocity is given in ft/sec, then the elapsed time must be given or expressed in seconds and the distance traveled in feet.

SAMPLE PROBLEM 1–13

What is the distance traveled by a locomotive moving at a speed of 3 miles per hour after the first 10 minutes?

SOLUTION:

Since speed is given in miles per hour, we should convert time into hours, calling it 1/6 hour. Then

$$d = 3\frac{mi}{hr} \times \frac{1}{6}hr = \frac{1}{2}mi$$

Always convert numbers by multiplying by a ratio (fraction) that includes the conversion factor in such a way that the fraction equals "one" and, by canceling units, the result has the proper units. If conversions were set up as fractions, the table would look like this:

Inches to feet: $\frac{12 \text{ in}}{1 \text{ ft}}$ or $\frac{1 \text{ ft}}{12 \text{ in}}$

Pounds to newtons: $\frac{1 \text{ lb}}{4.448 \text{ N}}$ or $\frac{4.448 \text{ N}}{1 \text{ lb}}$

Feet to meters: $\frac{1 \text{ ft}}{.308 \text{ m}}$ or $\frac{.3048 \text{ m}}{1 \text{ ft}}$

Using the fraction technique of conversion, the height and weight of a 6-foot tall, 50-newton man would be:

$$6 \text{ ft} = 6 \text{ ft} \times \frac{.3048 \text{ m}}{1 \text{ ft}} = 50 \text{ N} = 1.3 \text{ m}$$

$$50 \text{ N} \times \frac{1 \text{ lb}}{4.448 \text{ N}} = 11.1 \text{ lb}$$

CHAPTER **1**

PROBLEMS

1. Solve the following quadratic equations:
   a. $x^2 - 2x + 3 = 0$
   b. $x^2 + 5x - 5.5 = 0$
   c. $x^2 = 3.5x + 3.5$
   d. $2.5 + x^2 - 4.25 + 3x = 5.65x + 1$
   e. $3.4x - x^2 = 2$

2. Solve the following simultaneous equations for x and y:
   a. $2x + 3 = 0$
      $x + y = 6$
   b. $y + 3x = 2$
      $2x + y = 1$
   c. $2y - 4x = 1$
      $-y + x = 4$
   d. $y + 1 = x$
      $y - 1 = 2x$
   e. $xy = 3$
      $y + 3 = x$

3. Which of the following are direct proportion relationships?
   a. $x = 3y$
   b. $\dfrac{x}{y} = 3$
   c. Relationship between distance traveled and timed for a constant velocity vehicle.
   d. Relationship between the number of hours worked versus the amount of the paycheck.
   e. Relationship between the size of a piece of pie versus the number of people who will share.

4. Which of the following are relationships of inverse proportion?
   a. $y = 3x$
   b. $xy = 3$
   c. $y = \dfrac{3}{x}$
   d. The number of fish in a pond and the size of the pond.
   e. Your share of the profits and the number of shareholders.

5. Which of the following is an equation of joint proportion?
   a. $y = 3x$
   b. $y = \dfrac{3}{x}$
   c. $y = \dfrac{3z}{x}$
   d. $y = 3z + x$
   e. $xy = z$

6. Which of the following sets of equations are manipulations of each other?
   a. $x = ay$: $y = ax$
   b. $xy = z$: $\dfrac{z}{y} = x$
   c. $\dfrac{x}{y} = \dfrac{a}{z}$: $zx = ay$
   d. $\dfrac{x}{y} = \dfrac{z}{a}$: $\dfrac{1}{y} = \dfrac{zx}{a}$
   e. $\dfrac{x}{y} = \dfrac{z}{a}$: $\dfrac{y}{x} = \dfrac{a}{z}$

7. Find the missing exponents in the following simplifications:
   a. $x^3 \times x^2 = x^?$
   b. $\dfrac{x^3}{x^3} = x^?$
   c. $\dfrac{x^5}{x^2} = x^?$
   d. $(x^5)^2 = x^?$
   e. $\dfrac{(x^2)^3}{x^3} = x^?$

8. Solve the following equations for P_1, for example, simplify all equations to the form $P_1 =$
   a. $\dfrac{P_1}{P_2} = \left(\dfrac{V_1}{V_2}\right)^k$
   b. $\dfrac{P_2}{P_1} = \left(\dfrac{V_2}{V_1}\right)^k$
   c. $\left(\dfrac{P_1}{P_2}\right)^k = \dfrac{V_1}{V_2}$
   d. $\left(\dfrac{P_2}{P_1}\right)^{k-1} = \left(\dfrac{V_2}{V_1}\right)^k$

e. $\left(\dfrac{P_2}{P_1}\right)^{\frac{k}{k-1}} = \left(\dfrac{V_2}{V_1}\right)^{\frac{1}{k-1}}$

9. If x has units of feet and y has units of seconds, find the units of z in the following equations:

a. $xy = z$

b. $\dfrac{x}{y} = z$

c. $\dfrac{xz}{y} = 1$

d. $\dfrac{x}{y} + \dfrac{z}{x} = 4$

e. $zx + y = 2$

CHAPTER

2

Thermodynamic Properties

PREVIEW: The effect of heat on a substance is seen most clearly in a gas. A gas is a sea of roaming molecules with space between them. A gas exerts pressure and has mass and many other measurable properties.

OBJECTIVES:
- ☐ Describe temperature and pressure in terms of molecular motion.
- ☐ Introduce the properties of a substance that are important to thermodynamics.
- ☐ Solve problems using conservation of energy and the basic statement of the first law of thermodynamics.

2.1 INTRODUCTORY CONCEPTS

The study you are about to undertake deals with an area of basic science that has a variety of applications to practical problems in every engineering discipline, whether chemical, mechanical, thermal, industrial, civil, or electrical. In engineering and all other technical fields, thermodynamics is used to solve problems, design and maintain equipment, and perform specific tasks.

To begin, we offer a simple definition of *thermodynamics*, one that we can modify and expand as we learn more:

❖ **KEY TERM:** **Thermodynamics:** The study of heat, how it affects things, and how it can be used.

This definition is not perfectly useful, since it is not very explicit. It contains words whose meanings are yet unknown, such as *heat*. Furthermore, parts of it are very vague. What types of "things" is heat going to affect and why? And what does it mean by "using" heat?

Clarification of the definition will begin immediately through the investigation of the type of things that heat will affect. Heat (whatever it is) shows its effect on all materials: liquids, gases, and solids.

Whatever the material, whatever its shape, and regardless of how much of it there is, the total of the material will be called a *system*. This system, or quantity of matter, is contained within a boundary—a boundary that may be real or imaginary, movable or fixed. The system might be the air inside a balloon. The boundary then is real and movable; the skin of the rubber balloon. The boundary can be an imaginary one; for example, the system might be only a portion of the air within the balloon.

Regardless of the type of system being studied, matter must reside within the system. This matter is defined as the *working substance*. The function of the working substance is to receive, transport, or disperse heat, WORK, or energy. The working substance may be a gas, a liquid, or a solid or any mixture of these substances. A *closed system* is one in which mass cannot enter or leave the system (for example, a brick); an *open system* is one in which mass may enter and/or leave (for example, water in the bathroom sink). An *isolated system* is a *closed system* that cannot be influenced by anything that happens outside the system.

Everything outside the system is called the *surroundings*. In thermodynamics we shall be concerned with interactions between the system and its surroundings. These interactions can effect a change in the system. For example, in our system the brick might get hot from the sun, a part of its surroundings.

The *state* of a system is specified by its characteristic properties. The number of properties necessary to completely specify it will depend on the system; obviously the more complicated a system is, the more properties are needed to fully describe it.

Consider the air in a tank as the system. Temperature, pressure, and volume are properties of this system. The *state* of the air is completely specified by the properties temperature, pressure, and volume.

2.2 PROPERTY: MOLECULAR MASS

All working substances consist of molecules (Figure 2–1). If we were to look closely enough at any material to see molecules, they would look very much alike. They would appear like a collection of highly active molecules. In some materials the molecules would actually be moving; some would move fast and some would move slowly. In others, the molecules might be fixed in space but be vibrating or rotating. All molecules would look identical. Between the

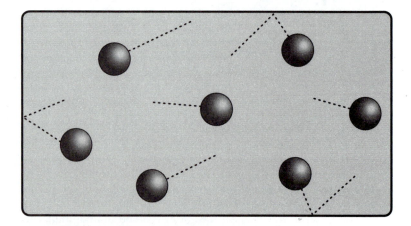

Figure 2–1. Materials are a collection of molecules.

molecules there would be empty space, as all materials include empty space—gases more than others.

Being this close to the material, it is very difficult to describe or classify it. It is very difficult to tell what type of substance it is. To do this we have to describe some properties of the substance that distinguish it from others.

One property that might distinguish a material from all others is the mass of each molecule.

❖ **KEY TERM:**   **Molecular Mass:** The term used to describe the mass of one molecule.

Measuring one molecule's mass is very difficult. It is a job for chemists with extremely precise instruments. Molecular mass is expressed in terms of *atomic mass units (AMU)*. Table 2–1 shows the molecular mass of several gases. The most important result from this table is the *relative* masses of different molecules. A molecule whose molecular mass is 32 weighs sixteen times as much as a molecule with a molecular mass of 2.

Table 2–1

Gas	Molecular Formula	Molecular Mass (AMU)
Hydrogen	H_2	2
Carbon	C	12
Water	H_2O	18
Oxygen	O_2	32
Nitrogen	N_2	28
Acetylene	C_2H_2	26

2.3 PROPERTY: MASS AND WEIGHT

Whereas molecular weight is often left to the chemists, the properties of *weight* and *mass* are used by all technicians, and sometimes confused by them as well. The difference between the two can best be described historically. Even back in the caveman era, the concept of weight was understood. Some things were heavy, others light. But in the middle 1700s, Newton demonstrated that weight depended on not just one object, but two; the object that is being weighed and the object that is providing the gravitational force, such as the earth. The weight of an object has to do with how much force it takes to lift it, and for a specific object, that force changes depending on the planet on which the object exists.

The idea that objects are lighter on planets other than earth doesn't mean that a 280-pound linebacker who weighs 65 pounds on the moon is less muscular, or less massive there. Something about the physical size of this individual is the same, whether he is on the moon or the earth. This is called *mass*. Every object can be described according to its mass, and this mass does not change no matter what gravitational force it is under.

The mass of an object is found by weighing it at sea level on the surface of the earth. This amount, in pounds, is assigned to the object as pounds of mass (designated lb$_m$). Although an object may weigh just 3 lb$_f$ (pounds of gravitational force) on a moon of Jupiter, it may have a mass of 280 lb$_m$ and be just as "big" as an earth-based linebacker.

❖ KEY TERM: **lb$_{mass}$ and lb$_{force}$:** Numerically the same on the surface of the earth.

In SI units, the mass unit is kilograms.

$$.4536 \text{ kg} = 1 \text{ lb}_m$$

or

$$2.2 \text{ kg} = 1 \text{ lb}$$

and the weight unit is newtons (N).

This discussion of weight and mass might sound like double-talk, but some actions depend on the *weight* of an object and some actions depend on the *mass* of that object. For example, does a person's ability to pick up an object depend on the object's weight or its mass? Definitely it depends on weight. Rocks are easy to move on the moon and can be thrown fantastic distances. If a large individual eats more than a small individual, however, is that a property of weight or mass? This time it is mass. Given the same amount of physical exertion, individuals on the moon would eat as much as they do on earth. How much they eat is dependent on their mass.

SAMPLE PROBLEM 2–1

Determine whether the quantities below depend on the mass or weight of an object. (Hint: Ask yourself whether they would be the same on the moon.)

a. Slumped shoulders aging (weight or mass of individual)
b. Amount of blue dye needed to tint a pond (weight or mass of the pond water)
c. Amount of fuel needed to move a vehicle 100 miles (weight or mass of vehicle)
d. Amount of water needed to make cement (weight or mass of cement)

SOLUTION:
a. weight
b. mass
c. weight
d. mass

SAMPLE PROBLEM 2-2

If you were an interstellar pilot and carried a license stating so, what would be printed on your license to help identify you, mass or weight?

SOLUTION:

Mass, of course. Your weight is changing as you fly through space; the police officer who pulled you over for speeding would have little idea what your weight meant unless the officer asked what celestial body you were weighed on.

In SI, the unit of mass is kilogram (kg) and of weight is the newton (N).

2.4 PROPERTY: VOLUME

The space occupied by a substance is called its *volume*. Volume might be considered to be a property; however, as you can see from Figure 2–2, the volume of a gas or liquid is determined more by the container that holds the fluid than by the material itself. Still, the volume occupied by a working substance plays an important role in its description.

The unit of volume in the English system is cubic feet (ft^3); in SI, it is the cubic meter (m^3). The cubic meter is used to measure large volumes, since it is about 37 times bigger than a cubic foot.

Thermodynamicists have defined a different type of volume that has dimensions of cubic feet per lb$_m$ or cubic meter per kg. This is called specific volume and will often be more valuable in our discussions than the traditional definition of volume.

❖ **KEY TERM:**

SPECIFIC VOLUME (v): The volume (V) occupied by one lb$_m$ of a substance (ft^3/lb$_m$) or the number of cubic meters comprising one kg of material (m^3/kg) in SI. $v = V/m$

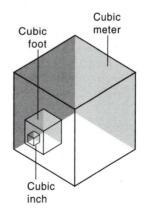

Figure 2–2. Comparison of volume units.

Notice the distinction of capital V for volume and small v for specific volume. Specific volume is usually used in reference to a gas. As you might expect, gases that are very dense have a low specific volume; one pound occupies very little space.

SAMPLE PROBLEM 2–3

Does air weigh anything? How much air would it take to weigh the same as, say, two apples? Assume that two apples weigh one pound and that the specific volume of air in your classroom is $v = 13.4 \text{ ft}^3/\text{lb}_m$. What is the volume of air that would weigh the equivalent of the two apples? Is it closer to the volume of a loaf of bread, a five-gallon aquarium, a refrigerator, or your whole classroom? Take a guess first, then solve the problem.

SOLUTION:

Most students would guess that the whole classroom contains one pound of air. But with

$$v = 13.4 \frac{\text{ft}^3}{\text{lb}} = \frac{V}{m}$$

and

$$m = 1 \text{ lb}_m$$
$$V = 13.4 \times 1 = 13.4 \text{ ft}^3$$

A loaf of bread has an approximate volume of $1/2 \times 1/2 \times 1 = 1/4 \text{ ft}^3$. A five-gallon aquarium has an approximate volume of $3/4 \times 1 \times 1 = 3/4 \text{ ft}^3$. A large refrigerator has an advertised internal volume of 15 ft^3. Your classroom has an approximate volume of $50 \times 50 \times 10 = 2500 \text{ ft}^3$.

One pound of atmospheric air is about enough to fill a refrigerator.

2.5 PROPERTY: DENSITY

Another property of a substance is its *density*. This is much easier to measure compared to the molecular weight, but not as easy to find as the actual mass.

❖ **KEY TERM:**

Density (ρ): The ratio of mass of an amount of a substance divided by its volume (lb_m/ft^3 or kg/m^3).

$$\rho = \frac{m}{V}$$

If the mass of the molecules in one cubic foot of a gas were measured and totaled, this would be the density of the gas.

As a consequence of their definitions, note the simple relationship between specific volume and density:

$$v = \frac{1}{\rho} \text{ or } \rho = \frac{1}{v}$$

In words, this says that the specific volume is the reciprocal of the density. Throughout this text, density and specific volume will be used interchangeably, according to the above equation.

SAMPLE PROBLEM 2–4

What is V and v for a substance that has a mass of 12 lb_m and occupies a volume of 3 ft^3? What is its density?

SOLUTION:

By the equation definition of specific volume, $v = V/m$ with $V = 3$ ft^3 and $m = 12$ lb_m

$$v = \frac{3 \text{ ft}^3}{12 \text{ lb}_m} = .25 \frac{\text{ft}^3}{\text{lb}_m}$$

Since the density is the reciprocal of specific volume $\rho = 1/v$

$$\rho = \frac{1}{.25 \dfrac{\text{ft}^3}{\text{lb}_m}} = 4 \frac{\text{lb}_m}{\text{ft}^3}$$

SAMPLE PROBLEM 2–5

A one m^3 container is filled with a gas that weighs .01 kg. What is the density of the gas?

SOLUTION:

$m = .01$ kg

Then by the definition of density $\rho = m/V$

$$\rho = \frac{.01 \text{ kg}}{1 \text{ m}^3} = .01 \frac{\text{kg}}{\text{m}^3}$$

SAMPLE PROBLEM 2–6

Density is a concept that is appropriate not only for gases, but for solids and liquids as well. Consider a block of cast iron with overall dimensions 5″ × 7″ × 12″. If its mass is 120 lb_m, what is the density of cast iron?

SOLUTION:

Inches should be converted to feet so that density can be found in terms of lb_m/ft^3.

Volume of cube = height × depth × length =

$$\frac{5}{12} \text{ ft} \times \frac{7}{12} \text{ ft} \times \frac{12}{12} \text{ ft} = \frac{420}{1748} = .24 \text{ ft}^3$$

By the definition of density $\rho = m/V$

$$\rho = \frac{120 \text{ lb}_m}{.24 \text{ ft}^3} = 500 \frac{\text{lb}_m}{\text{ft}^3}$$

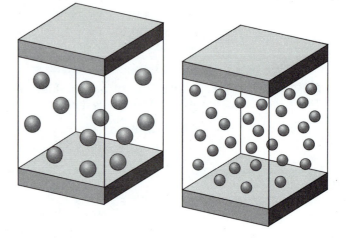

Figure 2–3. Two different gases may have the same density.

Volume, specific volume, and density are great illustrations of two types of thermodynamic properties: extensive and intensive. *Extensive properties,* such as mass and total volume, *depend* on the total mass of the substance present, while *intensive properties* are *independent* of the amount of substance considered. Intensive properties are definable at a point in a substance: if a substance is uniform and homogenous, the value of the intensive property will be the same at each point in the substance. Temperature and pressure are intensive properties. *Specific properties* are defined as properties per unit mass, and hence they are intensive.

The volume of a lake is very large and depends on what lake is being described. Volume is an extensive property. The specific volume of a lake is the same for all lakes (freshwater, at least), and therefore specific volume is an intensive property.

Some properties of substances describe exactly what type of material it is because no other substance can have the same value for the property. For instance, no two types of pure substance have exactly the same molecular weight. If the molecular weight is known, then the material is identified completely. In contrast, the density of two different substances can be precisely the same. One gas may have a molecular weight of only one-half that of another gas but have twice as many molecules stuffed into a cubic foot to give the same density (Figure 2–3). The density of a gas in no way describes what the gas is.

2.6 PROPERTY: PRESSURE

Continuing in the search for properties, notice that the molecules of a gas strike the side of the container and rebound (Figure 2–4). The molecules have a change in direction that is felt by the container in which they are enclosed. It is felt in terms of a force that is trying to push the container wall outward. The

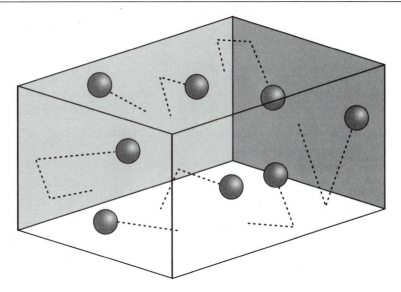

Figure 2–4. Rebounding molecules cause a force.

force pushing on the wall by one molecule is very difficult to measure, but the force from all the molecules is easy to determine using a force gauge.

The total force exerted on the container wall depends on:

- the velocity of the molecules
- the number of strikes against the wall
- the mass of the molecules
- the area of the walls

Total force exerted by a gas can be considered a property of the substance, but a more significant property might be the average force on one small portion of the wall. This property will tell what potential the working substance has to create a force on any size wall or system boundary.

❖ **KEY TERM:**   **PRESSURE (P):** The force on one square foot of a container, measured in psf (lb_f/ft^2) or in metric (N/m^2).

In practice, the pressure in a container is much easier to measure than the total force on the container walls. If we need to know the total force, then

$$Force = Pressure \times Area$$

or

$$F = P \times A$$

Pressure can also be measured in terms of pounds per square inch (psi). In this case we are interested in recording the total force on one square inch of the area of the endwall. Although inch is not one of the fundamental units we shall use for this text, in actual engineering work pressure is so often measured in pounds per square inch that we must become familiar with it here. In using it in problem solving, we almost always convert it immediately to psf.

SAMPLE PROBLEM 2-7

Wind is blowing against a wall of a house that has an area of 300 square feet. The total force of the wind is 600 pounds. What is the pressure on the wall in psf and psi?

SOLUTION:

Manipulating the equation $F = P \times A$ for pressure

$$P = \frac{F}{A} = \frac{600 \text{ lb}_f}{300 \text{ ft}^2} = 2 \text{ psf}$$

Use the conversion $1 \text{ ft}^2 = 144 \text{ in}^2$ to change to psi (see Section 1–10)

$$P = 2 \frac{\text{lb}_f}{\text{ft}^2} \times \frac{1 \text{ ft}^2}{144 \text{ in}^2} = .014 \text{ psi}$$

Pressure is measured in many ways. The most popular method is with an instrument called Bourdon tube gauge (Figure 2–5). This gauge does not measure the total force on one square inch of endwall; instead, it captures a sample of pressurized gas in a coiled cylindrical tube. The tube attempts to straighten out much like a limp balloon being blown up. The end of the tube is connected by gears to a needle that records the pressure on the dial face.

The Bourdon tube gauge is one of the most popular devices available to measure pressure. However, in special applications, instruments such as manometers and transducers may be used.

In SI, N/m^3 is such a small quantity (it is called a pascal) that measurements are usually made in kilopascals (kPa).

2.7 PROPERTY: TEMPERATURE

The final property of a substance that is of importance to this study deals with the average speed of the molecules of the substance. All the molecules of the material do not have the same velocity; therefore, when we speak of molecule velocity, we must clarify that we mean average velocity or average speed. Of course, the speed may have an effect on the pressure, as discussed previously, but more importantly, it is very significant in its own right, in a very spectacular way.

❖ KEY TERM:

TEMPERATURE (T): The indicator of the average motion of the molecules of a substance.

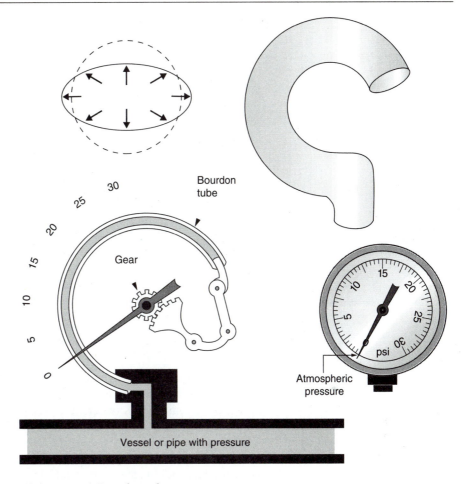

Figure 2–5. A Bourdon tube pressure gauge.

Molecules of a gas with a relatively low average velocity record a low temperature. Molecules strike a surface and transfer their energy to the surface. If the surface happens to be your hand, the energy is transferred to molecules in your hand, called temperature sensors, and your brain records the information as "hot." The mercury in a thermometer is excited, on a molecular level, by the fast strikes of surrounding molecules, and expands to record the increasing temperature.

Molecules of a liquid or a solid do not have velocities like the molecules of a gas, since these molecules are bound together. Nevertheless, solids and liquids have temperature, indicating that their motion is either vibrational or rotational. This vibration and rotation make up their temperature component (Figure 2–6).

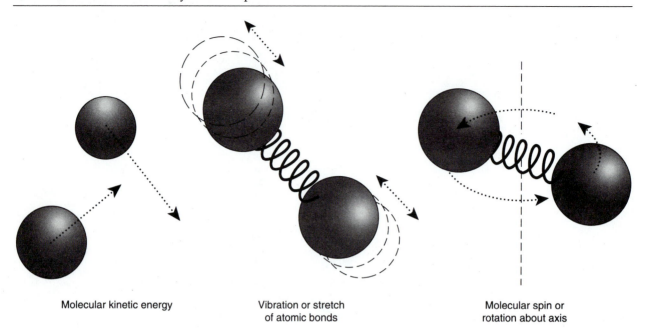

| Molecular kinetic energy | Vibration or stretch
of atomic bonds | Molecular spin or
rotation about axis |

Figure 2–6. Molecules in motion have temperature.

The basic SI unit for temperature is degrees Celsius (°C). The United States and other countries measure temperatures primarily in degrees Fahrenheit (°F). On the Celsius scale, water freezes at 0° and boils at 100°. Each degree Celsius is 1/100 of the difference between the boiling temperature and the freezing temperature of water. On the Fahrenheit scale, freezing is 32° and boiling is 212°; each degree is 1/180 of the difference between freezing and boiling, making it 5/9 smaller than a Celsius degree. Figure 2–7 shows some approximate temperature readings in degrees Celsius and Fahrenheit for some recognizable thermodynamic happenings.

The formulas for changing from degrees Celsius and degrees Fahrenheit are:

$$C = \frac{5}{9}(F - 32)$$

$$F = \frac{9}{5}C + 32$$

Equation 2–1

**SAMPLE
PROBLEM 2–8**   Change 68°F to degrees Celsius

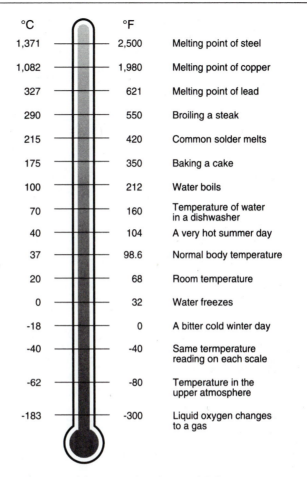

°C	°F	
1,371	2,500	Melting point of steel
1,082	1,980	Melting point of copper
327	621	Melting point of lead
290	550	Broiling a steak
215	420	Common solder melts
175	350	Baking a cake
100	212	Water boils
70	160	Temperature of water in a dishwasher
40	104	A very hot summer day
37	98.6	Normal body temperature
20	68	Room temperature
0	32	Water freezes
-18	0	A bitter cold winter day
-40	-40	Same termperature reading on each scale
-62	-80	Temperature in the upper atmosphere
-183	-300	Liquid oxygen changes to a gas

Figure 2–7. Comparison of degrees Fahrenheit and Celsius.

SOLUTION:

$$C = 5/9(F - 32°)$$
$$C = 5/9(68° - 32°) = 5/9(36°) = 20°$$

Thus, $68°F = 20°C$.

Temperature is usually recorded by a mercury thermometer. When a hot substance comes in contact with the thermometer, the mercury molecules speed up, causing the liquid mercury volume to increase slightly. This expansion causes the rise of mercury in the thermometer stem.

Another temperature indicator is the bimetallic strip, which is simply two dissimilar metal bars that are sandwiched together (Figure 2–8). When subjected to high temperatures, the molecules in each metal begin to vibrate faster

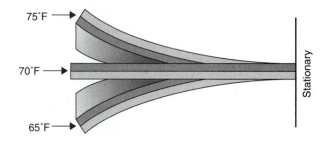

Figure 2–8. A bimetal thermometer.

and the metals expand. Each expands at a different rate, causing the bimetallic strip to bend. The amount of bending indicates the temperature.

2.8 PROPERTY: ENERGY

Energy is a term with which we are familiar. In mechanics, energy is defined as "the ability to do WORK." Mechanics deals with the motion of solid bodies, and a typical "system" of mechanics is a cannonball. A cannonball has the ability to do WORK; it can knock over a tree with its velocity. This type of energy is called kinetic energy. The cannonball also has the ability to knock over the tree if it is dropped out of a helicopter, if the cannonball has some elevation. Elevation is an indication of energy—potential energy.

The cannonball may be made of lead and acid and, in fact, be a battery, fully charged. The battery can be put into a bulldozer and provide just enough energy to make one pass at the tree and knock it down. This cannonball has electrical energy.

Or the cannonball may be very hot, at a high temperature, then put into a vat of water, a "boiler," to create steam. The steam can be fed to a steam-powered locomotive in an amount that is just sufficient to generate one pass at the tree and knock it down. This cannonball has thermal energy.

If a cannonball were shot out of a cannon, had velocity and elevation, and smashed into and knocked down part of a tree, at which time a charge of dynamite inside the cannonball exploded, knocking down more of the tree and getting the material so hot it could be dropped in a boiler to power a locomotive to knock down even more of the tree, and if when it was cold, it was connected to a battery-operated bulldozer to drain all the electrical energy from it and finish the task, it would be most interesting to find how much energy the cannonball originally had in total. Energy is a property or characteristic of a system. At any time, a system has a total capability to do WORK; and therefore, it has energy that is measurable.

A cannonball streaking through the sky with dynamite and electrical energy on board has a certain capability to do WORK, and that is often measured in foot-pounds or newton-meters. The principle of conservation of energy states that energy is a unique property, different from the others. It states that even if the cannonball exploded early, in midair, and all the chemical energy were exhausted, the energy of the system would not decrease. The cannonball, or the cannonball fragments, would speed up because of the explosion, or they would go higher in elevation, or they would increase in temperature, so that the chemical energy that was released would be converted into other forms of energy, and the capability of the system to do WORK would not change. Without any outside interference, *the energy of a system does not change.*

SAMPLE PROBLEM 2-9

In an attempt to break the existing speed record of Mach 2, Major Chuck Yeager flew the X-15 experimental rocket airplane subsonically at 90,000 feet over the Mohave Desert. The energy of the airplane system, including kinetic, potential, stored fuel (chemical), battery, and thermal energy, was precisely 3,605,301 ft-lb$_f$. At one point he turned on his auxiliary thrusters and burned 2180 lb$_m$ of solid rocket fuel, which was the equivalent of 146,206 ft-lb$_f$ of chemical energy. Assuming that the burn took place at level flight, and that the increase in temperature of the combustion chamber and rocket nozzle absorbed 3,459 ft-lb$_f$ of the chemical energy, how much kinetic energy was added to the plane? How much total energy was added to the system?

SOLUTION:

Since energy is conserved, there was no increase in the energy of the system.

The conversion of chemical energy was to thermal energy and kinetic energy, since potential energy did not change (level flight), and since we can assume that no batteries were charged or other energies were involved.

Therefore, the 3,605,301 ft-lb$_f$ of chemical energy was converted to 3,459 ft-lb$_f$ of thermal energy and 3,601,842 ft-lb$_f$ of kinetic energy.

Energy is conserved only if energy does not cross the system boundary, but in many instances energy does exactly that. For example, if we are beaming microwaves to the cannonball to charge on-board batteries, then the total energy of the system is increasing. Similarly, if the cannonball is hot and the atmosphere around it is cool, some of the thermal energy of the system is being transferred to the atmosphere and the total energy of the system is decreasing. Energy can flow in, and energy can flow out of the system.

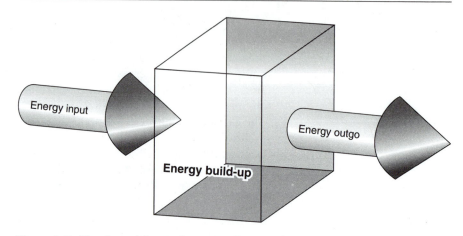

Figure 2–9. First law of thermodynamics illustrated.

The total energy of a system can change, then, by energy flow. A rule that governs the energy flow to a system is that the energy buildup of a system is equal to the net energy added to the system less the net energy given off by the system. In mathematical terms, it is stated:

$$Energy\ input - Energy\ outgo = Energy\ buildup$$

Equation 2–2

The name given to this principle is the *first law of thermodynamics*. This appears to be a simple, straightforward relationship, but in Chapter 5 we will see how important it can be.

Figure 2–9 indicates the simplicity of the first law. In effect, it simply states that whatever energy is put into a system and does not come out must remain in the system as in increase in total energy.

Although the unit of energy is often considered to be foot-pound or newton-meter, there are other equally important units that can be used. Electrical energy is often given in watt-hours, and the conversion from watt-hr to ft-lb$_f$ is 2656 ft-lb$_f$ = 1 watt-hr, and to newton-meters is 3601 N-m = 1 watt-hr. Thermal energy is often given in British Thermal Units (see Chapter 5), and 1 Btu = 778 ft-lb$_f$. In SI the thermal unit of energy is the joule = 1 N-m.

One way to transfer energy to the cannonball is to throw it into a fire. The total energy of the system increases due to the action of the fire according to the first law. The energy of the fire is created from a fuel, and different fuels generate different amounts of thermal energy. Table 2–2 lists the energy values of a variety of fuels.

TABLE 2–2

ENERGY VALUES (HEATING VALUES) OF POPULAR FUELS

Fuel	Heating Value	Heating Value (joules/kg)
Natural gas	1005 Btu/ft3	5.58×10^7
Fuel oil	140,000 Btu/gal	4.42×10^7
Wood	5000 Btu/lb$_m$	1.16×10^7
Coal	14,000 Btu/lb	3.24×10^7

SAMPLE PROBLEM 2–10

In a hot water boiler, 1000 cubic feet of natural gas is burned and 200,000 Btu's of the heat generated is lost up the chimney. What was the energy buildup of the water in the boiler?

SOLUTION:

Energy input = 1000 cu ft × 1005 Btu/ft^3 = 1,005,000 Btu

   (see Table 2–2)

Energy outgo (lost) = 200,000 Btu

Using the first law of thermodynamics:

Energy buildup = Energy input – Energy outgo

Energy buildup (added to water) = 1,005,000 – 200,000 = 805,000 Btu

To summarize the important aspects of energy conservation:

- An object has many types of energy associated with it: kinetic (KE), potential (PE), electrical (EE), chemical (CE), thermal (TE), and so on.
- Unless there is outside interaction (with the surroundings), an object's individual energies may change, but the total will not.
- If there are outside interactions, they can be energy additions or energy losses, but the net increase of energy is reflected in the net changes in the sum of the individual energies:

$$E_{in} - E_{out} = KE + PE + EE + CE + TE$$

- To determine the changes of energies in an object or system, the properties of the system must be measured before and after the energy changes occur.

2.9 EQUILIBRIUM: THE ZERO^TH LAW OF THERMODYNAMICS

Temperature plays an important role in energy flows. If two bodies at different temperatures are brought into contact, thermal energy will flow from the body of higher temperature (warmer body) to the body of lower temperature (cooler body). Further, if two or more bodies remain in contact over a sufficiently long period of time, they will reach the same temperature. This is so fundamental to thermodynamics that it is called the zero^th law.

For example, if a container of hot water and one of ice water are touched together, we say one feels hot and the other cold relative to the temperature of our body. The reason is that the hot water releases energy while the ice water extracts energy. The temperature of a body is a thermodynamic property but is also a measure of the ability of the body to transfer thermal energy to another body. When two bodies in contact have the same temperatures, no energy transfer will take place. This statement is called the "zero^th law of thermodynamics" and implies that the bodies are in *equilibrium*.

Equilibrium exists in a system when its properties do not change with time, that is, when they are constant. Thermodynamics deals chiefly with systems in equilibrium. In order to specify the temperature, pressure, or any other property of the system, the property cannot be changing. A fixed value cannot be assigned to a changing property—the property would be a function of time and the system would not be in equilibrium. When values are assigned to the properties of a system, these values are for the "equilibrium state" of the system.

CHAPTER **2**

PROBLEMS

1. Two similar vessels are filled with different gases, one with water vapor and the other with acetylene. If both vessels contain the same number of molecules, how many times more does the acetylene weigh than the water vapor?

2. If two similar vessels are to contain the same weight of two different gases, water vapor and nitrogen, how many times more water vapor molecules are needed than nitrogen?

3. Natural gas is a fuel consisting primarily of a mixture of two gases, methane and ethane. If a 5-cu-ft vessel contains 140 lb_m of methane and 60 lb_m of ethane, what is the density of the natural gas? What is the specific volume of the gas?

4. For each case below, fill in the missing quantities:
   a. $m = 3$ lb_m $V = 6$ ft^3 $v =$ ____ $\rho =$ ____ lb_m/ft^3
   b. $m = 4$ lb_m $V =$ ____ $v = .33$ ft^3/lb_m $\rho =$ ____ lb_m/ft^3
   c. $m =$ ____ $V = 2$ ft^3 $v = .10$ ft^3/lb_m $\rho =$ ____ lb_m/ft^3

5. For each case below, fill in the missing quantity:
   a. *Force* = 400 lb_f *Area* = 14 ft^2 *Pressure* = ____
   b. *Force* = ____   *Area* = 12 ft^2 *Pressure* = 100 lb_f/ft^2
   c. *Force* = 10 lb_f *Area* = ____   *Pressure* = 60 lb_f/ft^2

6. If you weigh yourself at home at 145 lb, is this weight or mass? What is your weight, including units? What is your mass, including units?

7. Convert the following temperatures to °C using Equation 2–1. Check Figure 2–7 to see if your answer is correct.
   a. 350°F
   b. 32°F
   c. –40°F
   d. 200°F

8. Convert the following temperatures to °F using Equation 2–1. Check Figure 2–7 to see if your answer is correct.
   a. 100°C
   b. 40°C
   c. 1400°C
   d. –40°C

9. Just like a cannonball in flight, water in a boiler has a certain amount of energy associated with it as a property. Before the water is heated, which of the following types of energy does it have?
   a. kinetic
   b. potential
   c. chemical
   d. thermal
   e. electrical

10. In a hot water boiler, 6 gal of oil is burned and 200,000 Btu's of the heat generated is lost up the chimney. What is the energy buildup in the water of the boiler?

11. If a missile in level flight is powered by natural gas and burns 225 ft³ of the fuel during its flight, by how much does the total energy of the missile change? By how much does the kinetic energy of the missile change?

12. If 2,000,000 watt-hr of energy is beamed up to a missile during flight, does its total energy change in ft-lb$_f$? If the microwaves are used to increase the speed of the missile, what is the increase of kinetic energy (in ft-lb$_f$)?

13. Consider a coal-fired power plant as a "system" with the following energy flows:

    1 ton coal input per hour (2000 lb$_m$ per ton)

    50,000,000 watts output per hour

    plus cooling water to maintain no energy buildup in the plant itself. In one hour, how many Btu's of thermal energy must be taken away in the cooling water?

CHAPTER
3

A Molecular Point of View

PREVIEW: Two gray steel bars are sitting on a table. Both are identical to the eye; they came from the same batch of metal. Touch one and your finger almost freezes tight to it. Touch the other and your finger recoils with the pain of a burn. One bar is very cold; the other is very hot.

If we took a microscope, could we see which bar was hot? Some scientists have claimed the heat would show up as blue fluid between the steel molecules. In 1904, James Jeans, a teacher at both Cambridge University in England and Princeton University in the United States proposed another idea: the molecular theory of heat. The theory is fascinating.

OBJECTIVES:
- ❑ Define the "state" of a gas.
- ❑ Describe the changes in gas properties in terms of the actions of individual molecules.

3.1 EXPERIMENTING WITH GAS PROPERTIES

The properties of materials that were defined in Chapter 2—molecular weight, density, volume, pressure, and temperature—are the building blocks of thermodynamics. As was noted, these properties are interrelated. For instance, the speed of gas molecules is measured as temperature, but is also an important contributor to pressure.

There are some precise relationships between gas properties that can be discovered by experimenting with a hypothetical molecular gas. We will investigate these relationships with four such experiments. As we get into the experiments, each property of the gas will be discussed in detail. The following symbols will be attached to the properties:

μ Molecular weight
m   Mass
ρ Density
V   Total volume
v Specific volume
P   Pressure
T   Temperature

The first experiment deals with a rectangular box that has a volume of one cubic foot and contains a gas consisting of four molecules. This gas has a certain density (for simplicity it can be said to be four molecules per cubic foot), and exerts a pressure on the container that can be measured with a Bourdon tube pressure gauge.

After the density and pressure of the gas has been recorded, a fifth molecule is injected into the chamber at the same average velocity of the other four. This means the temperature of the gas has not increased; however, the density is greater—five molecules per cubic foot. What will happen to the pressure? (Figure 3–1) Instead of having four molecules striking the wall, there will be more. More strikes against the wall means that the pressure will increase. The conclusion of this experiment is that if the *density* of a gas increases while the temperature of the gas remains the same, the *pressure* will similarly increase.

How can we write an equation to describe this experiment? As described in the first chapter, when two variables increase in the same proportion they are directly proportional and the equation relating them is:

$$P = a\rho$$

where a is the constant of proportionality.

This equation says that if the density increases, the pressure will also increase. Conversely, it states that if the pressure increases, the density must also increase. Now stop, go back, and read the last sentence again. Do you believe it?

Is the only way to increase pressure in the vessel by increasing density (by adding more gas)? You may be able to think of other ways. It is very important to remember that there are other ways to increase pressure but none if the

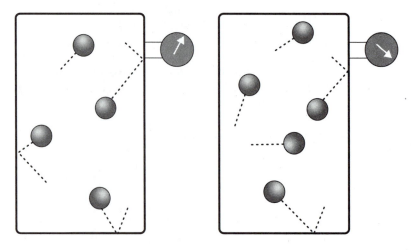

Figure 3–1. Experiment 1: The effect of density on pressure.

temperature remains the same. This experiment takes place with a constant temperature.

SAMPLE PROBLEM 3–1

Perform the same experiment but increase the size (mass) of the molecule instead of the number of molecules.

SOLUTION:

If the weight and mass of each molecule increases but the number of molecules remains the same, the density of the gas also must increase. Four heavy molecules now weigh the same as five of the original molecules. Similarly, since each molecule is heavier, its recoil off the wall will have a greater effect on the wall than previously. This effect is measured as a greater force or pressure. In this experiment, the density increased and the corresponding pressure increased, exactly the same results as the previous experiment. Therefore the same equation governs this experiment as the first:

$$P = a\rho$$

The fact that the pressure rises with an increase in density was demonstrated in the lab by Jacques Charles in 1810 and is called Charles' Law.

3.3 EXPERIMENT 2

The second experiment begins as the first one did with four molecules in a one-cubic-foot box. For this experiment, we add a thermometer next to the pressure gauge to record the temperature of the gas. Begin by recording the temperature and the pressure from the gauges and measuring the density. Then the average temperature of the molecules is increased by giving each one a push. (Figure 3–2) The thermometer records this increase by showing a greater temperature. What happens to the density? It remains the same, since we still have only four molecules in the box. What about the pressure? The pressure depends on the size of the molecules striking the side, the number of them, and the speed at which they strike. Since the speed is increased, the pressure is similarly increased.

Again for this experiment we have a direct proportion, this time between pressure and temperature. The equation for the experiment is:

$$P = bT$$

The equation says that if the temperature of the gas goes up, the pressure of the gas also will increase. Similarly, it states that an increase in the pressure of a gas will cause an increase in the temperature of the gas. These results are only true if the **density of the gas remains the same.**

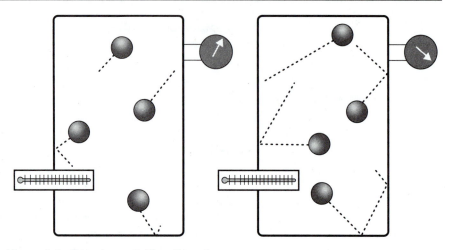

Figure 3–2. Experiment 2: The effect of temperature on pressure.

The fact that pressure rises when temperature increases was demonstrated in the lab by Robert Boyle in 1662 and is called Boyle's Law.

3.4 EXPERIMENT 3

The third experiment begins with the four molecules in the box that includes a pressure gauge and thermometer. As in the last experiment, we increase the average temperature of the molecules by giving each of them a push. This caused an increase in pressure before, but now we ask, "What can we do to make the pressure stay the same?" Remember that the pressure depends on the mass of the molecules, the number of strikes of the molecules against the wall, and the speed of the striking molecules. The speed of the average molecule has been increased in this experiment; therefore, to keep the pressure the same, we must either reduce the molecular weight of the gas or reduce the number of molecule strikes. Reducing the molecular weight of the gas is very difficult, since it would mean cutting away a little of each molecule. Reducing the number of strikes can be achieved by letting one of the molecules out of the box. Whichever of the two methods is used, the result is that the **density** of the gas is reduced. (Figure 3–3) Therefore, if the temperature of the gas is increased, the density of the gas in the box must be reduced in order to keep the gas pressure the same. Said in a different way, when the **temperature of a gas is increased, the density of the gas must be decreased if the pressure is to remain a constant.**

The equation to describe this experiment is the inverse proportion equation of Chapter 1. The equation governing this experiment, then, takes the form:

$$T = \frac{c}{\rho} \quad \text{or} \quad \rho = \frac{c}{T}$$

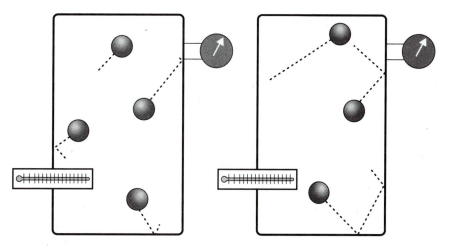

Figure 3–3. Experiment 3: The effect of temperature on density.

It was suggested that an easy way to change density was to let some molecules out of the container. Another way is simply to change the size of the container. Density is changed by keeping the same amount of gas encaptured and by increasing or decreasing the space it occupies. Figure 3–4 shows a vessel that can change size. It is a solid cylinder that has an expandable diaphragm that expands to maintain a constant pressure. Sample Problem 3–2 shows how to put the equation above to good use.

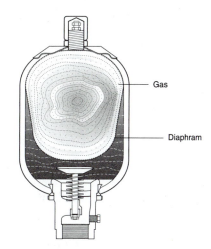

Gas

Diaphram

Figure 3–4. A variable volume vessel.

SAMPLE PROBLEM 3-2

If the temperature of the gas in the chamber shown in Figure 3–4 is increased by 10%, what change will occur in the volume of the cylinder to keep the pressure inside constant?

SOLUTION:

If the vessel expands to maintain a constant pressure, the temperature in the vessel will rise according to the equation

$$T = \frac{c}{\rho}$$

or

$$T\rho = c$$

This is true before the gas begins its change

$$T_1\rho_1 = c$$

and afterward

$$T_2\rho_2 = c$$

Since the left-hand sides of both equations are equal to c, they must be equal to each other.

$$T_1\rho_1 = T_2\rho_2$$

or

$$\frac{T_2}{T_1} = \frac{\rho_1}{\rho_2} = \frac{\dfrac{m}{V_1}}{\dfrac{m}{V_2}}$$

But since the mass of the gas does not change

$$\frac{T_2}{T_1} = \frac{V_2}{V_1}$$

but

$$\frac{T_2}{T_1} = 1.10$$

so

$$\frac{V_2}{V_1} = 1.10$$

which indicates that the volume increases by 10% also.

**3.5
THE PERFECT GAS
LAW**

The three experiments that have been described and the equations that have resulted from each have historical significance steeped in the tradition of thermodynamics: just envision a British college professor in his office or laboratory, working in the dead of night to prove his hypothesis about the relationship between pressure, temperature, and density, and eventually having a thermodynamic principle named after him.

Although each of the experiments was performed independently of the others, all three have much in common. For one thing, they all deal with the three most important properties of a gas: T, ρ, and P. Thermodynamically speaking, these three properties describe completely the situation the gas is in. They describe the "life signs" of the gas. If they are known, along with the name of the gas involved, then everything of importance about the gas is known. These three important properties collectively identify the "state" of the gas.

❖ **KEY TERM:**   **State of a Gas:** The properties P, v (or ρ), and T of a gas.

SAMPLE PROBLEM 3–3

Freon 134a (R-134a) is a gas that keeps a refrigerator cool. This gas is called refrigerant. Chemically, this refrigerant is called a hydrofluorocarbon. It is colorless, odorless, and usually stored at 70°F, is lighter than air with a density of 1.82 lb$_m$/ft³, and comes packaged in 145-lb containers under a pressure of 80 psia. What is the state of this R-134a?

SOLUTION:

$$P = 80 \text{ psia}, \quad \rho = 1.82 \frac{\text{lb}_m}{\text{ft}^3}, \quad T = 70°\text{F}$$

The three experiments helped define relationships between the state variables. Table 3–1 collects the equations involved in each and the conditions on which the experiment depended.

Table 3–1

Equation	Property That Is Maintained a Constant	Description of the Process
$P = a\rho$ (Equation 3–1)	T	Constant temperature
$P = bT$ (Equation 3–2)	ρ (or v)	Constant density or specific volume
$T = c/\rho$ (Equation 3–3)	P	Constant pressure

These equations would be much more convenient to use if they could be combined into one equation that was good for any of the three processes. That is easy to do. In Table 3–1, Equation 3–1 says that pressure is directly proportional to temperature, while Equation 3–2 says that pressure is directly proportional to density. Taken as a joint proportion this implies:

$$P = R\rho T$$

Equation 3–4

where R is the constant or proportionality. This new equation describes both Experiments 1 and 2. Conveniently enough, it also replaces Equation 3–3. This becomes clear if you solve the equation for T:

$$T = \frac{P}{R\rho} = \frac{\frac{P}{R}}{\rho}$$

and notice that it has the same form as Equation 3–3 in Table 3–1, except that c is replaced by P/R. But P/R is a constant for the conditions of Experiment 3, since P does not change. The equation $P = \rho RT$ can be called the "super" equation of Table 3–1, but properly it is called the *perfect gas law*.

❖ **KEY TERM**   **Perfect Gas Law:** The simple relationship between the state variables, $P = \rho RT$.

3.6
EXPERIMENT 4:
ADIABATIC

With the perfect gas law now available to find pressures, temperatures, and density, investigate one final experiment. Figure 3–5 shows a gas confined to a volume. But that volume can be enlarged, since one wall of the container is moveable. If something causes the wall to move, either the pressure of the gas

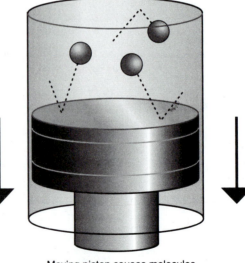

Moving piston causes molecules
to accelerate when rebounding

Figure 3–5. Experiment 4: Adiabatic. Moving boundaries tend to cause temperature changes.

or a force pulling on the piston rod, the volume obviously increases while the pressure decreases.

What happens to temperature? What happens to the average speed of molecules? It may appear that unless some heating or cooling occurs, the speed will not change. But looking at Figure 3–5 indicates that as molecules strike the receding wall, they rebound off the wall more slowly than when they went on. The average speed and temperature, therefore, drops. Similarly, if the piston is moving into the gas like a ping-pong paddle against a ball, the molecule velocities increase. The temperature changes by the mechanical action of the piston. Compression has a tendency to increase temperatures. Expansion has a tendency to decrease temperatures.

This Experiment 4 could be called Experiment Adiabatic—a funny word we'll see again in Chapter 9. For now, it describes *a mechanical method of increasing or decreasing gas temperature by compressing or expanding the gas.*

3.7 KINETIC THEORY OF GASES

We have described the properties of a gas as direct consequence of the actions of its molecules: how heavy they are, how much volume they occupy, and how fast they travel. This description of temperature and pressure and density is the one presented by James Jeans over a hundred years ago and is called the *molecular theory of heat* or the *kinetic theory of gases. Kinetic* means motion, so it is a good word to use for our discussion of thermodynamics, because thermodynamic properties imply molecular motion.

In the experiments of this chapter, we have imagined that the molecules act like round, isolated molecules. They move in straight lines and are unaffected by the other molecules in the gas. Is this the way a "real" gas behaves?

Strictly speaking, the answer is "no." Gases do have some intermolecular forces between them, which mean they may orbit around each other. But this

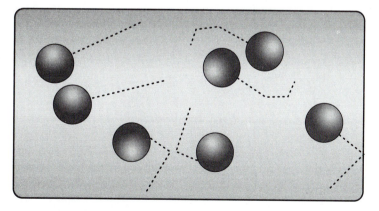

Figure 3–6. Molecules may collide or be attracted by each other, affecting the perfect gas law.

effect is minor, and the perfect gas law is valid for real gases, too, except for one very important situation. When a gas is near its *condensation* point, speeds of the molecules slow down to the point that the intermolecular forces start to take over, then the perfect gas law predicts inaccurately. (Figure 3–6) We will investigate this phenomenon in Chapter 18.

CHAPTER **3**

PROBLEMS

1. Write the perfect gas law $P = \rho RT$ in terms of v instead of ρ, then solve it for T.

2. When the density of oxygen in a closed vessel is doubled, how much does the pressure change?

3. When the temperature of nitrogen in a closed vessel drops by one-half, how much does the pressure change?

4. When 1 lb of atmospheric air in a sealed chamber is heated to twice its temperature:
   a. How much change in density does it show?
   b. How much change in pressure does it show?

5. Safety precautions must be made when filling a compressed air cylinder, since the pressure inside may cause the cylinder to break. Furthermore, if the temperature of the gas should increase after it is filled and while it is being stored, the pressure of the gas will increase according to Experiment 2. Regulations might suggest that cylinders must be filled so that even with a 20% increase in temperature ($T_2/T_1 = 1.20$), the pressure will not exceed P max.
   a. If the maximum pressure is 2000 psia, to what pressure can the cylinder originally be filled?
   b. If the temperature increases by 20%, will there be an increase in density?

6. A gas is contained in a vessel with the following state:
   $T = 500°R \qquad P = 50 \text{ psi} \qquad v = 20 \text{ ft}^3/\text{lb}_m$
   Determine the constants of proportionality for each of the experiments in this chapter. Determine a, b, and c for this gas.

7. The amount (weight) of air in Baby Jane's balloon is doubled due to the wind capability of Uncle Jeff. If the specific volume of the air in the balloon did not change, by how much did Uncle Jeff increase the size of the balloon?

8. Figure 3–7 shows the apparatus necessary to fill cylinders with compressed air. If the cylinder initially has some air in it, and the compressor is turned on until the amount (mass) of air is doubled:
   a. How does the density of air in the cylinder compare with its initial density?
   b. How does the specific volume of the air compare with its initial specific volume?
   c. If the temperature of the air did not change during the process, how much did the pressure change?

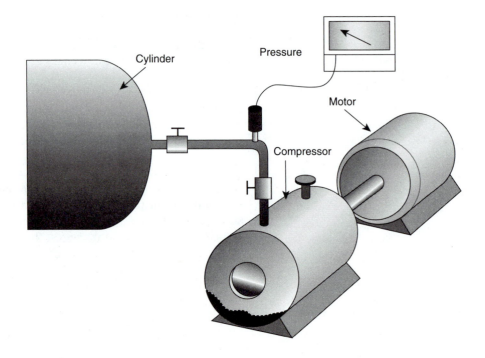

Cylinder

Pressure

Motor

Compressor

Figure 3–7. An air compressor.

9. Figure 3–8 shows a system arranged to maintain the pressure inside the vessel at a specified pressure, no matter what temperatures the gas may reach. Initially the vessel is filled to the prescribed pressure, but as the temperature changes, the pressure in the vessel may change according to Experiment 2. The constant pressure valve at the top either opens or closes to maintain the original pressure.

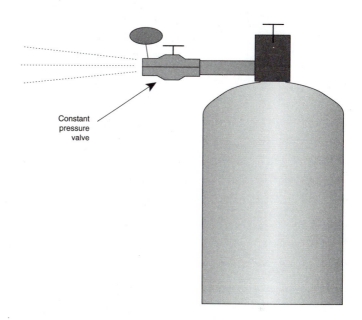

Figure 3–8. Pressure relief valve.

    a. If the temperature increases by 15%, how does the valve correct for the change in pressure? What percent of the original air is lost?

    b. If the temperature increases by 25%, what does the valve do and how much air is involved?

10. Which experiment of Sections 3–2 through 3–4 best describes each of the following thermodynamic processes (Experiment 1, 2, or 3)?

    a. Air escaping from a balloon.

    b. Hot air rises.

    c. Pop bottle explodes its top when left in car trunk.

    d. When canning fruit, sealed bottle top depresses when fruit cools.

11. Two compressed gas cylinders are connected by a hose and isolation valve. Chamber #1 has a volume of 5 ft^3 and contains 2 lb$_m$. Chamber #2 has a volume of 10 ft^3 and contains 1 lb$_m$. What is the specific volume of the gas in chamber #1? What is the specific volume of the gas in chamber #1? What is the specific volume of the gas in chamber #2? After the valve is opened, what is the specific volume of the gas? Also find the density of this gas.

12. Figure 3–9 shows some molecule experiments that appear to be different from the experiments in Sections 3–2 through 3–4 in this chapter. Upon further investigation you will find that each one is, in fact, identical (the results are the same) to one of the experiments in Sections 3–2 through 3–4. For each case, determine if it is equivalent to Experiment 1, 2, or 3:

   a. Each container has:
      Same molecular speed
      Same molecular weight
      Same number of molecules
      Different volume

   b. Each container has:
      Different molecular speed
      Same number of molecules
      Same molecular weight
      Different volume
      Same pressure

   c. Each container has:
      Same molecular speed
      Different molecular weight
      Same number of molecules

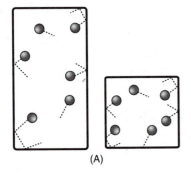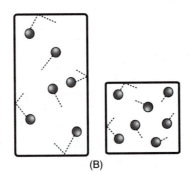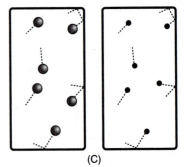

(A)   (B)   (C)

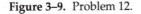

Figure 3–9. Problem 12.

13. The temperature of 4.83 lb of oxygen occupying 8 cu ft is changed from 570°F to 660°F, while the pressure remains constant at 115 psia. If there is no additional gas added to the expandable container, determine:

   a. the final volume
   b. the initial density and final density

Problems in SI Units

14. A gas is contained in a vessel with the following state:
$T = 300°K$ $P = 350\ kPa$ $v = 1.2\ m^3/kg$
Determine the constants of proportionality for each of the experiments in this chapter. Determine:
    a. final volume
    b. initial density
    c. final density

15. The temperature of 2 kg of oxygen occupying .25 m^3 is changed from 300°K to 400°K while the pressure remains constant at 800 kPa. If there is no additional gas added to the expandable container, determine:
    a. final volume
    b. initial density
    c. final density

CHAPTER
4

Applying the Perfect Gas Law

PREVIEW: The perfect gas law derived in Chapter 3 is one of the basic tools of thermodynamics. It is deceptively simple yet can be very powerful if used to its fullest. In this chapter we will see that this law (a) is actually two formulas in one, (b) can help us understand in detail how to describe the property of a gas, (c) can be written in at least ten different forms depending on what it is being used for, (d) can help predict the lowest temperature possibly achieved in laboratory conditions, and (e) can solve many significant engineering problems.

OBJECTIVES:
- ❑ Calculate the value of R in the perfect gas law for any gas.
- ❑ Solve many significant engineering problems using forms of the perfect gas equation.
- ❑ Perform an experiment that researchers have spent millions of dollars to duplicate; find absolute zero temperature.

4.1 THE GAS CONSTANT

In Chapter 3, the perfect gas equation was derived. It is one of the most important tools we use in thermodynamics. The equation

$$P = \rho RT$$

or

$$Pv = RT$$

Equation 4–1

came from a series of experiments. Each experiment had its own individual equation to describe it, but the perfect gas equation was one equation that could be used for all the experiments. The perfect gas equation simplifies thermodynamics not only because it makes all these experiments come under one general law, but also because it predicts the results for any thermodynamic experiment, not just the four we tried.

In Equation 4–1, the symbol R is a constant for each gas. It doesn't matter if the gas is hot or cold, pressurized, dense, or rarified; R will always be the same. But the constant is unique for each type of gas.

❖ **KEY TERM:**   **R:** Gas constant, different for each gas.

Table 4–1 gives the gas constant for many familiar gases. This table is an abbreviation of more complete lists of gases and gas constants available in handbooks on engineering of materials.

Table 4–1

GAS CONSTANTS FOR POPULAR GASES

Gas	$R\left(\dfrac{ft-lb_f}{°R-lb_m}\right)$	$R\left(\dfrac{kPa-m^3}{kg-K}\right)$
Acetylene	59.39	.320
Air	53.36	.287
Argon	38.73	.208
Carbon dioxide	35.12	.189
Hydrogen	766.53	4.157
Methane	96.40	.520
Nitrogen	55.15	.297
Oxygen	48.29	.260
Water vapor	85.60	.462
Natural gas	98.0	.520

If the tables are not available, the gas constant for a certain gas can be calculated, provided the molecular weight of the gas, m, is known. Do this using the equation

$$R = \frac{1545}{\mu}\left[\frac{(ft-lb_f)}{(lb_m -°R)}\right]$$

The number 1545 (ft-lb$_f$)/(lb$_m$–°R) or 8314 J/kg – °K is called the *universal gas constant* and is denoted by R. The temperature scale denoted by °R and °K will be explained in Section 4–3.

Dividing the universal gas constant by the molecular weight gives the gas constant for a specific gas. An illustration of this equation is to consider oxygen. It has a molecular weight of 32. The equation above predicts that the gas constant will be:

$$R = \frac{1545}{32} = 48.29\frac{(ft-lb_f)}{(lb_m -°R)}$$

Table 4–1 bears out that this is the proper value.
In SI,

$$R = 8.314 \frac{\text{kPa} - \text{m}^3}{\text{kg} - K}$$

For molecular weight of 32 (oxygen), this results in

$$R = \frac{8.314}{32} = .260$$

which is also found in Table 4–1.

4.2 USES OF THE PERFECT GAS EQUATION

The perfect gas equation is an important and useful tool and is particularly important for:

- defining a consistent set of dimensions or units for pressure, temperature, and specific weight
- determining the existing pressure, temperature, and density of any gas at any time (the "state" of the gas)
- determining the state of the gas after changes in the state have been made

4.3 UNITS OF P, T, AND ρ

As determined in Chapter 2, the most popular units of pressure are pounds (force) per square inch (lb_f/in^2, or psi) or pounds (force) per square foot (lb_f/ft^2, or psf). But these units may not be complete enough. It is important to consider what zero pressure actually means. From the discussion of the molecule gas, zero pressure would mean no molecular strikes against the wall; either there are no molecules in the container or they have no motion. This is called *absolute zero pressure* or *zero pressure absolute*. As the pressure increases by creating some motion, a positive pressure will be recorded in psi. Since this pressure is referenced to absolute zero pressure, it is called *absolute pressure*.

There is a predictable amount of atmospheric air that strikes our bodies as we stand at sea level. Typically we are subjected to 14.7 psi absolute pressure. Sometimes we wish to record pressures over and above that which is normally felt by our bodies. For example, the air pressure inside of an automobile tire is always described as additional to the atmospheric pressure: 25 psi air pressure in a tire means 25 psi over normal pressure. Absolutely, there is 25 + 14.7 = 39.7 psi total, or absolute, pressure. Since the 25 psi is the pressure read on a tire gauge, this measure of pressure is called *gauge pressure*. We distinguish the different measurements of pressure by using 39.7 psi*a* (absolute) or 25 psi*g* (gauge) (Figure 4–1).

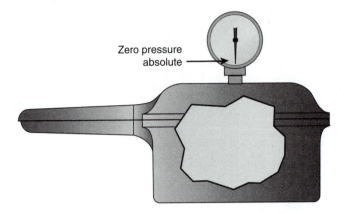

Zero pressure
absolute

Figure 4–1. Gauge and absolute pressure.

There is a simple equation that relates absolute and gauge pressure:

$$P\ absolute = P\ gauge + P\ atmospheric$$

where the atmospheric pressure at sea level is considered to be 14.7 psia, 2116 psfa, or 100 kPa.

It is important to note that the perfect gas law requires the absolute pressure measurement to be used in order to be correct. Either psia or psfa may be used, although psfa is preferred, as will be shown later. Forgetting to use absolute pressure is one of the major causes of errors in using the perfect gas equation! (Figure 4–2)

Temperature is measured popularly in degrees Fahrenheit or degrees Celsius (Centigrade). The two systems are related by the equation:

$$°F = \frac{9}{5}°C + 32°$$

Equation 4–2

Consider whether these temperature scales are absolute measurements—that is, whether zero temperature is the beginning of the measuring scale. Zero temperature means that the average speed of the gas molecules is zero. All molecules must be motionless! Zero degrees Fahrenheit certainly is not zero temperature because many parts of the world get "below zero" in the dead of winter. Zero degrees Celsius is even warmer than that, so it can't be absolute zero either.

You may well imagine that the perfect gas equation will not work unless *T* is in absolute units. Neither degrees Fahrenheit nor degrees Centigrade will work. An absolute temperature scale must be constructed. Scientists have shown that temperatures below –460°F cannot be achieved. This is absolute

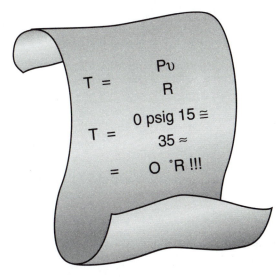

Figure 4–2. The major cause of errors using the perfect gas law equation is not using absolute units for pressure and temperature.

zero temperature Fahrenheit; there is no molecular movement. We can build an absolute temperature scale by starting it at –460°F and calling this point *zero*. This temperature scale is called the *Rankine scale* and is related to Fahrenheit by

$$°R = °F + 460°$$

If you convert –460°F into °C according to Equation 4–2, absolute zero temperature would be –273°C. The centigrade absolute temperature scale is called the *Kelvin scale*, and °K is found by:

$$°K = °C + 273°$$

When using the perfect gas law, temperatures must be given in terms of °R or °K.

Units of density are usually lb_m per cubic foot or lb_m per cubic inch. Fortunately both of these scales are absolute. Zero volume means that there can be no lower volume. Therefore, volume and density need not be changed, since they are already in absolute units, which is all that is required by the perfect gas equation.

The perfect gas equation is actually two equations in one. It says that the numerical value of the pressure must equal the numerical value of the product of the values of the gas constant, density, and temperature. It also says the units of pressure must equal the multiplication of the units of the gas constant, density and temperature. (Figure 4–3) As an example, if the units of pressure

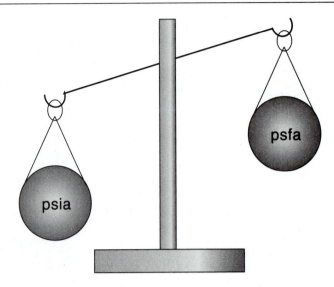

Figure 4–3. Units in the perfect gas equation must be consistent.

are lb_f/ft^2 absolute, units of temperature are degrees Rankine, and density is in lb_m/ft^3, then the perfect gas law unit equation states that

$$R = \frac{P}{\rho T}$$

and the units on the left side of the equation must be the same as the units on the right side. The units of R must be the same as the units of

$$\text{Units:} \quad \frac{P}{\rho T} = \frac{lb_f/ft^3}{lb_m/ft^3 - {}^\circ R} = \frac{(ft - lb_f)}{({}^\circ R - lb_m)}$$

The units of R are $(ft\text{-}lb_f)/({}^\circ R - lb_m)$.

Summarizing, the perfect gas law will work only if absolute units are used. Similarly, the results from the equation will make no sense unless consistent units are used. To be safe, in English units, always use quantities with distance dimensions in *ft*, time in *sec*, force in lb_f, and mass in lb_m. Therefore, the preferred units for the state variables are ${}^\circ R$, psfa, and lb_m/ft^3. In SI, use state variable units of ${}^\circ K$, kPa, and kg/m^3.

4.4
USING THE
PERFECT GAS LAW
TO DETERMINE
THE STATE

The second manner in which the perfect gas law is useful is that it helps to define the situation of the gas if nothing is happening to the gas and nothing is changing in the gas—that is, if the "state" of the gas is remaining the same. If the state of the gas is not changing and two of the three state properties are known, the third may be calculated by the perfect gas law. For instance, if we had a cylinder full of oxygen gas at a pressure of 300 lb per ft² absolute and a temperature of 560°R, the perfect gas equation will tell us that the density

$$\rho = \frac{P}{RT} = 300 \ \frac{lb_f}{ft^2} \left[48.29 \ \frac{(ft-lb_f)}{°R-lb_m} \times 560°R \right]$$

$$= .011 \ lb_m/ft^3$$

SAMPLE
PROBLEM 4–1

In an old style locomotive, steam was heated by coal in a boiler until it reached 300°F. The specific volume of the steam at this point was 6.46 cubic feet per pound. What is the steam pressure in the boiler?

SOLUTION:

The perfect gas law says:

$$P = R\rho T$$

where $R = 85.60 \ \dfrac{ft-lb_f}{°R-lb_m}$ (see Table 4–1)

$$T = 300 + 460 = 760°R$$

$$v = 6.46 \ ft^3/lb_m$$

But $\rho = \dfrac{1}{v} = \dfrac{1}{6.46 \ lb/ft^3} = .155 \ lb/ft^3$

therefore

$$P = 85.60 \left(\frac{ft-lb_f}{°R-lb_m} \right) \times 760°R \times .155 \frac{lb_m}{ft^3}$$

$$= 10083 \ lb_f/ft^2 \ (psfg)$$

which is an absolute pressure. In gauge, subtract 2116 lb_f/ft^2

$$P = 7967 \ psfg$$

The perfect gas equation can be used in this manner if the following are known: the kind of gas with which you are dealing, the temperature of the gas, and the pressure of the gas. Then you don't have to measure the density of the gas because the perfect gas law can calculate it for you. Similarly, if you know the kind of gas and its density and the temperature of the gas is measured with a thermometer, then the perfect gas law will tell what the pressure of the gas is: $P = R\rho T$

SAMPLE PROBLEM 4–2

It is necessary to store oxygen under a pressure of 50 psig and maintain a density of .2 lb_m/ft^3.
a. What storage temperature does this require?
b. If the gas were hydrogen, what temperature would be necessary?

SOLUTION:

a. $P = 50$ psig $= 64.7$ psia $= 9317$ psfa

$$\rho = .2 \; lb_m\big/ft^3$$

$$R = \frac{1545}{\mu} = \frac{1545}{32} = \frac{48.29 \; ft-lb_f}{°R-lb_m} \quad (\mu = 32 \text{ from Table 2–1})$$

Solving the perfect gas law for T yields $T = P/R\rho$

$$T = \frac{P}{R\rho} = \frac{9317 \; lb_f\big/ft^2}{48.29 \dfrac{ft-lb_f}{(°R-lb_m)} \times .2 \; lb_m\big/ft^2} = 964°R = 504°F$$

b.   If $\mu = 2$ (Table 2–1)

$$R = \frac{1545}{2} = 767.5 \frac{ft-lb_f}{(°R-lb_m)}$$

$$T = \frac{9317}{(767.5 \times .2)} = 60.7°R = -399.3°F$$

In the next sample problem, a new term is introduced. If a gas is stored in a lab, its temperature is often the same as the building, usually 70°F. If the gas is stored in a flexible container like a sack, then its pressure is atmospheric pressure, 14.7 psia. If a gas has conditions $T = 70°F$ and $P = 14.7$ psia $= 2116$ psfa, we say that it is at *standard temperature and pressure.*

❖ KEY TERM:

Standard Temperature and Pressure: Temperature of 70°F (21°C) and pressure of 14.7 psia (100 kPa).

SAMPLE PROBLEM 4-3

What is the density of acetylene at standard temperature and pressure? Solve this problem in English units and in SI units.

SOLUTION:

$$P = 14.7 \text{ psia} = 2116 \text{ psfa} \qquad P = 100 \text{ kPa}$$

$$T = 70°F = 530°R \qquad\qquad T = 21°C = 294°K$$

$$R = \frac{1545}{26} = 59.4 \frac{\text{ft-lb}_f}{°R\text{-lb}_m} \qquad R = \frac{8.314}{26} = .32 \frac{\text{kPa-m}^3}{\text{kg-}k}$$

Then the perfect gas law states

$$\rho = \frac{P}{RT} = \frac{2116}{(59.4 \times 530)} = .067 \frac{\text{lb}_m}{\text{ft}^3} \qquad \rho = \frac{P}{RT} = \frac{100}{(.32 \times 294)} = 1.100 \frac{\text{kg}}{\text{m}^3}$$

4.5 CALCULATING CHANGING STATE PROPERTIES

The third way to use the perfect gas law is for defining the properties of a gas after a change has been made to the gas. There are many ways a gas can be changed, for instance, by heating it. To predict the outcome of the state after the change, begin by measuring the gas properties before they change. Record the properties T, P, and ρ and assign them the subscript 1 (T_1, P_1, ρ_1) to indicate that they refer to properties before the change. From the discussion in the previous section, these properties satisfy the perfect gas law:

$$P_1 = \rho_1 RT_1$$

After measuring the properties of the gas, a change is made to the gas—by compressing it, heating it, adding more gas, or in any other way. After the change, the properties of the gas are measured and referred to this time as P_2, T_2, and ρ_2. Again they satisfy the perfect gas law:

$$P_2 = \rho_2 RT_2$$

Rather than measure all these conditions after the change, can the perfect gas law *predict* some of these conditions? The answer is "yes, in some situations." We do this by first dividing equation 1 by equation 2 resulting in:

$$\frac{P_1}{P_2} = \frac{\rho_1 RT_1}{\rho_2 RT_2} = \frac{\rho_1 T_1}{P_2 T_2}$$

Equation 4-3

This can be called the *perfect gas law for changing situations*. In some cases, this equation can be used to predict conditions after the change by knowing those before the change. For example, if we know that the density of the gas does not change during the process, then $\rho_1 = \rho_2$ and the perfect gas equation becomes:

$$\frac{P_1}{P_2} = \frac{T_1}{T_2}$$

With this equation, the pressure after the change, P_2, can be found by:

$$P_2 = \frac{P_1 T_2}{T_1}$$

If the properties P_1 and T_1 are known before and the temperature of the gas, T_2, is known after, the resulting pressure can be determined without measurement. Similarly, if the final pressure is known but not the final temperature, it may be found by the equation:

$$T_2 = \frac{T_1 P_2}{P_1}$$

SAMPLE PROBLEM 4–4

A sealed rigid vessel contains a gas at a temperature of 70°F and pressure of 20 psia. If the vessel is heated to 140°F, what will the pressure of the gas be?

SOLUTION:

If we were to try to use the perfect gas law, $P = \rho R T$, to find the final pressure, we would need to know the gas density. The density might be calculated by the basic relation $\rho = m/V$; however, in this problem, neither m nor V are given. Nevertheless, it is clear that neither m nor V changes throughout the heating. Therefore, ρ does not change during the process either. Therefore, the perfect gas for a changing situation (Equation 4–3) becomes

$$P_2 = \frac{P_1 T_2}{T_1}$$

But

$P_1 = 20 \text{ psia} = 2880 \text{ psfa}$

$T_1 = 70°F = 530°R$

$T_2 = 140°F = 600°R$

So

$$P_2 = 2880 \times \frac{600}{530} = 3260 \text{ psfa}$$

Or converting into psia

$$P_2 = 22.64 \text{ psia}$$

SAMPLE PROBLEM 4–5

At the Old Salt Seafood Restaurant, steamed clams are the specialty. (Figure 4–4) The chef has a large steaming vessel in which he places the clams on a bed of seaweed. The vessel is heated until the steam temperature is 260°F, then sealed shut. The chef then heats the vessel up to 350°F. He wants to know how much pressure the vessel must withstand.

SOLUTION:

This is the second time a sample problem has used steam as the gas. The properties of steam are listed under "water vapor" in Table 4–1. Steam is an often-used gas in industrial thermodynamic processes. It is, therefore, a gas we want to be familiar with; however, it has one peculiarity compared to the other gases in Table 4–1. It has a high vaporization temperature. In fact, steam does not exist at standard temperature and pressure. The material is, instead, a liquid—water. Since steam is generated at atmospheric pressures with temperatures of 212°F (boiling), and since in this problem the steam is at 260°F, its temperature is close to its condensation or boiling temperature. Section 4–6 will warn that when this occurs, gases

Figure 4–4. The chef performs a thermodynamic process.

do not act like "perfect gases," and the perfect gas law is somewhat inaccurate. To better solve this problem, we will have to wait for Chapter 18, but for now we assume that steam at this temperature acts like a perfect gas. Define the properties of the steam before heating with subscript 1. The properties are:

$T_1 = 260°F = 720°R$

$P_1 = 14.7$ psia $= 2116$ psfa

Then list the given properties for the steam after it has been heated:

$T_2 = 350°F = 810°R$

Finally, identify the quantity that is to be determined:

$P_2 = ?$

Since the density of the steam did not change (how could it?), the perfect gas equation for a constant P states:

$$\frac{P_1}{P_2} = \frac{T_1}{T_2}$$

or

$$P_2 = P_1 \times \frac{T_2}{T_1} = 2116 \times \frac{810}{720} = 2380 \text{ psfa} = 16.5 \text{ psia}$$

In a similar fashion, if the change in the gas occurs in such a manner that the pressure does not change, Equation 4–3 becomes:

$$\frac{P_2}{P_1} = 1 = \frac{\rho_1 T_1}{\rho_2 T_2}$$

$$\frac{\rho_2}{\rho_1} = \frac{T_1}{T_2}$$

This equation comes in handy when the conditions before the change are known and either ρ_2 or T_2 is measured after the change. The equation will then calculate the unknown property after the change.

SAMPLE PROBLEM 4–6

The chef at the Old Salt Seafood Restaurant bought a new clam pot that was better than the old one. (Figure 4–5) This pot had a pressure relief valve on the lid that would not allow dangerously high pressures to build up. The chef set the valve to maintain 100 kPa psia in the pot as the steam was heated from 130°C. What was the density of the steam after the pot was brought up to 180°C?

Figure 4–5. The chef changes the thermodynamic process.

SOLUTION:

Again list the known conditions before heat was added:

$T_1 = 130°C = 403$ K

$P_1 = 100$ kPa

The known conditions after heating:

$T_2 = 180°C = 453$ K

$P_2 = 100$ kPa

The problem asks for ρ_2. The appropriate equation to use is:

$$\frac{\rho_1}{\rho_2} = \frac{T_2}{T_1}$$

Unfortunately ρ_1 is unknown. However, it can be calculated from the perfect gas law:

$$\rho_1 = \frac{P_1}{RT_1} = 100\,(.462 \times 403) = .537 \text{ kg}/\text{m}^3$$

Therefore,

$$\rho_2 = \frac{\rho_1 T_1}{T_2} = .537 \text{ kg}/\text{m}^3 \times \frac{403\,\text{K}}{453\,\text{K}} = .477 \text{ kg}/\text{m}^3$$

Finally, the perfect gas law can help predict the state of the gas after a change has been made if the temperature during the change is constant. Then the perfect gas law from Equation 4–3 becomes:

$$\frac{P_1}{P_2} = \frac{\rho_1}{\rho_2}$$

which indicates that if the pressure of a gas is known before and after a change during which the temperature is constant, and the density is known before, then the final density can be calculated.

SAMPLE PROBLEM 4–7

A gas is pressurized from 5 psig to 20 psig without changing its temperature. If its final density is .2 lb_m/ft^3, what was its density originally?

SOLUTION:

For a constant temperature change, the perfect gas law states:

$$\frac{\rho_1}{\rho_2} = \frac{P_1}{P_2}$$

For this example, the following conditions are known:

$P_1 = 5$ psig = 19.7 psia = 2836.80 psfa

$P_2 = 20$ psig = 34.7 psia = 4996.80 psfa

Then by applying the perfect gas law above:

$$\rho_1 = \rho_2 \times \frac{P_1}{P_2} = .2\ lb_m/ft^3 \times \frac{19.7}{34.7} = .11\ lb_m/ft^3$$

Interestingly, we have used pressures in psia to solve this problem, although the preferred units are psfa. If you try the psfa units, you will find that the result comes out the same. This is because the equation we were using, $\rho_1 = \rho_2 \times P_1/P_2$, has the pressures as a ratio of each other. Therefore the units cancel out. It is correct to use either psia or psfa in this situation, but do not try in calculations with the perfect gas law that do not have the ratio of pressures.

Since either psia or psfa was correct in this solution, you may be tempted to use gauge pressures instead of absolute ones. After all, the units are going to cancel out. We can illustrate the danger of not converting to absolute pressure by using gauge pressure in this calculation:

$$\rho_1 = .2\ lb_m/ft^3 \times \frac{5}{10} = .05\ lb_m/ft^3$$

This result is completely different from the first answer and is, moreover, wrong.

These equations:

$$\frac{P_1}{P_2} = \frac{T_1}{T_2}$$

$$\frac{\rho_1}{\rho_2} = \frac{T_2}{T_1}$$

and

$$\frac{P_1}{P_2} = \frac{\rho_1}{\rho_2}$$

can be called the perfect gas law for a changing situation. Each was derived from Equation 4–3, and each is for a different special situation. Equation 4–3 also lends itself to other changing situations. For example, if the temperature and volume do not change, then pressure and mass have a relationship found from the following:

$$\frac{P_1}{P_2} = \frac{\rho_1}{\rho_2} \; (\text{for } T_1 = T_2)$$

Substituting the definition of density:

$$\rho_1 = \frac{m_1}{V_1}$$

and

$$\rho_2 = \frac{m_2}{V_2}$$

then:

$$\frac{P_1}{P_2} = \frac{m_1/V_1}{m_2/V_2} = \frac{m_1}{m_2} \text{ if } V_1 = V_2$$

or

$$\frac{P_1}{m_1} = \frac{P_2}{m_2}$$

When faced with a problem dealing with the perfect gas law, you should always read the problem through carefully. Make a diagram of the problem if possible. Each of the known quantities should be written down, and each should be defined with a letter, for example, T_1, T_2, p_2, and so on. Then define the unknown property that is to be found by giving it an appropriate name or variable. Determine whether the problem can be solved using one of the perfect gas equations, and, if so, solve it algebraically by substituting the known quantities into the equation. The procedure can be summarized as follows:

- Make a drawing of the problem.
- Write down the known properties.
- Identify the unknown and give it a name.
- Decide what properties are constant and write down the appropriate perfect gas law.
- Calculate to solve the equation for the unknown property.

SAMPLE PROBLEM 4–8

How many cubic feet of oxygen can be drawn from a 10-ft^3 cylinder filled to 200 psig at 70°F? The gas expands into the atmosphere (14.7 psia) at a constant temperature.

SOLUTION:

Often the solution to perfect gas law problems begins with a vision of the problem. (Figure 4–6) For example, consider the gas to initially occupy the space of the cylinder and to finally occupy the space of the cylinder plus a

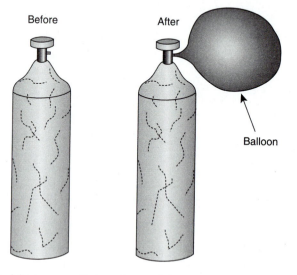

Before   After

Balloon

Figure 4–6. Envisioning a perfect gas law problem.

hypothetical balloon that holds the expanded air at 0 psig.

a.   Make a drawing.

b.   List the known properties:

$P_1 = 200$ psig $= 30916$ psfa

$V_1 = 10$ ft^3

$T_1 = 70°F = 530°R$

$P_2 = 0$ psig $= 2116$ psfa

$T_2 = 70°F = 530°R$

c.   Name the unknown:

$V_2 = ?$

d.   Select the appropriate perfect gas law equation. Since the temperature is constant:

$$\frac{P_2}{P_1} = \frac{\rho_1}{\rho_2} = \frac{\dfrac{m_1}{V_1}}{\dfrac{m_2}{V_2}} = \frac{V_1}{V_2}$$

e.   Solving for V_2, substituting and calculating:

$$V_2 = V_1 \times \frac{P_1}{P_2} = 10\ \text{ft}^3 \times \frac{30916}{2116} = 146\ \text{ft}^3$$

Therefore, a total of $V_2 - V_1 = 136$ cu ft can be drawn from the cylinder. This last step is important, since not all of the gas can be drawn from the cylinder. V_2 represents the volume of the gas after the expansion. Figure 4–6 clearly shows that volume includes the volume of the original vessel.

SAMPLE PROBLEM 4–9

How many cubic meters of oxygen can be drawn from a .5-m^3 cylinder filled to a pressure of 10^3 kPa absolute at 25°C? The gas expands into the atmosphere (100 kPa absolute) at a constant temperature.

SOLUTION:

$P_1 = 1000$ kPa

$V_1 = .5$ m^3

$T_1 = 25°C = 298$ K

$P_2 = 100$ kPa

$T_2 = 25°C = 298$ K

$$V_2 = V_1 \times \frac{P_1}{P_2} = 5\ \text{m}^3$$

Therefore, a total of $V_2 - V_1 = 4.5$ m^3 can be drawn from the cylinder.

4.6 APPLICATION: MIXTURES OF GASES

Oftentimes two or more gases are mixed to create one composite gas with desirable properties. For example, methane and ethane are mixed to give a clean, even-burning fuel known as natural gas. Oxygen and nitrogen are mixed in nature to become air.

Are mixtures of gases governed by the perfect gas law? Do we consider them as one gas whose gas constant R must be determined, or do we analyze them by each individual component, whose values of R can be found in Table 4–1?

Figure 4–7 gives the answer. Two gases, a and b, are mixed. All of the molecules are independent of each other; therefore, the two gases occupy the same space but otherwise are independent, except that the average speeds of both gases eventually will be the same. This is the principle of equilibrium discussed in Section 2–9. The gases can be considered as separated into two vessels of equal size:

$$V_a = V_b = V$$

and at the same temperature:

$$T_a = T_b = T$$

Therefore,

$$P_a = \frac{m_a R_a T}{V} \qquad P_b = \frac{m_b R_b T}{V}$$

are the pressures in each vessel, and when the gases come together in the same container, the strikes against the walls of each gas add together.

$$P_{total} = P_a + P_b$$

Equation 4–4

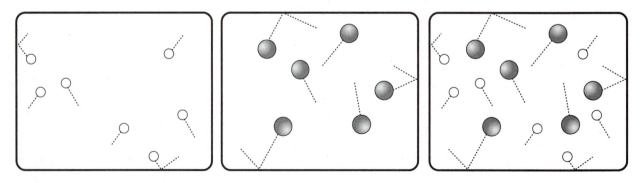

Figure 4–7. Gases that are mixed behave as though they were separate.

Equation 4–4 is formally called *Dalton's law of partial pressures*, and partial pressure is the name given to P_a and P_b defined above.

SAMPLE PROBLEM 4-10

One pound of hydrogen is mixed with 9 pounds of acetylene to give a hot brazing fuel. If the mixture is stored in a 20-ft³ container at 70°F, what are the partial pressures of each gas, and what is the pressure in the container?

SOLUTION:

$$V = 20\ \text{ft}^3, \quad T = 530°R, \quad m_{a(\text{hydrogen})} = 1\ \text{lb}_m, \quad m_{b(\text{acetylene})} = 9\ \text{lb}_m,$$

$$R_a = 766.53\,\frac{(\text{ft}-\text{lb}_f)}{(°F-\text{lb}_m)}, \quad R_b = 59.39\,\frac{(\text{ft}-\text{lb}_f)}{(°R-\text{lb}_m)} \quad \text{(Table 4–1)}$$

Then

$$P_a = \frac{m_a R_a T}{V} = 1 \times 766.53 \times \frac{530}{20} = 20313\ \text{psfa} = 141\ \text{psia}$$

$$P_b = \frac{m_b R_b T}{V} = 9 \times 59.39 \times \frac{530}{20} = 14164\ \text{psfa} = 98\ \text{psia}$$

Finally

$$P_{\text{total}} = P_a + P_b = 141 + 98 = 239\ \text{psia}$$

The important concept to remember is that the partial pressures of the mixture do not average together to give the total pressure, but instead add together. The reason for this is clearly seen in Figure 4–7.

4.7 WARNING: WHEN NOT TO USE THE PERFECT GAS LAW

The perfect gas equation is so useful in such a multitude of cases that you may get careless and use it in situations to which it does not apply. For example, do not use it on solids. When you heat a block of steel from 70°F to 600°F under atmospheric pressure, the perfect gas law says the block will double in size ($V_2 = V_1 \times T_2/T_1 = 2\ V_1$). If this were true, you might grow by about 10% if you flew from the Midwest during winter and landed in Miami. Do not apply the perfect gas law to liquids either. It is a *gas* law.

Another important point is, do not use the perfect gas law for a gas when there is a liquid in the same container. The density of the steam in a vessel that contains boiling water (Figure 4–8) cannot be predicted by measuring the temperature and pressure of the vessel and applying the equation

$$\rho = \frac{P}{RT}$$

This problem of water and the vapors it creates will be investigated in Chapter 18.

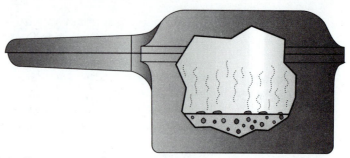

Figure 4–8. Gases near condensation cannot be treated as perfect gases.

TRY THIS EXPERIMENT

USING THE PERFECT GAS LAW TO FIND ABSOLUTE ZERO TEMPERATURE

How cold is absolute zero temperature? Researchers have been trying to achieve absolute zero for many years. It takes a great deal of scientific equipment to even try it, and to date no one has reached absolute zero.

Then how do we know that there is an absolute zero? How do we know what it is? A simple experiment you can do in your own home will allow you to determine the value of absolute zero. It takes very little equipment, plus an understanding of the molecular view of thermodynamics from Chapter 3. Your instructor may have you do this experiment in class, or you may just imagine it being done.

The experiment requires a small tight vessel that can be filled with a gas. The vessel must be fitted with a pressure gauge and a thermometer. (Such a vessel is available through the Broadhead-Garrett Co., Mansfield, Ohio. Ask for Refrigerant Demonstrator 9201.)

Fill the vessel with a gas, any gas. Air is acceptable; refrigerant gas may be even better. Record the pressure and temperature readings.

Place the cylinder into the oven and heat it to about 200°F. Notice that the pressure rises as the temperature rises. Why? If you don't know the answer, consider reading Chapter 3 again. Record the temperature and pressure.

Now remove the vessel from the oven, allow it to cool, then put it into the freezer. Once it gets very cold, record the temperature again. The table below shows some typical results for air and refrigerant R-134a.

Air		R-134a	
Pressure psia	Temperature °F	Pressure psia	Temperature °F
14.7	70	65	70
18.3	200	80.9	200
12.75	0	56.4	0

On a graph, plot the data of pressure versus temperature for each of the experiments above. This has been done for you in Figure 4–9.

What does this graph mean? A line has been drawn between the three points for each experiment and extended to negative temperatures. Using this line, we could predict what the pressure would be in the cylinder for any temperature from 200°F down to below –400°F. The graph predicts that molecules will stop moving (indicated by the pressure dropping to zero) when the temperature inside reaches –460°F. We have determined absolute zero temperature!

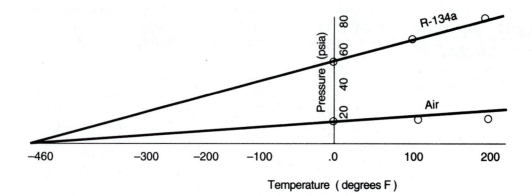

Figure 4–9. Results of an experiment to determine absolute zero temperature.

PNEUMATIC CONTROL DESIGNER

The technicians at Johnson Controls Company are very familiar with the benefits of the new designs for energy-efficient office buildings. After all, they install and service the control systems that make them work. Energy-efficient buildings are *tighter* than buildings have ever been before; they don't leak. You can't open windows in energy-efficient buildings. Exterior doorways are revolving, or include a foyer, so that air will not rush in when doors are opened. Leaky buildings make for cold drafts in the winter and muggy conditions in the summer, and the heating and air-conditioning systems have to work overtime to make the corrections.

But buildings are occupied by people who need fresh air. Tight buildings don't supply fresh air; therefore, it must be introduced into the hot-air and cold-air ducts in a regulated amount. This is called *outside air* and typically represents ten percent of the air that is recirculated through the warm-air and cold-air duct system. Johnson Controls is an international company that specializes in the pneumatic controls that regulate the outside air supply. Pneumatics, as it applies to outside air control, opens and closes dampers in the outside air ducts in a variable manner to conserve the energy necessary to condition the air. The dampers move by the action of a pushrod attached to an air-operated piston and cylinder, called a damper motor, that are opposed by a spring so that the greater the pressure in the cylinder, the further the rod will open the damper. The technicians that make up the workforce of the company in Johnson Controls' locations in major cities are often engineering technicians who are well versed in the field of thermodynamics.

In Johnson's systems, thermometers use air to measure temperature. In a pneumatic thermometer (Figure 4–10), which could be easily converted to a pneumatic thermostat, the thermometer does not record the temperature of a substance directly, as would a mercury-filled thermometer. Rather, the sensed temperature creates a unique pressure in an air-filled sensing bulb that is displayed on a pressure gauge. As the temperature of the air increases due to the placement of the probe in a hot environment, the pressure builds to move a needle on a pressure gauge. The pressure gauge is calibrated directly in terms of temperature.

**SAMPLE
PROBLEM 4–11**

One of Johnson Controls' most versatile thermometers is factory filled with 1 psig of 70°F air.
a. What is the density of the air in the bulb after it has been filled?
b. What will the pressure be when the bulb is immersed in a 300°F environment?
c. The sensitivity of the thermometer is measured as the amount of pressure

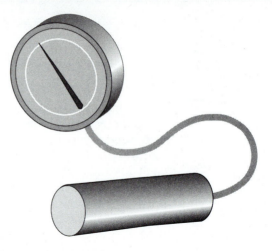

Figure 4–10. A pneumatic thermometer.

changes for every degree of change in temperature. What is the sensitivity of this thermometer in psi/°R?

SOLUTION:

a. From the perfect gas law, $\rho = P/RT$ where $P = 1 + 14.7$ psia $= 15.7 \times 144$ in^2/ft^2 = 2260 psfa, R = 53.3, and T = 530°R. Then

$$\rho = \frac{2260}{53.3} \times 530 = .074 \text{ lb}_m/\text{ft}^3$$

b. Assuming that the pressure gauge is a constant volume chamber (see Figure 2–5), then the thermodynamic process in the thermometer is constant volume. Therefore

$$P_2 = P_1 \times \frac{T_2}{T_1}$$

or

$$P_2 = 2260 \times \frac{760}{530} = 3240 \text{ psfa} = 22.5 \text{ psia} = 7.8 \text{ psig}$$

c. $$\text{Sensitivity} = \frac{\Delta P}{\Delta T} = \frac{6.8}{230} = .03 \frac{\text{psi}}{°\text{F}}$$

The types of circuits that a Johnson technician designs, installs, and services are much more complicated than the simple outside air damper controller described above. There are many more functions controlled by pneumatics, such as the start-stop of the air-conditioning or heating system, the control of humidity in the room, and the control of air supply to individual rooms. There are many other types of controls besides thermometers, and more types of actuators besides pneumatic cylinders.

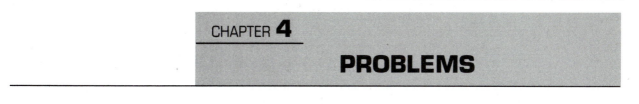

CHAPTER **4**

PROBLEMS

1. Determine the specific volume of air having a temperature of 70°F at normal atmospheric pressure.

2. What is the density of carbon dioxide that is stored at 15 psia and 90°F?

3. Carbon dioxide is held in a cylinder at an initial pressure of 2750 psig and a temperature of 460°F.
   a. What is its specific volume?
   b. If there are three cubic feet of gas in the cylinder, what is the mass of the gas?

4. Find the state of a gas with $R = 96$ (ft-lb$_f$)/(°R-lb$_m$) where $T = 240$°F and $v = 10$ ft^3/lb.

5. Use Table 4–1 to calculate the molecular weight of steam (water vapor).

6. Four pounds of oxygen have a volume of three cubic feet. If the pressure is 300 psia, what is the temperature of the oxygen?

7. During a future space mission to Philo, one of the moons of Mars, the astronauts may bring back a .01-lb$_m$ sample of its atmosphere in a 16-cu-in stainless steel sealed container. The container is designed with a window to see the gas, a pressure gauge, electrodes to test for conductivity, and a valve to allow the gas to react with other chemicals. Scientists at the Lunar Receiving Laboratory (LRC) inspect the container and notice that the gas has a brown tinge to it. The pressure gauge reads 255 psia. Wanting to find the molecular weight of the gas, they propose to perform a mass spectrometry experiment. It will cost the government $100,000 and take four weeks to complete. Can you save the taxpayers money and time by finding an easier way to measure μ, something the scientists cannot think of because the air-conditioning at the LRC has gone out and the room is a sweltering 92°F?

8. A steel cylinder weighs 75 lb empty and has an internal volume of 4 cu ft. The cylinder is filled with oxygen to 2200 psig at 70°F temperature.
   a. What will the cylinder weigh when it is filled with oxygen?
   b. How many cubic feet of oxygen can be drawn at 20 psig and 70°F temperature? (How many cubic feet would the cylinder occupy if the same amount of gas was kept at 20 psig and 70°F?)

9. Acetylene is a fuel gas that is used for flame cutting of metals. It is packaged in a steel cylinder that has a bursting pressure of 400 psig. At room temperature (70°F), the pressure in the cylinder is 300 psig. What temperature would the cylinder have to be warmed to before it would burst (°F)?

10. A gas having an initial volume of 5 ft^3 at a temperature of 1000°R is cooled at constant pressure until the volume decreases to 3 ft^3. Find the final temperature of the gas.

11. Two pounds of air are compressed maintaining constant temperature. The volume is compressed from 2.6 ft^3 to .89 ft^3. If the initial pressure is 35 psia, what is the final pressure?

12. Two cubic feet of gas expands from an initial pressure of 95 psia to a final pressure of 18 psia while maintaining a constant temperature. What is the final volume of the gas?

13. Air confined in a tank has an initial temperature of 150°F and an initial pressure of 50 psia. If the air is cooled until the pressure is 32 psia, what is the temperature of the gas?

14. With the new clam pot at the Old Salt Seafood Restaurant, the chef heats steam from 260°F to 350°F while maintaining a constant pressure. In order to do this, what percentage of the original steam will have to be removed?

15. J. Grisley is on trial for criminal negligence in the explosion of a steam boiler. It was Mr. Grisley's job to watch the pressure on the boiler to make sure it did not go above the safe limit of 1000 psig. Mr. Grisley testifies that just before the explosion, his control panel showed the pressure was 850 psig, the temperature was 750°F, and the specific volume was .6 cu ft per lb. All gauges were tested and found to be functional in the investigation following the explosion. The prosecutor says that Grisley fell asleep. As a technical witness, what do you say?

16. Thermodynamics Fable #1
    a. How many cubic feet will 44 lb of CO_2 (molecular weight = 44) occupy at 14.7 psia and 70°F (standard temperature and pressure, or STP)?
    b. How many cubic feet will 2 lb of H (molecular weight = 2) occupy at STP?
    c. How many cubic feet will 18 lb of H_2O (steam) occupy at STP (assuming water would vaporize at 32°F)?
    d. Moral: (pick one)

(1) Different gases occupy different volumes at similar conditions.
(2) All gases are at the same temperature at STP.
(3) The amount of any gas equal in pounds to its molecular weight (a mole of gas) occupies the same volume, no matter what gas it is.

17. Thermodynamics Fable #2—Avogadro's law
    a. How much will one cubic foot of CO_2 weigh at STP?
    b. How much will one cubic foot of H_2 weigh at STP?
    c. How much will one cubic foot of H_2O weigh at STP?
    d. How does the ratio of the weights of one cubic foot of CO_2 and H_2 compare with the ratio of molecular weights of the two?
    e. How does the ratio of the weights of one cubic foot of CO_2 and H_2O compare with the ratio of molecular weights of the two?
    f. How does the ratio of the weights of one cubic foot of H_2O and H_2 compare with the ratio of molecular weight of the two?
    g. Moral:
       Equal volumes of all perfect gases under the same conditions of T and P will contain the same number of molecules.

18. A metal sphere of 6-in internal diameter is weighed on a beam balance when it is empty (all gas evacuated before weighing). It is then filled with an unknown gas to a pressure of 110 psig and a temperature of 127°F and weighed again. The difference in the two weights was .05 lb. What is the gas?

19. What is the weight of the air in a 10-in wide automobile tire that has a 16-in hub diameter and a 32-in tire diameter if the air is contained at 70°F temperature and 30 psig pressure? Molecular weight of air is 28.97.

20. The pressure in a compressed air tank is originally at 20 psia when the temperature of the gas is 560°R.
    a. Find the temperature of the gas in °R when the gas pressure is increased to 25 psia, 30 psia, 50 psia, and 100 psia.
    b. What is the equation that relates P and T?
    c. On a piece of graph paper, make a careful plot of P versus T using the points mentioned above.

21. A refrigeration compressor takes in 15 cu in of gas on each stroke of the piston and compresses it to 5 cu in before discharging it. The cold gas enters the cylinder at 35°F and 32 psig and leaves the compressor at 175 psig. What is the temperature of the refrigerant as it leaves the compressor?

22. Cold outside air is drawn through a furnace and heated from 15°C to 50°C. If the pressure remains constant, by what factor will the volume of one pound of air be increased as it goes through the furnaces?

23. An air pump with a 4-in piston and a 12-in stroke compresses atmospheric air into a 5-ft^3 evacuated tank. Assume no temperature change during the pumping and find the gauge pressure in the tank after 100 strokes of the piston.

24. Solar panels are simple devices consisting of an enclosed box with a series of water-filled copper tubing loops inside. (Figure 4–11) In the manufacture of the tubing loops, many copper fittings are soldered into positions to make the turns. To check for leaks in these brazed joints, the panels are pressurized with an inexpensive, synthetic gas called R-22 (refrigerant with molecular weight = 92). A specially designed leak detector can be used to sniff for leaks of the R-22 through the solder joints. If the panels shown in the drawing are pressurized to 60 psig at a gas temperature of 70°F, estimate the weight of R-22 used for each panel.

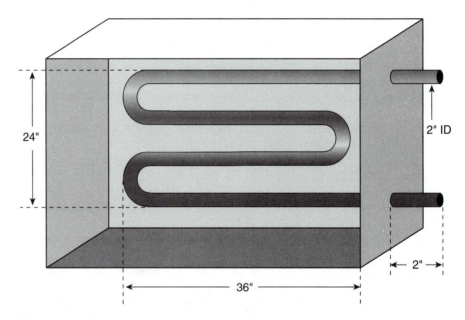

Figure 4–11. Solar panel.

25. Four tanks are sitting on a welder's bench. All are identical and are sized at 2.5 cubic feet. One is empty, another has hydrogen in it, another has oxygen, and another contains methane gas. It is unknown which cylinder has which gas, but each has a weight (gross weight of cylinder and gas) and pressure given below. Find what is in each cylinder ($T = 530°F$).

Tank No.	1	2	3	4
Weight (lb$_f$)	12	10	12	12
Pressure (psfg)	323,008	−2116	18375	38774

26. One drop of water as a liquid occupies approximately .005 in^3 and weighs .0016 lb$_f$. How much volume does it occupy when it is evaporated to a vapor ($T = 212°F$, $P = 2116$ psfa)? Calculate in cubic feet, then convert to cubic inches. How many times increase in volume does this represent?

27. Below are several statements of the perfect gas law. Each is true for the thermodynamic process of a special type of situation. Identify each with the number that describes the special situation. Each number may be used more than once or not at all.

    a. $Pv = RT$ (1) constant temperature

    b. $\dfrac{P_1}{T_1} = \dfrac{P_2}{T_2}$ (2) constant pressure and weight

    c. $P_1 V_1 = P_2 V_2$ (3) constant density

    d. $P = \rho RT$ (4) correct for all cases

    e. $\dfrac{P_1}{\rho_1 T_1} = \dfrac{P_2}{\rho_2 T_2}$ (5) incorrect for all cases

    f. $\dfrac{P_1}{P_2} = \dfrac{m_1 T_1}{m_2 T_2}$ (6) constant volume

    g. $P = b\rho$ (7) constant P and m

    h. $T = \dfrac{Pv}{R}$ (8) constant P

    i. $\dfrac{V_1}{V_2} = \dfrac{T_2}{T_1}$ (9) constant T and V

                                                      (10) constant T and m

28. A compressed air tank contains 1 pound of air at 150 psig and 70°F. How many 2-cu-ft balloons will it fill to 3 psig pressure at 70°F?

29. What is the specific volume of a gas at 180 psia and 90°F when its density is 0.0446 lb$_m$/ft^3 at 14.7 psia and 32°F? Calculate its gas constant and approximate molecular weight.

30. The NASA space shuttle re-enters the atmosphere at 200,000 feet. (Figure 4–12) A shockwave is created below the vehicle consisting of air at a density of .002 lb_m/ft^3 and a temperature of 2000°F. The shock wave acts as a container. What is the pressure of this air behind the wave?

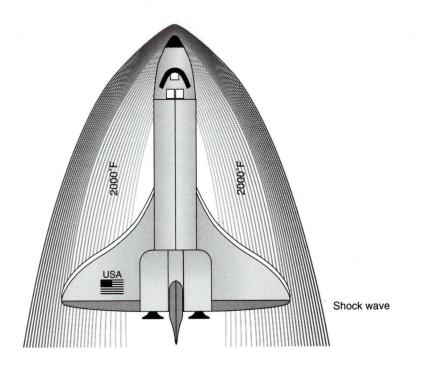

Figure 4–12. Re-entry of the NASA space shuttle.

31. A gas at 10 psig and 40°F is held in a rigid, sealed tank. The gas is heated to 540°F, at which temperature the density is measured to be .1 lb_m/ft^3. What is the weight of the gas in the container before and after the heating? Find the value of R and μ for this gas. Also find the final pressure.

32. A closed vessel A contains 3 cubic feet of air at a pressure of 500 psia and a temperature of 120°F. (Figure 4–13) This vessel connects with vessel B, which contains an unknown volume of air at 15 psia and a temperature of 50°F. After a valve separating the two vessels is opened, the resulting pressure is 200 psia and temperature is 70°F. What is the volume of vessel B?

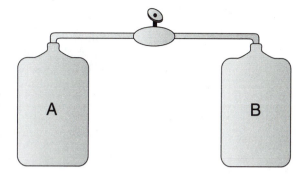

Figure 4–13. Two-sided vessel.

33. The specific volume of hydrogen at 80°F and 14.7 psia is 198 ft³/lb$_m$. If the gas is cooled under constant pressure conditions to 30°F, the specific volume changes to 178.6 ft³/lb$_m$. From this information, what do you estimate to be the temperature of absolute zero?

34. The temperature of an ideal gas remains constant while the absolute pressure changes from 103.4 psia to 827 psia.
    a. If the initial specific volume is .3 ft³/lb$_m$, what is the final specific volume?
    b. If the initial temperature is 250°F, what is the molecular weight of the gas?

35. In an Indy Grand Prix car race, tires are filled to 50 psi before the race. (Do not ask whether this is psia or psig!) At this time the tires are typically at 70°F, but during the race the tire temperature often gets to 300°F. Give an estimate of the pressure in the tires at this point. What assumptions did you make? What state property does not change?

36. Two spherical drums 6 ft in diameter each contain oxygen at 80°F. The drums are connected by a valve, which is initially closed. One sphere contains 3.75 lb, and the other 1.25 lb. What is the common pressure of the two vessels after the valve is opened (assuming constant T)?

37. A 3-ft³ oxygen tank used for welding is filled to 45 psi on a typical 70°F day. After using the tank for two days, the pressure reads 30 psi. How many lb$_m$ of oxygen were taken out of the tank?

38. .5 lb$_m$ of nitrogen and 2 lb$_m$ of helium are mixed at 70°F in a 3-ft³ cylinder.
    a. What is the partial pressure of nitrogen?
    b. What is the partial pressure of helium?
    c. What does the pressure gauge on top of the cylinder read?

39. Natural gas is a futuristic fuel for automobiles and buses. One concept to store fuel on-board a bus is to carry it in a large sack on top of the vehicle. For small automobiles, the gas must be compressed into smaller containers. If the sack on the bus (25 ft^3 and 1 psig) is compressed to a 1-ft^3 vessel (constant temperature), what would be the pressure in the smaller vessel?

Problems in SI Units

40. Find the volume of 1500 kg of propane stored at 42°C and 450 kPa (absolute).

41. A 3-m^3 rigid tank contains helium at 800 kPa and 30°C. A balloon is filled with the helium until the pressure in the tank falls to 200 kPa (constant T). What is the mass of helium removed from the tank?

42. A 10-m^3 tank at 150°C and 100 kPa absolute is evacuated to 10 kPa. If the final temperature is 130°C, determine the mass of gas remaining in the cylinder.

43. A 5-m^3 tank contains 32 kg of oxygen and 15 kg of nitrogen at 20°C. Find the partial pressure of each gas and the total pressure in the tank.

Computer Problems

44. Do Lesson 1 on your computer disk.

45. Do Lesson 2 on your computer disk.

46. Do Lesson 3 on your computer disk.

CHAPTER
5

Heat Is Not Temperature: Introducing the First Law of Thermodynamics

PREVIEW: Thermodynamics is the study of heat, and the tools of the thermodynamics technician are equations. Equations predict how heat affects matter and ways in which heat can be used.

In Chapters 2, 3, and 4, the text hinted at situations that involve "heating." Problems were solved concerning temperature change, but the question of how these temperature changes came about was not addressed.

How can heat be described in an equation? In what terms do we measure heat? These questions are answered in this chapter as we derive the first equation involving heat: the "first law of thermodynamics."

OBJECTIVES:
❑ Define Btu and joule, units of heat.
❑ Describe what happens when a substance is heated: in terms of internal energy, latent heat, and WORK.

5.1 DEFINING HEAT

In Chapter 2 we introduced a simple definition of *thermodynamics* in terms of "heat," then proceeded to discover something about heat. Chapters 2, 3, and 4, however, dealt primarily with the properties of a substance that heat might affect. It is time to continue to look for the meaning of *heat*.

Each of us has an idea of what the word *heat* means. It is part of our everyday vocabulary. Unfortunately, everyone may have a different understanding of this word. To some, heat is temperature, and we hear the phrase "turn up the heat." In a similar way, heat is warmth; we say "it's very hot." These definitions, however, are vague. *Heat* is a quantity that is much more precise than these phrases suggest. Heat is much more than just "warmth." It is measurable. It is predictable.

Are heat and temperature the same thing? If this is true, then (1) every time a substance is heated, it will show an increase in temperature, and (2) every time the same amount of heat is put into two different substances, they will show an increase in temperature of the same amount. Are these statements true? Several simple experiments can be performed to find out exactly what heat is, and in particular to show whether heat is the same thing as temperature.

The first experiment calls for the use of a hot plate. A hot plate is just a convenient device that gives off heat. We are using the hot plate in this experiment because it gives off a constant supply of heat. On the top of the hot plate is a beaker filled with 16 oz of water. Immersed in the beaker is a thermometer that will give a precise record of the water temperature at any time during the experiment, which is measured by a stop watch. This experiment begins by turning on the hot plate and recording the temperature of the water at certain time intervals as it is "heated." The result is that as the hot plate, or the heat generator, gives off a constant amount of heat, the temperature of the water increases. This seems to indicate that heat and temperature are the same thing.

Anytime you add heat, you will increase the temperature of the substance (water). In fact, this experiment can be used to define *a unit of heat*. It can be said that one unit of heat is equivalent to the heat that the generator must give off in order to raise the temperature of the water one degree Fahrenheit. This being agreed upon, we can name the unit of heat anything we like. However, the proper term for this amount of heat is a British Thermal Unit, or Btu. (Figure 5–1)

❖ **KEY TERM:**   **BTU:** The amount of heat required to raise one pound (mass) of water one degree Fahrenheit.

By this definition, it takes 10 Btu's to raise the temperature of a pound of water 10°F, and to raise $1/2$ lb_m of water 1°F takes $1/2$ Btu.

A joule or kilojoule is another unit of heat based on temperature changes in degrees centigrade or Kelvin. The amount of heat required to raise one kg

Figure 5–1. Experiment 1 defines a Btu.

of water one degree Kelvin is 4.18 kilojoules. (See Section 5–8 for a complete description.)

Now it is possible to describe the "heating capacity" of the heat generator. The generator that we are using (hot plate) might raise the temperature of one pound of water only a half of a degree a minute. If this is true, we can classify the heat generator according to our definition of Btu—we can say that it gives off one-half of a Btu every minute.

SAMPLE PROBLEM 5–1

A hot plate is rated to raise the temperature of one pound of water 4°F in one minute's time. How many Btu's are added to the water in one minute, and how many Btu's per minute does the hot plate deliver?

SOLUTION:

By definition of a Btu, four Btu's are added to the water. The hot plate delivers 4 Btu/min.

A subtle difference has crept into our vocabulary now in describing *heat* and a *heat generator*. In the definition of the measure of heat, no mention was made of how long it took to heat the water. A Btu of heat was added to the beaker when one pound of water was raised 1°F, no matter how long it took. Whether it takes a year or a second to add the heat to the vessel, the temperature will raise one degree. Rapid heating or slow heating will cause the same increase in temperature of the substance, if the same amount of Btu's are added. Yet in the measure of the output of the heat generator, the time it takes to generate the heat determines the size specification of the hot plate—the heating capacity.

SAMPLE PROBLEM 5–2

A hot plate is sized to heat the same amount of water as in Sample Problem 5–1 up to the same temperature, but does this in half the time. How much heat is added and what is the capacity of the hot plate?

SOLUTION:

Since the amount of water being heated and the number of degrees of increase in temperature are the same as in Sample Problem 5–1, the heat added is the same: 4 Btu. But now the rating is for

$$\frac{4 \text{ Btu}}{1/2 \text{ min}} = 8 \text{ Btu/min}$$

The hot plate has twice the capacity as that of the previous example.

SAMPLE PROBLEM 5–3

The beaker in the above experiment is placed in a refrigerator and the temperature is decreased by 10°F in 2 minutes. Was heat added to the container or taken away (rejected)? How much? Where did the heat go?

SOLUTION:

Since the temperature of the water dropped, heat was removed from the beaker. Because temperature decreased by 10°F, the amount of heat in the glass decreased by 10 Btu. Mathematically:

Change in temperature = –10°F (negative since there was a decrease)

therefore

Heat = –10 Btu

The amount of heat removed from the beaker of water did not disappear, but rather increased the temperature in the refrigerator itself.

SAMPLE PROBLEM 5–4

In Sample Problem 5–3, what is the cooling capacity (Btu/min) of the refrigerator?

SOLUTION:

$$Capacity\left(\frac{Btu}{min}\right) = \frac{10\ Btu}{2\ min} = \frac{5\ Btu}{min}$$

To write an equation that describes the amount of heat added compared to the increase in water temperature during this experiment is relatively simple. If there is 1°F increase in temperature, there must have been 1 Btu added. This suggests that numerically, the heat transferred is equal to the temperature change.

$$Q\ pe\ \Delta T$$

Equation 5–1

The "pe" instead of an equal sign means that this is a "potential equation"; that it is possibly correct in some sense and in some situations, but has not been checked for units and for universal application. Q will be the symbol used for heat. Therefore, the potential equation states that the amount of heat added to a substance equals the increase or decrease in the temperature of the substance. ΔT means $T_2 - T_1$, the difference in the temperature before and after the heating.

The potential equation above is a simple statement that heat and temperature are the same thing.

SAMPLE PROBLEM 5–5

A stove burner typically has a heat-generating capacity of 1000 Btu/hr while a slow-cooking crock pot may put out only 50 Btu/hr. If they both add ten Btu's of heat to a 1-lb$_m$ pot of water, how much will the temperature of the water increase in each case, and how long will it take?

SOLUTION:

Since Q pe ΔT and the amount of heat added in both cases is 10 Btu, then the increase in the temperature of the water will be 10°F in both cases. The stove will heat the water in 10 Btu/1000 Btu/hr = .01 hr or 36 sec. The crock pot will take 10 Btu/50 Btu/hr = .2 hr = 12 min.

5.3 EXPERIMENT 2

This experiment is to determine again whether heat and temperature are the same thing. This time two hot plates are used as generators, and again two beakers of water will be heated. In each beaker a thermometer is positioned as before. Both beakers are placed on the heat generators at the same time, and the temperature is noted at specific time intervals.

If heat is the same thing as temperature and both hot plates are identical and give off the same amount of heat every minute, we would expect the temperature of the two beakers to increase exactly the same amount. That is, when the same amount of heat is added to one object as to another object, the temperature increases are exactly the same. In this experiment, however, it is readily seen that the temperature of the 8-oz beaker of water increases much more rapidly than the temperature of the 16-oz beaker of water. (Figure 5–2)

Figure 5–2. Experiment 2 deals with different amounts of water.

The same amount of heat was absorbed by each beaker, but two different temperatures resulted. This gives us a suspicion that temperature and heat are not the same thing.

The definition of Btu gives a hint about why this would happen. The definition says a 1°F increase in temperature can be expected if 1 Btu is added to 1 lb of water. But, it seems to imply that if we had ½ lb of water, 1 Btu would be more concentrated in the water and create a larger temperature increase.

In this experiment, the ½ lb of water will increase its temperature by 1°F with only a ½ Btu addition of heat. The equation for Q in Section 5–2 does not give this result and must be modified somewhat to take into consideration the mass of the water. A better equation is:

$$Q \text{ pe } m \, \Delta T$$

Equation 5–2

With $m = ½$ lb, $T = 1°F$, it predicts a heat addition of ½ Btu.

SAMPLE PROBLEM 5–6

The average hot water heater holds 30 gallons of cold water which fills at 65°F. How many Btu's does it take to heat a tank full of water 1°F? (1 gallon of water contains 8.33 lb_m.)

SOLUTION:

Since one gallon of water weighs 8.3 lb_m:

$$m = 30 \text{ gal} \times 8.3 \frac{lb_m}{gal} = 250 \, lb_m$$

Also the temperature difference is given: $\Delta T = 1°F$. Then Equation 5–2 states:

$$Q = m \times \Delta T = 250 \times 1 = 250 \text{ Btu}$$

5.4 EXPERIMENT 3

To verify our latest equation, Equation 5–2, we ought to do another experiment. Again we need the two heat generators, and two 16-oz beakers. This time one beaker is filled with water and the other with oil. (Figure 5–3) Start the hot plates and watch the temperature of each material increase. Since we have the same amount (mass) of each substance, we would expect to see each have the same increase in temperature if they are heated for the same time. The thermometer readings do not bear this out. The water temperature increases more slowly than the oil temperature. Does this mean that the effect of heat on different substances is different? It certainly does!

Figure 5–3.   Experiment 3 deals with different liquids.

How can one measure the effect of heat on substances other than water? It is accomplished by keeping a table of all different types of materials and the amount of Btu's necessary to raise 1 lb_m of each substance 1°F. These values we define as the *specific heat of the material.* The specific heat of oil is .40 Btu/lb_m – °F. It takes only .40 Btu's of heat to raise one pound of oil 1°F. Table 5–1 is a list of specific heat for many familiar substances.

The equation for Q in the last paragraph does not give correct results for anything but water; therefore, we must further modify the equation. For this experiment the equation is:

$$Q \text{ pe } Cm\Delta T$$

Equation 5–3

where C is the specific heat. This equation takes into consideration that temperature and heat are not the same, and that the effect of heat on a substance depends on not only how much of that substance there is, but also what type of substance it is.

SAMPLE PROBLEM 5–7

Oil is used as a lubricant and a coolant on an automatic drill press. The oil is sprayed onto the drill bit at 80°F and rolls off at 115°F. How much heat does each ounce of oil pick up?

Table 5–1

SPECIFIC HEATS OF LIQUIDS, SOLIDS AND GASES

Material	c(Btu/lb$_m$ – °R)	c(kJ/kg – °K)
Solids		
Bronze	.104	.435
Brick	.22	.921
Copper	.092	.398
Glass	.199	.833
Granite (stone)	.195	.816
Humus	.44	1.842
Lead	.031	.129
Lithium	.79	3.307
Nickel	.105	.439
Quartz	.28	1.172
Silver	.056	.234
Steel	.12	.502
Liquids		
Aerosol	.19	.795
Alcohol	.58	2.428
Gasoline	.50	2.093
Machine oil	.40	1.674
Mercury	.033	.138
Sea water	.94	3.936
Sulphuric acid	.336	1.407
Gases*		
Air	.17	.712
Nitrogen	.175	.725
Steam	.36	1.507

* Specific heats for gases are discussed at length in Chapter 7.

SOLUTION:

$$Q \text{ pe } Cm\Delta T = .40 \, \frac{\text{Btu}}{\text{lb}_m - °F} \times \frac{1}{16} \, \text{lb}_m \, (115° - 80° \, F)$$

$$= .82 \text{ Btu's}$$

SAMPLE PROBLEM 5–8   How much heat (kJ) must be added to one kg of silver to raise its temperature 1°C?

SOLUTION:

When measuring the temperature of the silver before heating, its temperature in °C is different from its temperature in °K. But when heat is added, the temperature in both Celcius and Kelvin increases by one degree. This is because one Celcius degree is the same as one Kelvin degree, just like one Fahrenheit degree is the same as one Rankine degree. So the specific heat of silver can be written as .234 kJ/kg – °C or .234 kJ/kg – K. The answer to the question is:

$$Q \text{ pe } .234 \frac{kJ}{kg - °C} \times 1 \text{kg} \times 1°C = .234 \text{ kJ}$$

**5.5
EXPERIMENT 4**

Equation 5–3 is not completely correct, which can be illustrated by another experiment. Set up the two heat generators with 16-oz beakers filled with water. This time the temperature of the first beaker of water is 65°F, while that of the second beaker of water is 210°F. Turn on the generators and measure the temperature of each. From what we've seen in the previous paragraphs, there is no reason that the temperature increases of each beaker should not be the same.

Yet at a later time we may record a 20°F increase in the temperature of beaker 1, while beaker 2 is sitting at 212°F, a mere 2°F increase! Of course, what is happening is that the second beaker is boiling. (Figure 5–4) Water boils at 212°F in the open atmosphere, which means that it is changing phase from a

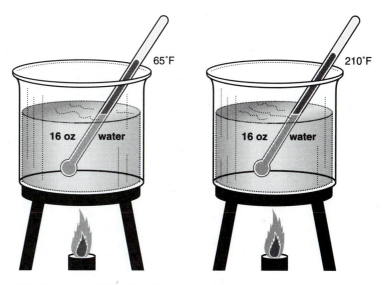

Figure 5–4. Experiment 4 deals with water boiling.

liquid to a gas. The change of phase is caused by heat being added. While this is going on, the temperature of the boiling liquid–gas combination remains the same.

❖ **KEY TERM:**   **Phase Change:** The change of a material from a liquid to a gas, or a solid to a liquid, or a gas to a liquid, or a liquid to a solid, or a solid to a gas.

The equation for Q from the last experiment does not predict this mere 2°F increase and must be further modified. To do so, we must examine the way heat affects a substance. Any heat that causes an increase in temperature of the substance is called *sensible heat*. Any heat addition that changes the phase of the substance is called *latent heat*. All the previous experiments dealt with sensible heat. For situations that deal with both sensible heat and latent heat, we must sum each one to get the total heat addition:

$$Q \text{ pe Sensible and Latent} = Cm\Delta T + \text{Latent Heat}$$

In order to measure latent heat, use the Btu that we have previously defined. For example, it takes 970 Btu of heat to completely evaporate one lb_m of water. If we define the latent heat per pound of a substance to be L, then the equation for heat addition is:

$$Q \text{ pe } Cm\Delta T + m_L L$$

Equation 5–4

where m_L is the mass of the substance that has gone through a phase change.

Table 5–2

LATENT HEAT OF MATERIALS		
Material	Latent Heat of Melting/ Freezing (Btu/lb_m)	Latent Heat of Vaporization/ Condensation (Btu/lb_m)
Water	144/–144	970/–970
Aerosol propellant	8/–8	68/–68
Nitrogen	——-	86/–86

Table 5–2 presents the latent heat of two substances in changing phase from both solid-liquid and liquid-gas. Note that the latent heat of freezing is the opposite of the latent heat of melting and carries a negative sign. Similarly, the latent heat of condensation is the opposite of the latent heat of vaporization and carries a negative sign.

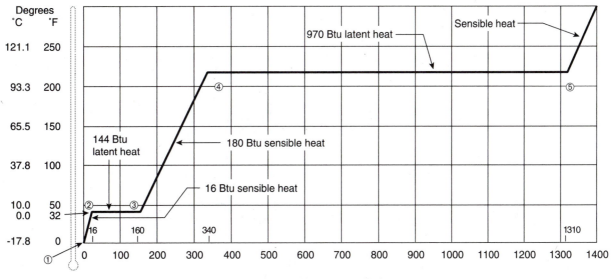

Figure 5–5. Plot of temperature during thawing and boiling. From Whitman & Johnson, *Regrigeration & Air Conditioning Technology,* 3rd ed. (Albany, NY: Delmar Publishers, 1995).

SAMPLE PROBLEM 5–9

A heat generator is giving off 1 Btu/min. A beaker containing 16 oz of 60°F water is placed on top of the burner. Plot the temperature of the water on a graph for the first 1200 minutes of heating. Show which portions represent latent heat and which represent sensible heat.

SOLUTION:

(See Figure 5–5.)

Equation 5–4 has evolved to a fairly sophisticated point. This equation should now apply to any situation and, therefore, is very general in nature. One more experiment, however, will show that it is not yet complete.

5.6 EXPERIMENT 5

This experiment requires more sophisticated equipment than we have used before. Only one heat generator is necessary, on top of which a strange-looking vessel is placed. The vessel is much like a pressure cooker in that it will hold 16 oz. of steam. However, the lid is made to move up and down in the vessel without losing any steam. We set this chamber on top of the burner and hold the lid at one point. As the steam is heated, the temperature rises. According to the perfect gas law of Chapter 4, if the gas temperature goes up, so does the

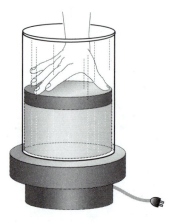

Figure 5–6. Experiment 5 Deals With WORK.

pressure. After a certain amount of heat is added, we can remove the Btu generator and be left with the high pressure-temperature gas in the cylinder.

The pressure inside can be relieved by moving the lid up to provide a larger volume for the gas. If the pressure is reduced, the gas temperature must simultaneously decrease. The lid can be pulled far enough out for the temperature to go back exactly to the original temperature!

Now we have a puzzling problem. Heat was added to the cylinder and the temperature rose. This indicated that heat was showing up as an increase in temperature. However, as the additional pressure was relieved by increasing the volume of the cylinder, the temperature decreased back to its original value. Where did the heat go? From the beginning to the end of this process, we have added heat, but we have no overall increase in temperature! Equation 5–4, then, predicts we must not have added any heat.

To discover what has happened to the heat, notice that the lid has moved to a new position. This motion was caused by increasing the gas pressure inside the lid, and allowing this pressure to push the lid up. This is an encouraging observation, since we could connect the lid to a piston and the piston to a gear and let it turn a set of tractor wheels, an electricity generator, or any one of a variety of other devices. As such, the heat we put into the cylinder would be performing useful work. Instead of increasing the temperature of a gas, the heat would be converted to WORK. (Figure 5–6)

Why can heat be converted to mechanical WORK? If we can answer this, we may have a clue to a definition of what heat really is.

WORK is a form of energy. Other forms of energy are chemical energy, electrical energy, thermal energy, and so on. Each form of energy can be turned into a different one. An electric motor turns electrical energy into mechanical

WORK. A power plant turns thermal energy into electrical energy. Heat can be converted to WORK, so it also must be a form of energy.

Since Experiment 5 has indicated that heat can be converted to WORK, the equations for Q previously given are not quite correct. To include the possibility of converting heat to WORK, a more correct equation is:

$$Q = Cm\Delta T + m_L L + WORK$$

Equation 5–5

This equation can be used for the previous experiments where no piston was involved and, therefore, no work was done. Simply said, $WORK = 0$ for those experiments. Note that this equation is valid for all cases. Therefore, the "pe" has now been replaced with an "=".

SAMPLE PROBLEM 5–10

A gas in a vessel similar to that in Figure 5–6 is heated using a hot plate with a 4-Btu/min capacity. After being heated for 3 minutes, the hot plate is shut off and the piston is allowed to rise until the temperature of the gas inside is reduced back to its original point. How much WORK, in terms of Btu's, is done on the piston to move it?

SOLUTION:

The total amount of heat put into the vessel is

$$Q = 4\,\frac{Btu}{min} \times 3\ min = 12\ Btu$$

Directly applying Equation 5–5

$$Q = Cm\Delta T + m_L L + WORK$$

where $\Delta T = 0$ for the overall process

$m_L = 0$ (no evaporation takes place)

Then $Q = 12\ Btu = WORK$

5.7 THE ENERGY EQUATION: WHERE HEAT GOES

Heat is a form of energy, but energy is a property of a system. In the experiments, the system was the water in the beaker. The water had no kinetic, potential, or electrical energy. The water did have thermal energy, though, since it had temperature. The thermal energy of the water was not conserved (constant), because the water received energy from the surroundings, the hot plate. Since there was energy input, the experiments are governed by Equation 2–2, the first law of thermodynamics:

Energy Input – Energy Outgo = Energy Buildup

The energy input is heat, Q. In these experiments there was no energy outgo. The energy buildup represents an increase in thermal energy of the system. Now we can see exactly what this thermal energy is because the experiments have shown that there is sensible thermal energy $Cm\Delta T$ and there is latent thermal energy $m_L L$. The experiments also have shown that the energy buildup of a system can be mechanical work done on or by the system, designated here as WORK.

The equation:

$$Q = Cm\Delta T + m_L L + WORK$$

is a statement of the first law of thermodynamics when the only energy flows are heat energy.

❖ **KEY TERM:**　　**HEAT:** A form of energy flow created by placing a high temperature substance next to a low temperature substance.

Sensible thermal energy is often called *internal energy*, or *stored energy*, which seems to imply that the thermal energy is embedded within the substance, and in fact it is, because the increased temperature implies that the molecules are vibrating or rotating faster due to the addition of the heat.

When heat is added to a substance, it may be used to provide the energy necessary to change the phase of the substance.

Whenever heat flows into a system, it does not just disappear. It goes somewhere. It appears as a form of thermal energy or mechanical WORK. Where it goes is predicted by the equation we have derived:

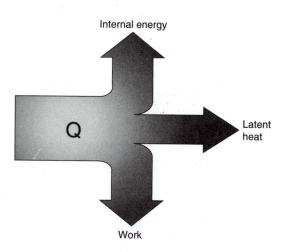

Figure 5–7. Once heat is absorbed by a system, it becomes internal energy, latent heat, or WORK.

$$Q = Internal\ Energy + Latent\ Heat + WORK$$

Equation 5–6

This equation is a form of the first law of thermodynamics, but we will refer to it as *the energy equation.*

The energy equation can be illustrated pictorially by Figure 5–7. This set of thick arrows shows the addition of heat Q to any substance and the split of heat energy into the forms predicted by the energy equation or the first law of thermodynamics.

SAMPLE PROBLEM 5–11

When 35 Btu's of heat was added to a substance, it was found that 5 Btu's were used to do WORK and 1 Btu was used for latent heat. (Figure 5–8) Determine how much heat was stored and show the energy diagram for the process.

SOLUTION:

$Q = Internal\ Energy + Latent\ Heat + WORK$

$35 = Internal\ Energy + 5 + 1$

$Internal\ Energy = 35 - 5 - 1 = 29\ Btu$

SAMPLE PROBLEM 5–12

A beaker similar to the one in Figure 5–6 is filled with 2 lbs of water. A 1000-Btu/min hot plate is turned on for two minutes; 650 Btu's of WORK is done (Chapter 7 will show how), and 1 lb of water is changed to steam. How much did the water temperature increase if you consider that 10% of the hot plate heat was lost by the beaker into the surrounding air?

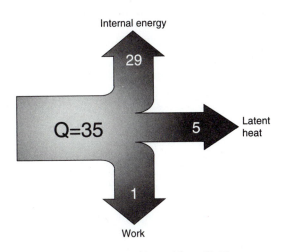

Figure 5–8. The thermal energies of Sample Problem 15–11.

SOLUTION:

Using the first law of thermodynamics:

Energy Buildup = Energy In − Energy Out

Energy In = 1000 Btu/min × 2 min = 2000 Btu

Energy Lost = 10% (2000) = 200 Btu

Energy Buildup = 2000 − 200 = 1800 Btu

This energy buildup was in the form of heat, $Q = 1800$ Btu, and the energy equation states:

$$Q = Cm\Delta T + m_L L + WORK$$

$$WORK = 650 \text{ Btu}$$

$$m = 2 \text{ lb}_m$$

$$m_L = 1 \text{ lb}_m$$

$$L = 970 \frac{\text{Btu}}{\text{lb}_m}, \quad C \text{ for water is } 1 \frac{\text{Btu}}{\text{lb}_m - °F}$$

Substituting these values into the energy equation:

$$1800 = 1\frac{\text{Btu}}{\text{lb}_m} - °R \times 2 \text{ lb}_m \times \Delta T + 1\text{lb}_m \times 970\frac{\text{Btu}}{\text{lb}_m} + 650 \text{ Btu}$$

$$\Delta T = \frac{(1800 - 1620)}{2} = \frac{180}{2} = 90°F \text{ (and } 90°R)$$

5.8 THE ENERGY EQUATION IN SI

One of the critical steps in deriving the first law of thermodynamics was the definition of a Btu. It is the basic unit of heat. The Btu is obviously not an SI unit. It involved one pound of water and one degree Fahrenheit of temperature.

In SI, the weight of the water will be in kg, the temperature change in K, and the basic unit of heat the kilojoule. In SI units, the first law of thermodynamics is written exactly as before:

$$Q = Cm\Delta T + m_L L + WORK$$

In SI, specific heat, C has units kJ/kg − K. Water has a specific heat of 4.18 kJ/kg − K. L, the latent of water, in terms of kJ/kg has a value of 2256 kJ/kg.

SAMPLE PROBLEM 5-13

A vessel fitted with an expansion diaphragm and including 10 kg of water is heated with 3000 kJ. When this happens, the water temperature rises by 20 K, and .1 kg of water is boiled to steam. How much WORK is done from this heating?

SOLUTION:

$$Q = Cm\Delta T + m_L L + WORK$$

$$C = 4187 \frac{kJ}{kg-K} \text{ (Table } 5-1\text{)}$$

$$Q = 3000 = 4.187 \frac{kJ}{kg-K} \times 10 \text{ kg} \times 20 \text{ K} + .1 \text{ kg} \times 2256 \frac{kJ}{kg} + WORK$$

$$1947 \text{ kJ} = WORK$$

TECHNICIANS IN THE FIELD

NUTRITIONIST

Technicians are found in a wide variety of jobs, many which do not appear to be in their field. Yet the principles of technology do not apply just to gas cylinders and chef's pots. One of the well-respected nutritionists on the campus of Penn State University, College Station, PA, got her beginning in engineering technology and claims that she uses her thermodynamic principles every day.

This nutritionist's days are split between being the trainer for the Penn State athletic teams and consulting to the medical college hospital for special patient diets. One of the principles that gives her a great perspective on body conditioning is that she looks at the human body as a large thermodynamic process. Energy goes into the body in the form of food and is burned there to create heat. In the absence of food, the body may burn stored fat to create heat. The heat energy is used to increase the internal energy of the body when it is cold, or to offset the loss of internal energy when the body is overheated. The fuel also goes to do the WORK the body performs, either walking, running, performing labor, or the simple metabolism of the body; powering the heart and moving the diaphragm; and overcoming heat loss from the skin.

There are special thermodynamic data tables for nutritionists such as Table 5–3. This table lists the rate at which WORK is performed by the body for a variety of activities. This nutritionist finds different terminology in her field than that used by traditional thermodynamicists: heat is given in terms of calories—expressed by a capital C, which is, in fact, a kilocalorie. Her field uses SI units exclusively; therefore, body weight is given in terms of kg, and WORK is in units of joules. With these modifications, she utilizes the first law of thermodynamics:

$$Q = IE + WORK + m_L L$$

With this information, she can give valuable advice.

When wrestlers start the season, they need to get into condition and make weight. The nutritionist advises them to control their weight before training because losing real weight, rather than temporary water weight, can take a long time.

SAMPLE PROBLEM 5–14

Assuming a person went on a total fast (no food at all), how long would it take to lose 5 kg (11 lb) if the person spent half the day playing golf and the other half sleeping (basal metabolism).

TABLE 5–3

ENERGY AND EXERCISE FACTS			
Output		**Input**	
Exercise	**Energy Required (kJ/hr)**	**Food Component**	**Energy Content (kJ/kg)**
Wrestling	3.400	Carbohydrates	4.200
Jogging	2.500	Protein	8.400
Shoveling snow	2.100	Fat	33.100
Volleyball	1.500		
Golf	1.000		
Gardening	.800		
Watching TV	.300		
Basal metabolism	.300		
Playing baseball	.200		

SOLUTION:

We will assume that the body temperature overall does not change ($T = 98.6°F$), therefore $IE = 0$. Also we will assume no sweating, therefore $m_L = 0$.

The amount of WORK done per hour is the average between golf (1.000 kJ/hr) and basal metabolism (.300 kJ/hr), which is .650 kJ/hr, or 15.6 kJ/day.

The Q comes strictly from burning body fat, since there is no food intake and body fat burns at a rate of 33.100 kJ/kg.

The first law of thermodynamics assures that

$$Q = 33.100 \frac{kJ}{kg} \times 5 \text{ kg} = WORK = 15.6 \frac{kJ}{day} \times \text{Number of Days}$$

or

Number of Days = 10.61

No wonder it takes so long to lose weight on a diet!

When the nutritionist is consulting at the hospital, the situation is sometimes turned around. Many times she is called on to plan a diet for a patient who needs to gain weight. There are other times when she is called on to solve a problem that no one else can solve. Recently, for example, a doctor called her about a problem he had during open heart surgery of a child patient. The operation took ten hours, during which time the chest cavity was completely open. After the surgery, the chest tissue remained blue for an extensive length of time, and there was some concern whether some of the tissue might die. After a period of consideration, the consultant believed she had determined the cause. The open chest cavity increased heat loss from the body, causing a significant drop in the internal energy of the body—a drop in body temperature. She figured that if the body stopped processing food under the anesthetic ($Q = 0$) and the BMR (WORK) was increased 200%, due to the increased heat loss, to .900 kJ/hr, the internal energy drop of the body would be

$$0 = IE + .900\frac{kJ}{hr}$$

or

$$IE = Cm\Delta T = -.900\frac{kJ}{hr}$$

For a 200-lb patient (80 kg) with a body specific heat = 4.187 kJ/kg – K (water), this implies a temperature drop of –.900 kJ/hr × 10 hr = 4.187 × 80 × ΔT or ΔT = –.02 K, an insignificant amount. For a 10-lb baby (4 kg), however, this meant a temperature drop of ΔT = –.900 × 10/4.187 × 4 = –.5 K, or about –1°F, a significant body temperature change.

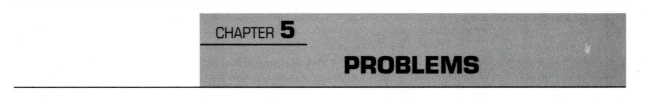

CHAPTER **5**

PROBLEMS

1. The term $m_L L$ in the equation of the first law of thermodynamics refers to
    a. sensible heat
    b. latent heat
    c. internal energy
    d. heat

2. The term "pe" in Chapter 5 refers to
    a. internal energy
    b. potential energy
    c. partial energy
    d. potential equation

3. The symbol ΔT means
    a. $T_2 - T_1$
    b. $/T$
    c. $\dfrac{T_2}{T_1}$
    d. $T_2 + T_1$

4. In the equation

    $$Q = IE + ? + \text{latent heat}$$

    ? refers to:
    a. heat
    b. thermal energy
    c. sensible heat
    d. WORK

5. Below are the specific heats for several materials, a, b, c, and d. Which material best matches the following description: This material shows the effect of heat readily; only a few Btu's are necessary to send its temperatures skyrocketing.
    a. c = .1
    b. c = .5
    c. c = 1.0
    d. c = 10

6. Which material (the specific heat of each is given below) best matches the following description: This material can store large amounts of heat

without showing much increase in temperature, and it takes a long time to heat up and cool down.

a. c = .1
b. c = .5
c. c = 1.0
d. c = 10

7. A picture of a complicated machine is shown in Figure 5–9. Although there are a lot of gears and pulleys, the thing that makes it all work is the substance trapped inside. This material goes through a thermodynamic process to make the gears work. The equation that governs this machine, therefore, is the energy equation. If 1000 Btu's of heat are put into the medium, of which 50 are stored and 350 are used to change the phase, how much WORK will the machine do?

Figure 5–9. A thermodynamic machine.

8. How much energy is stored in 55 gal (459 lb) of water when its temperature is raised from 100°F to 200°F? How much heat is stored when the temperature is raised from 560°R to 660°R?

9. In the days before electric blankets were available, people would place warm rocks under the covers of their beds before they retired. How much heat would a five-pound stone release as its temperature dropped from 85°F to 70°F?

10. A bronze metal annealing plant handles coils of metal 3 ft wide and 1/16 in thick. The bronze is passed over a natural gas flame and heated from 70°F to 600°F. (Figure 5–10) It travels at a rate of .5 ft/sec. The density of bronze is .31 lb_m/in^3.
    a. How many Btu's of heat must be added to one foot of metal?
    b. How many Btu/min must the flame provide?
    c. If natural gas is the fuel and has a heating value of 1050 Btu/cu ft, how many cu ft/hr of fuel is used?

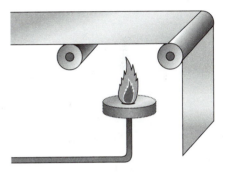

Figure 5–10. Strip metal heating.

11. By how many degrees could you raise one-half pound of water using the same amount of heat necessary to evaporate that same one-half pound?

12. A can of aerosol deodorant contains one pound of a scented liquid that will vaporize and spray out when the top button is pushed. When 1 oz of the substance (latent heat is 68 Btu/lb_m) is vaporized, the energy needed for the change in phase of the substance comes from the stored energy of the remaining deodorant. How many degrees does the temperature of this remaining substance drop when 1 oz is sprayed?

13. From the previous chapters we have learned that "temperature is a measure of the average speed of molecules" and "an indicator of stored heat is temperature." From this it can be deduced that "heat energy is stored in a substance by increasing the speed of the molecules of the substance." True or false?

14. The diagram in Figure 5–11 shows an industrial boiler consisting of a fuel flame, hot water heat exchanger, and flue stack. The first law of thermodynamics states that in any closed hypothetical volume, the "energy in" must equal the "energy out" plus the "energy buildup" in the volume. Use this law to obtain an algebraic relation between
    • Q *added* (heat generated by combustion)
    • Q *flue* (heat going up the chimney)
    • Q *water* (amount of heat the boiler water absorbs)

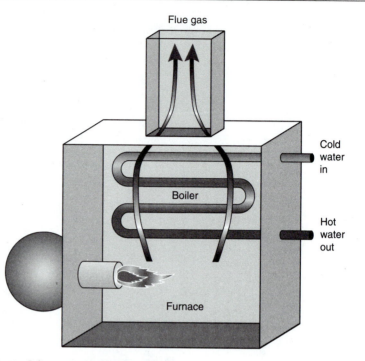

Figure 5–11. Schematic of a gas-fired boiler.

15. In a liquid nitrogen plant, the temperature of air is reduced from atmospheric 70°F to –196°F where it changes from gas to liquid. Assuming that all the air is nitrogen, how much heat must be removed (*Q*) to generate one lb of liquid nitrogen? (No WORK is done. See Tables 5–1 and 5–2.)

16. The LNG (liquid natural gas) tanker Natalia was built in Belgium as the largest tanker of its kind, carrying 65,000 tons of fuel cargo at tempera-

Figure 5–12. LNG tanker.

tures far below zero. (Figure 5–12) To keep both pressures and temperatures down during shipping, about .2% of the fuel is exhausted as a gas at atmospheric pressure and temperature. If the price of natural gas is $4/1000 cu ft, how much money is lost per day?

17. A steam boiler in the mechanical room of Allen County Hospital is rated to deliver 12,000 lbs of steam per hour. The water is supplied to the boiler at 100°F and is supplied to the hospital at 260°F steam.
    a. How many Btu's are required by the burner each hour to generate the 12,000 lbs of steam?
    b. If all the water is placed in the boiler at the beginning of the hour, plot a graph of the water temperature as it is heated throughout the hour.

18. Determine the change in internal energy of 2 lb of oxygen as its temperature changes from 70°F to 90°F.

19. Compute the quantity of heat necessary to raise the temperature of
    a. 10 lb of machine oil from 60°F to 150°F (in Btu)
    b. 5 lb of quartz from 30°C to 75°C (in joules)

20. An electrical power plant is a system governed by the first law of thermodynamics. Enough coal is fed to the plant every hour to generate 1,000,000 Btu of heat. The output from the plant is WORK, equivalent to 800,000 Btu in an hour. How much heat must be rejected from the plant each hour? Why?

21. A pressure cooker set to boil water at 230°F is filled with two cups of water (8 oz cups) and heated from 60°F to boiling, then heated more until half of it has turned to steam. If no WORK is done in the boiler, how many Btu's are required for this to happen?

22. What is the difference between heat and internal energy? Are heat and temperature the same thing? What is the relationship between heat and temperature?

23. Before Imperial Margarine, butter was churned from cream. Cream was loaded into a vessel outfitted with paddle wheels and a hand crank. It took about 45 minutes for a person to turn the crank and churn one pound of butter (from 20 lb of cream). If the churn vessel did not allow any heat to escape, and if a person typically invested one Btu/min of churning, how warm did the cream get during the operation?

24. Consider a coal-fired electrical power plant as a thermodynamic device that burns 4000 lb of coal in one hour (heating value 14,000 Btu/lb).
    a. During the hour, does the internal energy of the plant change? (Does the plant get a lot hotter inside?)

b. Where does the heat go?
c. If 30% of the heat is lost up the chimney, how much WORK is generated?

25. During a process, 200 Btu/lb of heat is added to a fluid, while 25,000 ft-lb/lb of WORK is extracted. If there are 4 lb of fluid, what is the change of internal energy? (778 ft-lb/lb = 1 Btu)

26. Nitrogen enters a compressor at 35°F. The compressor does 5 Btu's of WORK on each pound of the gas as it is compressed.
   a. In the equation

$$Q = Cm\Delta T + m_L L + WORK$$

   is WORK equal to +5 Btu or –5 Btu?
   b. What is m_L?
   c. What is Q?
   d. What is T_2 (temperature after compression)?

27. Air flows across a cooling coil at a rate of 250 cu ft/min while its temperature falls from 80°F to 55°F. Find the change of internal energy per lb_m during this process.

28. An electronic instrument is cooled by air blown through the case. Air is supplied at 60°F and is exhausted at 90°F. Pressure remains essentially atmospheric (14.7 psia). Find:
   a. change in internal energy (per lb_m) of air as it flows through
   b. change in specific volume (%)

Problems in SI Units

29. A healthy male adult has 19% body fat. If Jim weighs 60 kg and has a body fat percentage of 23, how many hours a day would he have to shovel snow to get his body fat down to average, assuming that his daily caloric intake matches his energy output when the shoveling is not considered.

30. During a process, 200.000 kJ/kg is added to a gas while 75.000 kJ/kg of WORK is done by the gas (taken out of the system). If there are two kg of gas, determine the change of internal energy of the gas.

31. In a liquid nitrogen plant, nitrogen gas is cooled from 21°C to –100°C where it changes to a liquid. How much heat must be removed (Q) to generate 5 kg of liquid nitrogen? (WORK = 0. *Latent heat of nitrogen* = 1450 kJ/kg. *Specific heat* = .725 kJ/kg – K.)

CHAPTER

6

Fundamental Processes: Constant Volume

PREVIEW: The energy equation, or first law of thermodynamics, was established using the most simple ideas and equipment, yet it can be used to immediately solve a host of important, meaningful engineering problems.

We stand before three tons of steel coils that have emerged from an industrial oven at 1300°F and are teetering over a giant vat of oil. In a moment the coils will be lowered into the oil to be rapidly cooled below 200°F. Is there enough oil in the tank? We can be confident there is if we have sized it according to the first law of thermodynamics.

OBJECTIVES:

❑ Use the first law of thermodynamics to analyze simple heating processes.
❑ Solve industrial quenching problems.
❑ Adapt the energy equation for situations of heating or cooling liquids in pipes (hydronic heating and cooling).

6.1 A MORE SPECIFIC ENERGY EQUATION

In the last chapter, an energy equation that accounts for all of the Btu's of heat that are added to a substance was derived. For any system, the energy equation is:

$$Q = Cm(T_2 - T_1) + WORK + m_L L$$

Equation 6–1

Q = heat added or removed from the system (substance)
C = specific heat of the substance
m = mass of the substance
$T_2 - T_1$ = temperature increase or decrease
$WORK$ = work done by the substance during the process
m_L = mass of the substance that changes phase during the heat process
L = Latent heat value of the substance

This equation and the perfect gas law are the two basic building blocks of thermodynamics.

Equation 6–1 isn't as specific as it might be, since it doesn't explicitly describe how to calculate the amount of WORK done. The equation has little

111

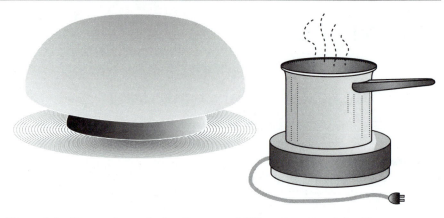

Figure 6–1. Thermodynamic situations are of different sizes and significance.

use until the term *WORK* is defined in terms of measurable quantities such as the state properties of the gas.

To describe how much WORK is done while a system is being heated or cooled is a very difficult task. There are many, many different ways in which a substance may be heated, and the amount of WORK that results depends on how the change in the thermodynamic properties of the system or process takes place.

❖ **KEY TERM:**   **Process:** The method by which a system has its thermodynamic properties changed.

In some processes, a great amount of WORK is done, while in others much heat is stored and very little WORK is done. (Figure 6–1) In Chapters 6 to 9, we will apply the energy equation to four standard processes that cover most of the ways by which a thermodynamic change can come about.

6.2 THE CONSTANT VOLUME PROCESS

The first process that will be investigated is the most fundamental of all. It is the process that occurs so that **the volume occupied by the gas does not change**. Witness what happens in the heating of food in a commonplace pressure cooker. The volume of the pressure cooker does not change even as the steam inside is heated and the pressure builds in an attempt to expand the pressure cooker.

❖ **KEY TERM:**   **Constant Volume Process:** Heating or cooling while keeping the volume of the material fixed.

The energy equation can be modified to apply specifically to constant volume processes by considering an example.

The cooking vessel of Figure 6–2 is one that is rigid, sealed, and supports

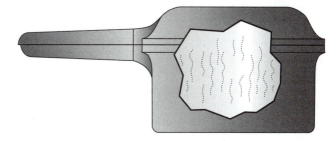

Figure 6–2. Constant volume processes do no WORK.

a constant volume process. Originally it contains 10 lb$_m$ of steam at 260°F and 15 psia. Then it is heated with 1000 Btu's. What are the thermodynamic properties of the steam after the heating process—that is, what is the pressure and temperature inside the vessel?

Equation 6–1 can be applied to this case by considering first the latent heat. Since the water started as steam and ended as steam, no change of phase took place. Therefore, the latent heat is zero. Anytime there is no change in phase, the latent heat is zero.

The next term to attack is *WORK*. How much WORK was done during the heating? To begin answering this question, we must first ask, "What is the nature of WORK?" In physics, it is said that WORK is a force pushing through a distance:

$$WORK = F \times d$$

In order for WORK to be done, not only must there be a force involved, but that force must be pressing against an object and that object **must move**. In the pressure cooker situation, there definitely is a force acting, since the pressure inside the vessel is great. However, nothing actually moves because the pressure cooker holds the steam pressure tight inside the vessel. Therefore, the WORK done is zero. **There is no WORK during a constant volume process.** (Figure 6–2)

Notice that the density of the gas does not change during this process either, since neither the mass nor the volume occupied by the gas changes. Therefore, constant volume problems are constant density problems, too.

In this pressure cooker example, the energy equation reduces simply to:

$$Q = Cm(T_2 - T_1)$$

All of the heat energy that is put into the cooker is transformed into internal energy, showing up as an increase in temperature. This simplified energy equation, along with the perfect gas law, will help solve the original problem. The energy equation can be used to find T_2 as follows: (use Table 5–1)

$$1000 \text{ Btu} = .36 \frac{\text{Btu}}{(\text{lb}_m - °F)} \times 10 \text{ lb}_m \times (T_2 - 720°R)$$

$$T_2 = \frac{1000}{3.6} + 720°R$$

$$T_2 = 998°R = 538°F$$

The perfect gas law now can be employed to find the final pressure. Since P_1, T_1, and T_2 are known, and since $\rho_1 = \rho_2$, the perfect gas law reduces to

$$P_2 = \frac{P_1 T_2}{T_1}$$

$$= 15 \frac{\text{lb}_f}{\text{in}^2} \times 144 \frac{\text{in}^2}{\text{ft}^2} \times \frac{998}{720} = 2994 \text{ psfa} = 20.8 \text{ psia}$$

6.3 APPLICATION: HEATING A ROOM

There are many applications of the constant volume process in nature and in the operation of familiar machines where all the heat energy goes to increasing the temperature of the system (none goes to latent heat or WORK). If we were to devise a machine that would convert heat into temperature at a maximum efficiency, we would do it using the principles of a constant volume process.

In what kind of situation is it important to convert heat into the maximum amount of temperature that it will provide? Immediately you think of the burning of fuel to heat a building for comfort, and a simple heating example points out this application.

Consider the problem of burning one pound of coal in an air-tight room that is $20' \times 50' \times 10'$ high. This room is originally at 70°F. At what temperature will the room be after the burning is completed?

This example clearly involves a thermodynamic process. (Figure 6–3) There is gas (air) that is enclosed in a volume (the room) and heat is added to the gas, causing the gas temperature to increase. This is a very simple thermodynamic process in which the gas temperature increases, the pressure increases, but the volume of gas remains the same.

The heat being added is produced by burning coal. One pound of coal will burn to provide approximately 14,000 Btu of heat, as seen from Table 2–2.

The heat added is $Q = 14,000$ Btu. The specific heat of air is .17 Btu/lb$_m$ – °R (see Table 5–1), and the temperature before burning the coal is 530°R. Therefore, the energy equation is:

$$14,000 = .17 \frac{\text{Btu}}{\text{lb}_m} -°R \times m \times (T_2 - 530°R)$$

Equation 6–2

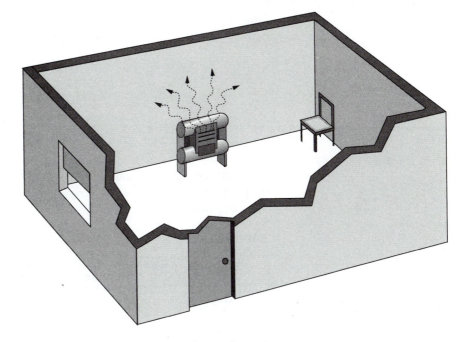

Figure 6–3. Heating a room is a thermodynamic process.

Before calculating the temperature after burning (T_2), the mass (m) of the air in the room must be determined. This is easy, since it is known that the volume of the room is:

$$V = 10{,}000 \text{ cu ft}$$

The weight of air occupied in this volume can be found in two ways. The easiest method is to know that density of room air at standard temperature and pressure is .075 lb_m/ft^3. With the density of the air and the volume of the room both known, the total mass of air in the room is found to be:

$$m = \rho_{air} \times V = .075 \, \frac{\text{lb}_m}{\text{ft}^3} \times 10{,}000 \text{ ft}^3 = 750 \text{ lb}$$

If the value of ρ_{air} is not readily available, it can be calculated using the perfect gas law $\rho_{air} = P/RT$. The pressure of the air is atmospheric pressure of 14.7 psia, or 2116 psfa. The gas constant (R) for air is 53.3 ft/°R (lb$_f$/lb$_m$), and the temperature (T) of the air before burning is 530°R.

$$\rho = \frac{P}{RT} = 2116 \, \frac{\text{lb}_m}{\text{ft}^2} \times \left(53.3 \, \frac{\text{ft}-\text{lb}}{\text{lb}-°\text{R}} \times 530° \text{R} \right) = .075 \, \frac{\text{lb}_m}{\text{ft}^3}$$

Now that the total mass of the air in the room has been found, Equation 6–2 can be used to find the final temperature of the room air.

$$\left(T_2 - 530°R\right) = \frac{14,000 \text{ Btu}}{\left(.17\dfrac{\text{Btu}}{\text{lb}_m} - °R \times 750 \text{ lb}\right)} = 110°R$$

$$T_2 = (110 + 530)°R = 640°R = 180°F$$

This is the constant volume situation: complete conversion of heat into temperature. The specific heats for common gases for the constant volume situation are shown in Table 6–1. This table is an extension of Table 5–1.

Table 6–1

SPECIFIC HEATS (C) OF COMMON GASES (FOR CONSTANT VOLUME SITUATIONS)		
Gas	C (Btu/lb$_m$ – °R)	C (kJ/kg – °K)
Air	.17	.712
Helium	.754	3.157
Hydrogen	2.435	10.195
Nitrogen	.175	.732
Oxygen	.15	.628
Propane	.36	1.507
Steam	.36	1.507

6.4 HEATING WITHOUT FUEL

Heat doesn't always have to come from burning or by using some external source. Consider the problem of a blacksmith who is shaping a horseshoe. (Figure 6–4) The horseshoe is made up of one pound of steel and it is heated to 800°F. In order to cool it after it is completed, the blacksmith thrusts it into a pail containing three pounds of water that is standing at 70°F. What is the final temperature of the horseshoe and the water?

We might ask first whether a thermodynamic process actually occurred. The answer is inevitably yes, since a thermodynamic property of the horseshoe (temperature) is changed. Furthermore, the temperature of the water is changed also. There are actually two thermodynamic processes involved here, the cooling of the horseshoe and the heating of the water in the bucket.

Consider the water in the bucket as a complete and closed system to which heat is being added. How much Q is added by the horseshoe? This is not specified in the problem, but we do know that **the heat added to the water is the same as the heat taken out of the horseshoe**.

To describe the heat addition to the water, which energy equation should be used if it is assumed that no vaporization occurred? Before the horseshoe

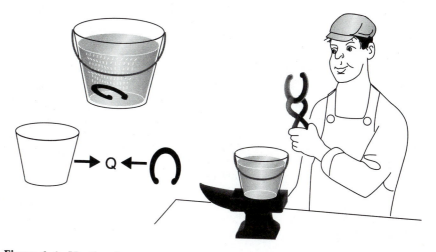

Figure 6–4. Heating does not require a flame or a hot plate.

hits the water, the water had a certain density and volume. After the interaction with the horseshoe, the density of the water will be basically the same as it was initially; therefore, the volume occupied by the water will remain the same. The heating process takes place with constant volume.

The energy equation for the process is:

$$Q_{\text{water}} = C \times 3\,\text{lb}_m \times (T_2 - 530°\text{R})$$

Since it is water whose temperature is being raised, we should use the specific heat of water, which is 1 Btu/°R-lb$_m$. The equation now becomes:

$$Q_{\text{water}} = 1\frac{\text{Btu}}{°\text{F}-\text{lb}_m} \times 3\,\text{lb}_m \times (T_2 - 530°\text{R})$$

Equation 6–3

This equation has two unknowns in it, Q and T_2, and cannot yet provide an answer.

Take notice of the other system being heated, the horseshoe. As the horseshoe loses heat, the density of the steel does not change appreciably. It is true that metal expands when heated, which could change the density, but this is a very minor effect. Since the volume occupied by the horseshoe does not change, the equation that describes the cooling is the constant volume energy equation.

❖ **KEY TERM:**   **Heating a Solid or a Liquid:** No matter how it is done, it is always a constant volume process. Solids and liquids do not expand significantly during heating.

Knowing the specific heat of steel (see Table 5–1), the energy equation for the horseshoe can be written as:

$$Q_{shoe} = .12 \frac{Btu}{lb_m - °F} \times 1\ lb_m \times (T_2 - 1260°R)$$

Equation 6–4

Again this equation has two unknowns and cannot be solved. However, we understand that the heat absorbed by the water is given up by the horseshoe:

$$Q_{water} = -Q_{shoe}$$

The negative sign in this relationship is very important. What one substance is absorbing (+), the other one is giving up (–). The values for Q_{water} and Q_{shoe} from Equations 6–3 and 6–4 can be substituted to give:

$$3(T_2 - 530°R) = -.12(T_2 - 1260°R)$$

Since the water and the horseshoe eventually reach the same temperature, both occurrences of T_2 represent the same temperature.
Solving the equation yields:

$$3\ T_2 = -.12T_2 + 151 + 1590$$

$$3.12\ T_2 = 151 + 1590$$

$$T_2 = \frac{1741°R}{3.12} = 558°R = 98°F$$

This problem of horseshoe and water is just one of many of the same type. Here is another problem that can be solved using two equations and two variables.

SAMPLE PROBLEM 6–1

Consider that there is a pan containing some water and several silver balls. Specifically, there is a half pound of water and one-half pound of silver balls. When we heat this combination with ten Btu's, what will be the resultant temperature of each of the materials?

SOLUTION:

Each substance has its own energy equation:

$$Q_{silver} = C_{silver}m_{silver}(T_{2silver} - T_{1silver})$$
$$Q_{water} = C_{water} \times m_{water} \times (T_{2water} - T_{1water})$$

Notice the $T_{2\text{silver}}$ and $T_{2\text{water}}$ in each equation are the same, since both substances are in contact with each other and achieve the same temperature. Therefore, we can just call them T_2. The same is true about $T_{1\text{silver}}$ and $T_{1\text{water}}$, and we can call them simply T_1. Furthermore, the total heat put into the pan must equal the total heat absorbed by each substance.

$$Q_{\text{total}} = Q_{\text{silver}} + Q_{\text{water}}$$

The energy equation for silver is:

$$Q_{\text{silver}} = .056 \times \frac{1}{2} \times (T_2 - 530°\text{R}) \text{ (see Table 5–1 for specific heat of silver)}$$

The energy equation for water is:

$$Q_{\text{water}} = 1.0 \times \frac{1}{2} \times (T_2 - 530°\text{R})$$

Now, since the total heat added is 10 Btu:

$$Q_{\text{total}} = 10 \text{ Btu} = .056 \times \frac{1}{2} \times (T_2 - 530°\text{R}) + 1 \times \frac{1}{2} \times (T_2 - 530°\text{R})$$

$$10 = .528 \, (T_2 - 530°\text{R}) \qquad T_2 = 548.9°\text{R} = 88.9°\text{F}$$

6.5 APPLICATION: QUENCHING PROCESSES

Situations similar to the horseshoe and water bucket are common in engineering and technology; where two materials at different temperatures are brought into close contact so that they transfer their heat to arrive at one common temperature. They are referred to as quenching processes.

❖ **KEY TERM:**

Quenching Process: When two materials are brought into close contact, and the heat from one is given up (Q_1) and absorbed by the other (Q_2) so that their temperatures become the same ($Q_1 = -Q_2$).

Quenching is a popular method of treating steel. Steel is not a basic metal (compound) but rather an alloy of iron and carbon. The percentages of the components and the way they are combined determine whether the steel is very soft, so that it is readily formed into products such as automobile fenders, or hard and tough, such as that used for gears or bulldozer blades. (Figure 6–5) Carbon and iron are combined by melting each and mixing them. How their molecules align themselves in the resulting alloy depends on how fast they were melted, at what temperature they are combined, and how fast the alloy is cooled. If the alloy is cooled very rapidly, the process is called quenching, and the resultant steel is tough, hard, strong, and water-resistant.

There is a variety of quenching methods, depending on the desired characteristics of steel. Besides direct quenching, which is the type described with the water bucket and horseshoe, there is the "time quench," where quenching occurs at changing temperatures in a slow cooling cycle so that the cooling rate changes as different components of the steel structure are solidi-

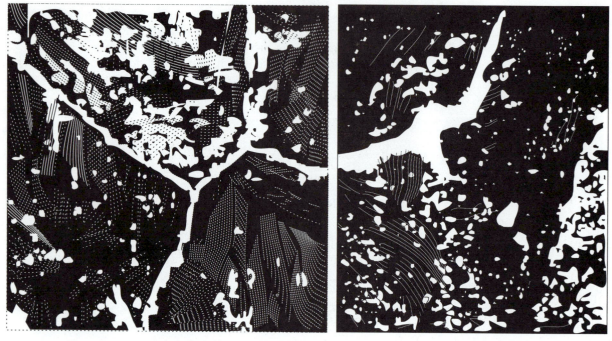

1.2% C   0.7% C

Figure 6–5. Microstructure of steel for different percentages of carbon and different mixing methods.

fying. "Selective quenching" is done on a finished part, where only a portion of the object is subjected to the severe cooling. Water is often used as the quenching medium, although it has a tendency to flash to steam near the part being quenched, sometimes causing undesirable results. To eliminate this problem, oil may be substituted as the quenchant.

SAMPLE PROBLEM 6–2

Oil is used as the quenchant in a continuous system where steel parts are fed to a quenching tank on a conveyor belt. (Figure 6–6) One hundred gallons of oil are maintained at a quenching temperature of 300°F, while 200 lb/hr of 1000°F parts are fed through. The quenching temperature of oil is kept constant by taking off 300°F oil and supplying 100°F oil from an outside air cooler. At what rate (gal/hr) must oil be fed? Assume no quenchant evaporation. Oil has a specific heat of .40 Btu/lb$_m$ – °F and weighs 7.6 lb$_m$/gal.

SOLUTION:

Consider a one-hour time period:

$$Q_{\text{steel}} = C_{\text{steel}} \, m \, (300° - 1000°)$$

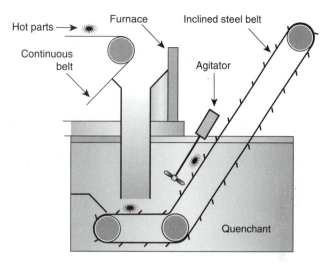

Figure 6–6. An industrial quenching process.

$$= .12 \times 200 \times (-700°) = -16,800 \frac{Btu}{hr}$$

$$Q_{quenchant} = .40 \times m_{oil} \times (300° - 100°)$$

$$= 80 \times m_{oil}$$

Now apply the "quenching rule" that

$$Q_{quenchant} = -Q_{steel}$$

$$80 \times m_{oil} = -(-16800)$$

$$m_{oil} = 210 \text{ lb}_m$$

Since this answer in fact represents the flow in one hour, this oil flow rate of lb_m/hr converts into gallons fed per hour by utilizing the conversion of 7.6 lb_m/gal for oil:

$$m_{oil} = 210 \frac{\text{lb}_m}{hr} \times \frac{1}{7.6 \frac{gal}{\text{lb}_m}} = 27.6 \frac{gal}{hr}$$

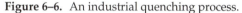

6.6 APPLICATION: HEATING WATER AS IT FLOWS

In the beaker-and-hot plate example in Chapter 4 and the horseshoe-and-pail example in this chapter, a specific amount of water was "captured" and heated. A different situation involves water that flows past a heat source and is heated continuously as it flows by.

For such a situation, the equation $Q = Cm\Delta T$ needs some specific interpretation. In particular, what is m if the water is flowing? We cannot say that one pound or three pounds of water are being heated. Instead, we say that a specific amount of water is being heated every minute, or every hour. We must

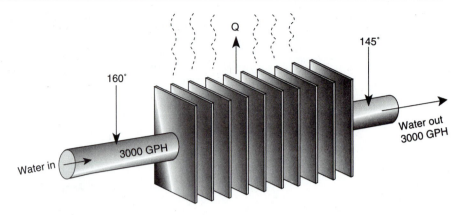

Figure 6–7. Heat given off in a radiator cools the water as it flows.

interpret the m in the heat equation as flow rate (lb$_m$/hr), and when the units are on a per time basis, we shall call the flow rate by the symbol $\dot{m}$. Then the term $\dot{Q}$ will be heat added or lost *per hour* and $\dot{Q} = Cm\Delta T$. Below is a sample problem illustrating this concept.

SAMPLE PROBLEM 6–3

A baseboard hot water radiator gives off heat to keep a room warm. If 3000 gallons per hour of hot water at 160°F are supplied to the radiator unit and 145°F water is returned from the unit (Figure 6–7),

a. how many lb$_m$/hr of water is flowing, and how many Btu/hr are given off by the radiator?

b. how many pounds of water flow per minute, and how many Btu's are given off per minute?

SOLUTION:

a. $\dot{m} = \text{Flow rate} = 3000\,\dfrac{\text{gal}}{\text{hr}} \times 8.3\,\dfrac{\text{lb}_m}{\text{gal}} = 24{,}900\,\dfrac{\text{lb}_m}{\text{hr}}$

$\dot{Q} = Cm\Delta T$

$= 1.0\,\dfrac{\text{Btu}}{\text{lb}-°\text{F}} \times 24{,}900\,\dfrac{\text{lb}_m}{\text{hr}} \times (605°\,\text{R} - 620°\,\text{R})$

$= -373{,}500\,\dfrac{\text{Btu}}{\text{hr}}$

b. $\dot{m} = 24{,}900\,\dfrac{\text{lb}_m}{\text{hr}} \times \dfrac{1}{60}\,\dfrac{\text{hr}}{\text{min}} = 415\,\dfrac{\text{lb}_m}{\text{min}}$

$\dot{Q} = Cm\Delta T$

$= 1.0\,\dfrac{\text{Btu}}{\text{lb}_m - °\text{F}} \times 415\,\dfrac{\text{lb}_m}{\text{min}} \times -15°\,\text{R}$

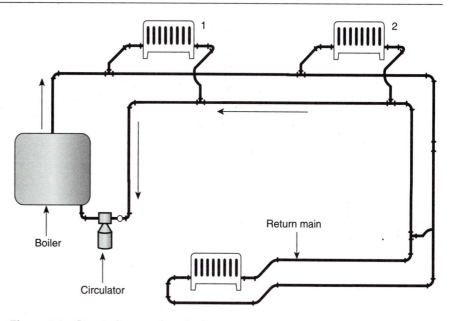

Figure 6–8. Circuit diagram for a hydronic system.

$$= -6225 \frac{\text{Btu}}{\text{min}}$$

As Sample Problem 6–3 suggests, heating rooms with hot water heat is a popular technique: it is called *hydronic heat*, since the word *hydronic* means "pertaining to water." Figure 6–8 illustrates a complete hydronic system, including boiler, water pump or circulator, piping, and terminal devices such as radiators, baseboard convectors, or unit heaters. Notice that the terminal devices are connected one after another in straightforward fashion. They are connected "in series."

Each terminal unit must be sized to give off the correct amount of heat for the room it is in. This depends on the water temperature in the device and its size. Baseboard radiators come in all lengths. Table 6–2 indicates the amount of heat given off per foot of convector for a variety of water temperatures.

Table 6–2

HEAT RADIATED FROM BASEBOARD RADIATORS (PER FOOT)								
Water Flow Rate (lb/hr)	**Heat Released (Btu/hr per lineal foot)**							
	140°F 150°F 160°F 170°F 180°F 190°F 200°F 210°F 220°F							
2000	335 400 460 530 590 670 730 800 870							
1000	310 370 430 500 560 630 690 760 820							

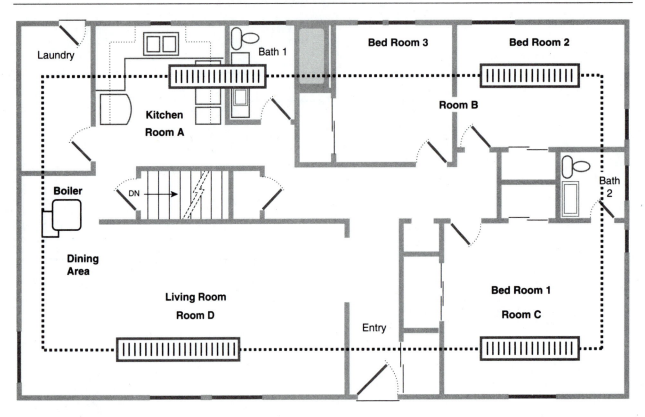

Figure 6–9. Floor plan.

Sample Problem 6–4 shows a typical design problem that can be solved using the first law of thermodynamics.

Figure 6–9 shows a building with four rooms that are to be heated using a baseboard hydronic system with a terminal unit (convector) in each room as shown. The heating required room-by-room is given below

     Room A   5000 Btu/hr
     Room B   4000 Btu/hr
     Room C   6000 Btu/hr
     Room D   3000 Btu/hr

The boiler provides 180°F water, and it is desired to have an overall temperature drop of water from boiler outlet to boiler return of 20°F. Find:

a.   the lb_m/hr and gal/hr of water circulating
b.   the temperature of water entering each radiator

c. the length of each radiator unit needed to provide the necessary heating for each room

SOLUTION:

a. $\dot{Q}$ (given off by water) = 5000 + 4000 + 6000 + 3000

$$= 18{,}000 \frac{\text{Btu}}{\text{hr}}$$

$$\dot{Q} = C\dot{m}\Delta T$$

$$C = 1 \frac{\text{Btu}}{\text{lb}_\text{m} - {}^\circ\text{F}} \text{ (water)}$$

$$\Delta T = 20{}^\circ\text{F} = 20{}^\circ\text{R}$$

$$18{,}000 \frac{\text{Btu}}{\text{hr}} = 1 \times \dot{m} \times 20{}^\circ\text{R}$$

$$\dot{m} = 900 \frac{\text{lb}_\text{m}}{\text{hr}}$$

$$or \quad \frac{900 \text{ lb}_\text{m}/\text{hr}}{8.3 \text{ lb}_\text{m}/\text{gal}} = 108.5 \frac{\text{gal}}{\text{hr}}$$

b. The supply water temperature to room A is 180°F. The temperature drop is

$$\dot{Q}_{\text{water in rad A}} = -5000 \frac{\text{Btu}}{\text{hr}} = C\dot{m}\Delta T = 1 \times 900 \times \Delta T \quad \text{(heat is given off}$$
by the water)

$$\Delta T = \frac{-5000}{900} = -5.6{}^\circ\text{F}$$

Temperature going into B is 180 – 5.6 = 174.4°F

$$\dot{Q}_{\text{water in rad B}} = -4000 \frac{\text{Btu}}{\text{hr}} = C\dot{m}\Delta T = 1 \times 900 \times \Delta T$$

$$\Delta T = \frac{-4000}{900} = -4.4{}^\circ\text{F}$$

Temperature going into C is 174.4 – 4.4 = 170°F

$$\dot{Q}_{\text{water in rad C}} = -6000 \frac{\text{Btu}}{\text{hr}} = C\dot{m}\Delta T = 1 \times 900 \times \Delta T$$

$$\Delta T = \frac{-6000}{900} = -6.6{}^\circ\text{F}$$

Temperature going into radiator D is 163.4°F

$$\dot{Q}_{\text{water in rad D}} = -3000 \frac{\text{Btu}}{\text{hr}} = C\dot{m}\Delta T = 1 \times 900 \times \Delta T$$

$$\Delta T = \frac{-3000}{900} = -3.4°\text{F}$$

Temperature returning to boiler is 160°F.

c. The length of baseboard radiation needed to satisfy each room is found with the help of Table 6–2. First we must find the average temperature of hot water in the radiator in the particular room. For example, in room A, the water in is 180°F and the water out is 174.4°F for an average of 177.2°F. Interpolating from Table 6–2, this implies the heat given off is 543 Btu/1 hr-ft. To satisfy the load of 5000 Btu, this requires:

$$Length_A = \frac{5000 \text{ Btu/hr}}{543 \text{ Btu/hr} - \text{ft}} = 9.21 \text{ ft}$$

In rooms B, C, and D:

$T_{Bavg} = 172.2°\text{F}$

$$Heat\ Radiated = 510\frac{\text{Btu}}{\text{hr} - \text{ft}}$$

$$Length_B = \frac{4000 \text{ Btu/hr}}{510 \text{ Btu/hr} - \text{ft}} = 7.84 \text{ ft}$$

$T_{Cavg} = 166.7°\text{F}$

$$Heat\ Radiated = 476\frac{\text{Btu}}{\text{hr} - \text{ft}}$$

$$Length_C = \frac{6000 \text{ Btu/hr}}{476 \text{ Btu/hr} - \text{ft}} = 12.6 \text{ ft}$$

$T_{Davg} = 161.7°\text{F}$

$$Heat\ Radiated = 436\frac{\text{Btu}}{\text{hr} - \text{ft}}$$

$$Length_D = \frac{3000 \text{ Btu/hr}}{436 \text{ Btu/hr} - \text{ft}} = 6.88 \text{ ft}$$

TECHNICIANS IN THE FIELD

SOLAR HEATING SYSTEM DESIGNER

There are not many companies in southern Ohio that will build a system that guarantees year-round solar heating, enough for the complete comfort heating of a house, but Solar Technologies, Inc. (STI), located in Rising Sun, Ohio (where else?) is one such company. A small company, STI has only four installers and one engineering technician who serves as designer, purchasing agent, salesperson, and owner. Seven years ago she was working in a manufacturing plant and building a home that she designed to be completely solar heated. Once she had moved in, gotten some experience operating her system, and made modifications based on that experience, she decided to start designing systems for other people.

This engineering technician's concept was a good one. She wanted to pattern all her designs after her successful initial one. She wanted to specify the same pumps, the same collectors, the same controls. (Figure 6–10) In this way, she would be using only components that she was confident in, and she would be minimizing the number of components she would have to stock as replacement parts to service the systems she installed.

To do this, though, she had to have the best components right from the start. In the case of the solar collectors, there were many options and many

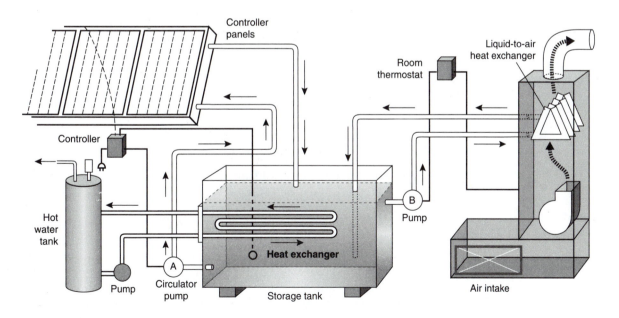

Figure 6–10. Complete solar heating system with storage.

price ranges. Some had thick insulation around the base to hold the heat in once it was captured. Some had none. Some had double glass, some single. She picked three models to test herself, outside in the dead of winter. The test she performed was to determine the efficiency of the panels: how much of the input solar radiation did the panel collect and transfer to the water flowing through? She outfitted her test panel setup with water flowmeters and electric thermometers, and put each model to the test.

SAMPLE PROBLEM 6–5

At the time of the panel testing, the local university reported that the solar incidence was 180 Btu/hr/ft². The first panel measured 36 in × 8 ft and was tested with 1.5 gal/min of water flowing through. The temperature rise of the water was 22°F. What was the efficiency of the panel that day?

SOLUTION:

The amount of input radiation into the panel can be calculated from the solar incidence:

$$\dot{Q}_{in} = 180\frac{Btu/hr}{ft^2} \times \frac{36}{12} \times 8 = 4320\frac{Btu}{hr}$$

The amount of heat taken out of the panel by the water is given by the energy equation:

$$\dot{Q}_{out} = C_{water} \times \dot{m} \times \Delta T = 1\frac{Btu}{lb-°F} \times .3\frac{gal}{min} \times 8.47\frac{lb}{gal} \times 22°F$$

$$= 55.9\frac{Btu}{min} = 3354\frac{Btu}{hr}$$

This yields an efficiency of $\eta = 3354/4320 = .76 = 76\%$

Another part of the design the designer wanted to firm up early on was the type of heat storage she would incorporate. Solar energy is collected during the day, but most of it is used at night. The designer would transfer the heat from the solar panel to a material in a storage bin in the basement of the house, then transfer the heat out at night. What type of material would store heat best? She looked at Table 5–1 to get a clue. The material with the highest specific heat would store the most heat for every degree of temperature rise in the material. This was water. But specific heat is based on a per pound basis, and the designer wanted to minimize the space necessary to store the heat, so she wanted the best material based on a cubic-foot basis. Therefore, she multiplied the specific heat times the density of the material. She found two that were the highest: granite rock (ρ = 180 lb/ft³) and water (ρ = 62.4 lb/ft³). Performing the calculations she found

$$\text{for rock: } C \times \rho = 180 \times .195 = 35 \text{ Btu}/\text{ft}^3 - {}^\circ\text{F}$$

$$\text{for water: } C \times \rho = 62.4 \times 1 = 62.4 \text{ Btu}/\text{ft}^3 - {}^\circ\text{F}$$

Therefore, on either basis, water was found to be the best thermal storage medium. All her designs are based on a basement thermal storage hot water tank.

❖ **KEY TERM:**   **Solar Technologies, Inc.:** An example of the many small businesses that are owned and operated by engineering technicians, often based on designs that they have created.

APPLICATIONS OF CALCULUS

We have been very successful in deriving and using equations of thermodynamics to solve significant and important problems. The mathematics we have used is algebra. The next level of mathematics is calculus. What can calculus do to increase our ability to solve thermodynamic problems?

In Section 6–2, we added 1000 Btu's of heat to 10 lb of steam and predicted that the final temperature would rise by 278°F. We did this based on a specific heat of .36 Btu/lb$_m$ – °F. But what if the specific heat of steam is changing during the process? What if the specific heat of the substance is not a constant but changes according to the temperature of the substance? Table 6–1 presents a value of .36 for the specific heat of steam, but in fact this is for 70°F. At 500°F, its specific heat is .385. The specific heat of steam is not constant but, instead, is a function of temperature; we can describe it as $c(T)$. If we must, we can disregard this deviation in specific heat and approximate it as constant, but calculus allows us to solve problems in which the specific heat of the material changes during the process.

It is not possible to write the constant volume energy equation as $Q = c(T)m(T_2 - T_1)$ because this equation gives us no idea of how to find a value for the function $c(T)$, since the T is everchanging from T_1 to T_2. If we stop the process somewhere in the middle of the heating, at a temperature designated as T, and examine what happens during the next small amount of heating, then the energy equation makes more sense. Instead of calling the next small amount of heat added Q, let's call it dQ, where the d means an intermediate and small amount. To further describe dQ, we can call it a differential amount of heat added. The result of this differential amount of heat is a differential amount of temperature change; call it dT. Now it is fair to write the constant volume energy equation as $dQ = c(T)mdT$. The equation implies that each differential amount of heat input causes a different amount of differential change in temperature, based on the changing value of $c(T)$.

The beauty of calculus is that it provides a formal manner by which to total all the differential amounts of heat additions to get the total heat addition. This process is integration—that is,

$$Q = \int dQ$$

Similarly, integrating both sides of the differential form of the energy equation above gives

$$Q = \int dQ = \int c(T)mdT = m\int c(T)dT$$

To try out the differential form above, take the case that c is a constant. Then

$$Q = mc\int dT = mc(T_2 - T_1)$$

The specific heat of steam changes slightly in the range of 70-400°F. It can be best described by the function $c_{steam}(T) = .36 + .000045T$. If we solved the heating problem of Section 6–2 using this varying specific heat, we would find

$$Q = 1000 = m\int\left(.36 + 4.5\times10^{-5}T\right)dT$$

$$Q = 10\left[.36(T_2 - 260) + \frac{4.45}{2}\times10^{-5}\left(T_2^2 - 260^2\right)\right]$$

simplifying

$$100 = .36T_2 + 2.25\times10^{-5}T_2^2 - 95.1$$

using the quadratic formula for T_2 yields

$$T_2 = 524.5°F$$

which is a 13.5 °F difference from the result found at the end of Section 6–2.

SAMPLE PROBLEM 6–7

Researchers have found that the specific heat of CO_2 varies greatly over the range of 0°F to 700°F, but that this change can be predicted by the function

$$c_{CO_2}(T) = .15 + .000109T - 2.5\times10^{-8}T^2$$

Find the amount of heat necessary to raise two-lb_m CO_2 from 70°F to 500°F.

SOLUTION:

$$Q = m\int c(T)dT = 2\,lb_m\left(.15(T_2 - T_1) + .000054\left(T_2^2 - T_1^2\right) - .8\times10^{-8}\left(T_2^3 - T_1^3\right)\right)$$

$$= 153\ Btu$$

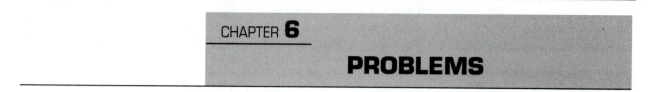

CHAPTER **6**

PROBLEMS

1. The formula for *WORK* in a constant volume process is:
   a. $P\dfrac{(V_2 - V_1)}{J}$
   b. 0
   c. ρRT
   d. $Cm\Delta T$

2. The equation that describes how heat is interchanged between two substances (1 and 2) that are initially at different temperatures but come into direct contact with each other is:
   a. $Q_1 = -Q_2$
   b. $Q_T = Q_1 + Q_2$
   c. $Q_1 = Q_2$

3. When a liquid or solid is heated, what fundamental process is always involved?
   a. constant V
   b. constant P
   c. constant T

4. When two materials are brought into contact so that heat given off from one is absorbed by the other, the process is called:
   a. heat exchange
   b. heating
   c. material cooling
   d. quenching

5. Heat has rapidly become a precious commodity. Someday you may check into a hotel room and be required to purchase heat by the Btu to warm the room. By how many degrees will you be able to heat a room with dimensions 12ft × 20ft × 8ft with 1000 Btu's?

6. What is the final temperature of the air in a 24ft × 12ft × 8ft insulated room if 100 lb of 500°F rock (granite) is placed inside? Originally the temperature was 70°F.

7. If a .01-lb$_m$ sample of a black, solid fuel is burned inside a sealed cannister that also contains 8 lb$_m$ of water, what is the heating value of the fuel

(Btu/lb$_m$) if the temperature of the water is raised from 60°F to 73°F during the burning?

8. One pound of silver at 500°F is plunged into four pounds of water at 60°F in a manner such that no water is evaporated. Calculate the final temperature of the silver and water.

9. An industrial quenching process sends three tons of hardened steel (C = 0.1 Btu/lb –°F) at 1500°F into a quenching tank of 4000 lb of 60°F water. What is the temperature of the steel when it comes out of the tank, and how much heat has it lost?

10. Oxygen at a pressure, temperature, and volume of 40 psia, 80°F, and 20 ft^3 undergoes a constant volume process to a pressure of 60 psia. Determine T_2, WORK done, and heat added during this process.

11. A rigid tank contains oxygen (O_2) gas at 30 psia and 60°F. Two pounds of O_2 are added and the contents are at 40 psia and 80°F. What is the volume of the tank?

12. A three-pound block of an unknown material at 300°F is dropped into a 10-ft^3 chamber filled with air at 2116 psfa and 70°F. The chamber is then sealed. If the temperature of the air rises to 260°F when it is in equilibrium with the block, what is the specific heat of the block material?

13. A roll of silver metal is quenched in a water bath from 1000°F to 200°F. (Figure 6–11) The quenching tank contained 80 gallons of water initially

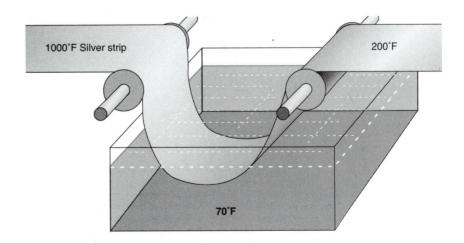

Figure 6–11. Continuous roll quench tank.

at 70°F. If the metal flows through at a rate of 2000 lb$_m$ every hour, how long will it take the water to heat up to 200°F? (Assume no evaporation.)

14. A refrigerated warehouse for produce receives a full stock of 300 lb of beets (C = 1.1 Btu/lb$_m$ – °R), 500 lb of spinach (C = .9 Btu/lb$_m$ – °R), and 1000 lb of cucumbers (C = .98 Btu/lb$_m$ – °R) every morning at 8:00. Typically the produce arrives at 80°F and must be refrigerated to 50°F by noon. How many Btu's must be removed each hour to achieve this temperature?

15. A refrigeration system that is rated to remove 6000 Btu/hr is used to cool a 55-gal water tank filled with water. By how many degrees will the unit be able to cool the water in one hour?

16. The drawing of a "Passive Solar House" is shown in Figure 6–12. Sun radiation comes through the overhang window and is captured in the sun room. In order to store the heat of the day, a cement (ρ = 50 lb$_m$/ft^3) wall painted black is included as part of the room. On a good day, radiation will strike the 14ft × 20ft wall for 8 hours at a rate of 50 Btu/hr-ft^3.
   a. If the wall is 4 in thick, how warm will it get by the end of the day?
   b. If the wall is 8 in thick, how warm will it get by the end of the day?

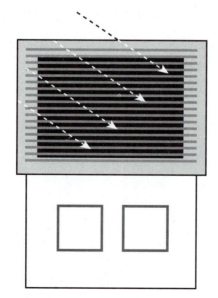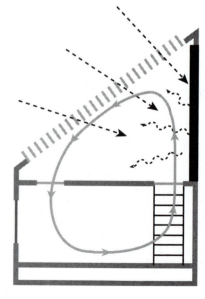

Figure 6–12. Rooftop solar panel.

17. Molten bronze and silver are mixed in a vat to make an alloy used for high-quality, low-resistance wire. If the vat contains 250 lb of silver and 750 lb of bronze at 1000°F, what will the temperature be after 1000 Btu's are added?

18. Air confined in a tank has an initial temperature of 170°F and an initial pressure of 60 psia. If the air is cooled until the pressure falls to 32 psia, what is the final temperature of the gas in degrees Fahrenheit? How much heat is removed per pound of air?

19. A tank in a manufacturing plant is filled with compressed air to a pressure of 120 psig and a temperature of 120°F. If the tank is left standing for several days, what will the pressure in the tank be? The temperature in the factory is 70°F.

20. Figure 6–13 shows a solar collector with 25 ft^2 of surface area that receives approximately 40 Btu/ft^2 per minute from the sun, but loses 25% of that amount to the surroundings. Water enters the top of the collector at 130°F and leaves the bottom at 160°F. What is the flow rate of water through the panel (in lb$_m$/min)?

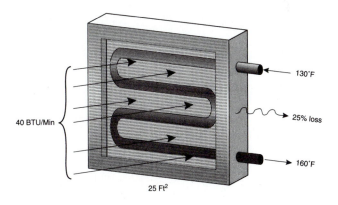

Figure 6–13. Solar collector.

21. If the capacity (Btu/hr output) of a baseboard radiator should be 10,000 Btu/hr, with a temperature drop of 12°F,
   a. what should the water flow rate be?
   b. what is the proper length of radiator (T_{in} = 175°F)?

Problems in SI Units

22. A refrigeration system that is rated to remove 6000 kJ/hr is used to cool a 2-m^3 water tank filled with water. By how many °C will the unit be able to cool the water in one hour?

23. Two kg of silver at 250°C is plunged into eight kg of water at 15°F in a manner such that no water is evaporated. Calculate the final temperature of the silver and water.

CHAPTER
7

Fundamental Processes: Constant Pressure

PREVIEW: This chapter describes a unique elevator that uses no pulleys, motors, or chains to make it move—only the principles of thermodynamics. This thermodynamic process is not a constant volume process, which means another special case of the energy equation must be described. Along the way, the concept of specific heat takes on two meanings.

OBJECTIVES:
- ❏ Solve problems for situations in which the volume of the gas changes during the process.
- ❏ Calculate the WORK done during a process and understand the units.
- ❏ Introduce a new type of specific heat and discuss its relationship to the specific heat previously defined.

7.1 THE CONSTANT PRESSURE PROCESS

All processes do not necessarily take place without changing the size of the system itself. Heating air in a balloon is a good example of a process during which the vessel does not maintain a constant size. Figure 7–1 illustrates a device that is designed so that the volume occupied by the gas may change. It looks similar to the pressure cooker of the previous chapter except that the lid slips into the cooker and supports a heavy weight on top. The lid is free to move much like a piston. As it moves, the volume occupied by the "captured"

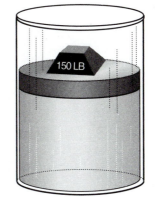

Figure 7–1. A vessel with free-floating piston.

substance (gas) changes considerably. This piston is called *free floating* since it is suspended in space by the very gas that it is confining. The pressure of the gas is sufficient to support the weight of the piston. If the weight of the piston is *W*, then the amount of pressure required to just support it is given by the equation:

$$P = \frac{W}{A}$$

where *A* is the cross-sectional area of the piston.

In this type of device, the piston will fall if the pressure in the gas is not sufficient to support it. As the piston falls, the gas is compressed until the pressure rises to the supporting level, $P = W/A$.

Similarly, if the pressure in the gas is greater than the supporting level, the piston will be pushed upward. As it goes up, the gas rarifies and the pressure drops until it once again reaches the supporting level. (Figure 7–2)

Therefore, the pressure of the gas in this vessel is determined solely by the weight of the piston and the area of the piston. The piston will always arrange itself so that $P = W/A$. This is a most important point and is critical to the understanding of the usefulness of this device. No matter what gas is used, no matter what the temperature of the gas, when it is placed inside this vessel and the lid is placed on and allowed to fall or rise, the pressure of the gas will always be the same. Therefore, the device can be called a *constant pressure vessel*.

❖ **KEY TERM:**   **Constant Pressure Vessel:** A canister fitted with a floating piston so that the pressure of the gas inside is determined by the area of the piston and its weight, and does not depend on any properties of the gas.

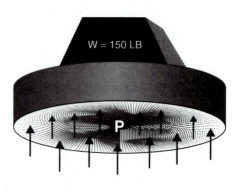

W = P*A

Figure 7–2. Upward-pushing forces equal downward forces.

SAMPLE PROBLEM 7–1

A constant pressure vessel contains acetylene gas. A 10-lb piston with an area of 10 sq in is placed over the gas and falls until the pressure of the acetylene is sufficient to support the weight. What pressure is this?

SOLUTION:

$$P = \frac{W}{A} = \frac{10 \text{ lb}_f}{10 \text{ in}^2} = 1 \text{ psi}$$

SAMPLE PROBLEM 7–2

What would be the supporting pressure for Sample Problem 7–1 if the gas in the vessel were steam?

SOLUTION:

$$P = \frac{W}{A} = \frac{10 \text{ lb}_f}{10 \text{ in}^2} = 1 \text{ psi}$$

It doesn't matter what gas is put into the cylinder, the area and weight of the piston dictate the gas pressure under it.

SAMPLE PROBLEM 7–3

What would be the supporting pressure for Sample Problem 7–1 if the piston surface were twice as large?

SOLUTION:

$$P = \frac{W}{A} = \frac{10 \text{ lb}_f}{20 \text{ in}^2} = \frac{1}{2} \text{psi}$$

The only ways to change the pressure inside the vessel are to change the weight of the piston or its area.

In the vessel described above, the pressure remains the same even if the gas is heated or some other thermodynamic changes take place. While the process is occurring, the gas is attempting to change its pressure, but the piston moves instead to maintain the pressure constant. This implies that the volume occupied by the gas changes during a thermodynamic process.

Since the volume of the gas in this vessel changes during the thermodynamic process, what is the resulting energy equation for this constant pressure process? The energy equation is the same as that developed in Chapter 5, which can be made specific for a constant pressure process.

First, it safely can be said that if the process starts with a gas and ends completely with a gas, the latent heat expended or absorbed during the process is zero. Therefore, we do not consider latent heat at this time.

Recall from Chapter 6 that WORK will be done on an object if there is a force on it, and if it indeed moves. For the constant pressure situation, there is a force due to the pressure of the enclosed gas, $F = P \times A$, and there is the motion of the piston. These two facts result in the conclusion that WORK is being done. The equation for WORK is:

$$WORK = F \times d = P \times A \times d$$

where d is the distance the piston moved. The product of $A \times d$ is the formula for the volume of the cylinder created by the moving piston. (Figure 7–3) This volume is also the additional volume that the gas will occupy as it is heated.

If the initial volume that the gas occupies is V_1 and the final volume that the gas occupies is V_2 then:

$$V_2 - V_1 = A \times d$$

The WORK done by the gas, therefore, can be related to the properties of the gas before and after the process by the equation:

$$WORK = P(V_2 - V_1)$$

The units of this equation are found to be:

$$(WORK) = \frac{lb_f}{ft^2} \times ft^3 = ft - lb_f$$

$$(WORK) = \frac{N}{m^2} \times m^3 = N - m \text{ (in SI)}$$

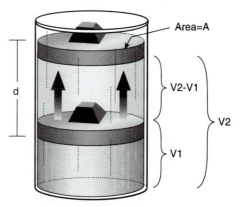

Figure 7–3. WORK means a change in volume.

It was previously established that a part of the heat can be converted into WORK, and vice-versa. But heat has units of Btu's and WORK has units of ft-lb$_f$, so a conversion factor from Btu's to ft-lb$_f$ is needed, and similarly from joules to N-m. The conversion factors are:

$$778 \text{ ft-lb}_f = 1 \text{ Btu}$$

$$1 \text{ N-m} = 1 \text{ joule}$$

These conversion factors are used so often that they are given a letter designation: J = 778 ft-lb/Btu; J = 1 N-m/joule.

If the equation derived in the previous paragraph were written in terms of Btu or joules, it would read:

$$WORK = \frac{P(V_2 - V_1)}{J}$$

Summarizing, the equation for energy change during a constant pressure process can be written as:

$$Q = Cm(T_2 - T_1) + \frac{P(V_2 - V_1)}{J}$$

Equation 7–1

SAMPLE PROBLEM 7–4

In Section 6–2, a problem was solved to find the temperature increase in a constant volume vessel filled with steam when 1000 Btu's of heat were added. Consider the same situation as that problem but find the temperature increase if the heating took place in a constant pressure vessel. (Figure 7–4) There are 10 lb$_m$ of steam in the vessel initially at 260°F and under 15 psia pressure.

SOLUTION:

Known conditions:

P_1 = 15 psia = 2160 psfa = P_2 (constant pressure process)

T_1 = 260°F = 720°R

$m_1 = m_2 = 10$ lb$_m$

Q = 1000 Btu

V_1 can be calculated from the perfect gas law for state 1:

$P_1 V_1 = m_1 R T_1$

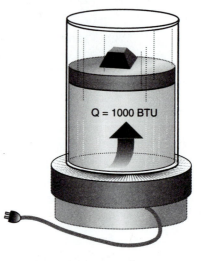

Figure 7–4. Heating in a constant pressure vessel.

or

$$V_1 = 10 \text{ lb}_m \times 85.6 \frac{\text{ft--lb}_f}{\text{lb}_m} - °R \times \frac{720° R}{2160 \text{ lb}/\text{ft}^2} = 285 \text{ ft}^3$$

To find T_2, consider the perfect gas law (at this temperature steam can be considered to be a perfect gas) (see Chapter 17):

$$P_2 V_2 = m_2 R T_2$$

or

$$\frac{P_2 V_2}{m_2} = R T_2$$

Notice that every quantity is known in the equation except P_2 and V_2. Substituting known values:

$$2160 \frac{V_2}{10} = 85.6 T_2$$

or

$$V_2 = .40 T_2$$

Equation 7–2

Equation 7–2 does not determine T_2, since V_2 is also unknown. This should be expected, since the perfect gas law does not include Q, the amount of heat added. After all, the heat added is what makes the temperature rise. The equation that does consider Q is the energy equation:

$$Q = Cm(T_2 - T_1) + \frac{P(V_2 - V_1)}{J}$$

Substituting known quantities into this equation yields:

$$1000 \text{ Btu} = .36 \times 10(T_2 - 720^\circ\text{R}) + 2160\frac{(V_2 - V_1)}{778}$$

Using the result above and simplifying the energy equation yields:

$$1000 = 3.6T_2 - 2592 + 2.78V_2 - 791$$

$$4383 = 3.6T_2 + 2.78V_2$$

Equation 7–3

This energy equation also does not reveal the value of T_2. However, this equation together with Equation 7–2, comprise a system of two equations and two unknowns that can be solved simultaneously. Substituting for V_2 in (from Equation 7–2) in Equation 7–3:

$$4383 = 3.6T_2 + 2.78\,(.40T_2)$$

$$4383 = 4.73T_2$$

$$926^\circ\text{R} = T_2$$

After the cylinder stops rising, the temperature of the gas will be 926°R, or 466°F.

Comparing the result of Sample Problem 7–4 with the problem in Section 6–2 points out the basic differences between constant volume processes and constant pressure processes. In the problem in Section 6–2, an addition of 1000 Btu's caused a temperature increase of 278°F. This resulted in an average of 1000 Btu/278°F or 3.6 Btu/°F, or an average of .36 Btu per 1°F per pound-mass of the gas (10 lb were involved). In the heating process of Sample Problem 7–4, 1000 Btu's added caused a temperature difference of 206°F, not nearly as large an increase as in the constant volume case. This computes to an average of 1000 Btu/206°F = 4.9 Btu/°F, or an average of .49 Btu per 1°F per pound-mass. More Btu's were required to raise the same amount of water vapor one degree Fahrenheit in the constant pressure heating example.

Isn't it unexpected that if we place one pound of steam into two different types of vessels, it will take more heat to raise the temperature in one than the other? And furthermore, if this is true, then what is the meaning of specific heat? If .36 Btu raises one pound of steam 1°F in one case, why did it take .49 Btu in another? Is the specific heat for the constant pressure situation .49 Btu/°F/lb_m? If so, why did we use specific heat of .36 Btu/°F/lb_m when we solved Sample Problem 7–4?

The confusion created by these two numbers can be eliminated by reviewing Equation 7–1 to see that the specific heat C is part of the internal energy term. To best describe its use, we should say that *when* .36 Btu is *stored* in one pound of steam, the temperature is raised 1°F. When the temperature of one pound of steam is raised 1°F, it is certain that .36 Btu of energy was stored. So the symbol *C* deals with internal energy and would best be called the *specific heat of internal energy or stored energy*, and can be designated as C_{IE} .

❖ **KEY TERM:**   **Specific Heat of Internal Energy (C_{IE}):** The amount of heat that must be stored in a substance to cause one pound to rise one degree Fahrenheit.

In the constant pressure case, a portion of the Btu's were used to do WORK and others were stored. In the constant volume case, all heat was used to raise the gas temperature, and none to do WORK, so it took fewer total Btu's to raise water vapor 1°F. We can be so precise as to say that in the second case, of the .49 Btu that was required to raise 1 lb of water vapor 1°F, .36 Btu actually went to increase the temperature while .13 Btu/°F-lb$_m$ were sacrificed to do WORK.

Sample Problem 7–4 indicates that there is another type of specific heat. This one, .49 Btu/lb$_m$ –°R, is an *overall specific heat dealing with the process.* For this situation, it can be called the overall specific heat of steam in a constant pressure process and termed c_p.

A substance has two specific heats, C_{IE} and c_p. (Figure 7–5) Which of these is the *C* that appears in Equation 7–1 when written for a constant pressure process? The answer is C_{IE}. That equation shows that the specific heat impacts only the internal energy term.

When, then, do we use the other specific heat, c_p? It allows us to write a shortcut form of Equation 7–1 when it applies only to a constant pressure process. The abbreviated energy equation is:

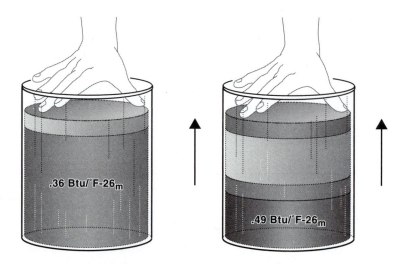

Figure 7–5. Gases have two specific heats.

$$Q = c_p m(T_2 - T_1)$$

This is not a replacement for Equation 7–1, just a restatement of it.

❖ **KEY TERM:**   **Overall Process Specific Heat:** The amount of heat that must be added to one pound of a substance to raise its temperature one degree Fahrenheit.

SAMPLE PROBLEM 7–5

Solve Sample Problem 7–4 again, this time using the shortcut form of the energy equation for a constant pressure process.

SOLUTION:

$$Q = c_p m(T_2 - T_1)$$
$$1000 = .49 \times 10 \times (T_2 - 720) \qquad T_2 = 926°R$$

Notice how easy this solution is compared to Sample Problem 7–4 and the same answer.

Why have we not seen the necessity of two specific heats before? For example, when we studied the constant volume process, the energy equation resulted in: $Q = C_{IE} m(T_2 - T_1)$. For that case, the overall specific heat and C_{IE} were the same. In a sense, the specific heat of internal energy can be determined by subjecting the gas to a constant volume process, c_v. Then $C_{IE} = c_v$.

❖ **KEY TERM:**   **Specific Heat for Constant Volume Processes (c_v):** The overall process specific heat if the heating takes place at constant volume. It has the same value as the specific heat of internal energy (C_{IE}).

❖ **KEY TERM:**   **Specific Heat for Constant Pressure Processes (c_p):** The overall process specific heat if the heating takes place at constant pressure. Numerically it is greater than the c_v for the substance.

Just as every gas has its own gas constant, each type of gas has its own characteristic specific heat at constant volume and specific heat at constant pressure. The values of these specific heats for any gas can be found in engineering handbooks. Specific heats for some of the more common gases are given in Table 7–1.

7.2 DETERMINING SPECIFIC HEATS

Laboratory analysis of a gas easily determines the specific heats of the gas—both c_v and c_p. A weighted sample of the gas is placed in a vessel whose interior volume is known. The vessel is fitted with an electric heating element. The initial temperature is measured very accurately, then the gas is heated by the electric heaters. The amount of heat added is known precisely by measuring the voltage and amperage to the heater. Multiplying the volts times amps yields watts of heating (.293 watt-hr is equivalent to one Btu).

SAMPLE PROBLEM 7–6

a. A .67-lb$_m$ sample of gas is placed in a .15-cu-in constant volume vessel. The temperature of the gas rises 2°F when the heater is connected to 30 volts and draws a total of .1 amp for one minute. What is the c_v of the gas?

b. The same specimen is placed in a similar constant pressure vessel where the gas temperature rises 2°F with .15 amp at 30 volts for one minute. What is the c_p of the gas?

Table 7–1

PHYSICAL PROPERTIES OF GASES*

Name of Gas	Molecular Weight	R Gas Constant $\frac{lb_f-ft}{lb_m-°R}\left(\frac{kPa-m^3}{kg-K}\right)$	Specific Heat $\frac{Btu}{lb_m-°R}\left(\frac{J}{g-K}\right)$		k $\frac{c_p}{c_v}$
			c_p	c_v or c_{IE}	
Acetylene	26.0	59.4 (.320)	.35 (1.46)	.27 (1.13)	1.30
Air	29.0	53.3 (.287)	.24 (1.00)	.17 (.71)	1.40
Ammonia	17.0	90.8 (.489)	.52 (2.18)	.40 (1.67)	1.32
Argon	40.0	38.7 (.208)	.12 (.50)	.07 (.29)	1.67
Carbon dioxide	44.0	35.1 (.189)	.25 (1.05)	.16 (.67)	1.30
Carbon monoxide	28.0	55.2 (.297)	.24 (1.00)	.17 (.71)	1.40
Chlorodifluoromethane (R-22)	93.0	17.5 (.089)	.21 (.88)	.18 (.75)	1.16
Dichlorodifluoromethane (R-12)	121.0	12.77 (.067)	.14 (.59)	.12 (.50)	1.13
Ethylene	28.0	55.19 (.297)	.40 (1.67)	.33 (1.38)	1.22
Helium	4.0	386. (2.08)	1.25 (5.23)	.754 (3.16)	1.66
Hydrochloric acid	36.0	42.4 (.231)	.19 (.79)	.136 (.57)	1.40
Hydrogen	2.0	767. (4.16)	3.42 (14.3)	2.435 (10.2)	1.41
Methane	16.0	96.4 (.520)	.59 (2.47)	.47 (1.97)	1.32
Methyl chloride	50.5	30.6 (.165)	.24 (1.00)	.20 (.84)	1.20
Nitrogen	28.0	55.2 (.297)	.25 (1.05)	.175 (.73)	1.41
Nitric oxide	30.0	51.5 (.277)	.23 (.96)	.165 (.69)	1.40
Nitrous oxide	44.0	35.1 (.189)	.22 (.92)	.175 (.73)	1.31
Oxygen	32.0	48.3 (.260)	.22 (.92)	.15 (.63)	1.40
Propane	44.0	34.8 (.189)	.40 (1.67)	.36 (1.51)	1.15
Steam	18.0	85.8 (.462)	.445 (1.86)	.335 (1.40)	1.33
Sulphur dioxide	64.0	24.1 (.130)	.15 (.63)	.12 (.50)	1.26

* Specific heats actually vary slightly with temperature. Values shown above are average values for temperatures between 32°F and 212°F.

SOLUTION:

a.   First find the number of watts and Btu's used in one hour:

$$Watt - hrs = 30 \text{ volts} \times .1 \text{ amp} \times \frac{1}{60} \text{ hr} = .05 = .17 \text{ Btu}$$

$$Q = c_v m(T_2 - T_1)$$
$$.17 = c_v \times .67 \times 2°$$
$$.127 = c_v$$

b. $Watts = 30 \times .15 \text{ amp} \times \dfrac{1}{60} \text{ hr} = .075 = .256 \text{ Btu}$

$$Q = c_p m(T_2 - T_1)$$
$$.256 = c_p \times .067 \times 2$$
$$.191 = c_p$$

The laboratory analysis of c_v and c_p may be even simpler than this, since there is a concrete mathematical relationship between them. For a constant pressure process, there are two equations for Q:

$$Q = c_p m(T_2 - T_1)$$

Equation 7–4

$$Q = c_v m(T_2 - T_1) + \frac{P(V_2 - V_1)}{J}$$

Equation 7–5

But notice that

$$P_2 V_2 = mRT_2$$

and

$$PV_1 = mRT_1$$

Then from Equation 7–5:

$$Q = c_v m(T_2 - T_1) + \frac{(mRT_2 - mRT_1)}{J}$$

$$= c_v m(T_2 - T_1) + \frac{mR}{J}(T_2 - T_1)$$

$$= \left(c_v + \frac{R}{J}\right) m(T_2 - T_1)$$

This compares with the "other" Q equation:

$$Q = c_p m \, (T_2 - T_1)$$

if

$$c_p = c_v + \frac{R}{J}$$

Equation 7–6

SAMPLE PROBLEM 7-7

Are the specific heats of ammonia from Table 7–1 consistent with Equation 7–6?

SOLUTION:

From Table 7–1:

$$c_{v \text{ammonia}} = .40, \quad R_{\text{ammonia}} = 90.8$$

then

$$c_p = .40 + \frac{90.8}{778} = .52$$

Notice that this calculated value corresponds to the value for c_p given in Table 7–1.

7.3 APPLICATION: A THERMODYNAMIC ELEVATOR

Throughout the day, you are confronted with the constant pressure process in the machines you use, the weather you feel, the prepared meals you eat, and other situations. Many of these situations are presented throughout the remainder of the book. One we will look at for illustrative purposes you may never have seen before. It is a "gas elevator."

A gas elevator is an experimental device for moving cargo from a freight dock up to the deck of a ship. It consists of an elevator car suspended in an elevator shaft built at the edge of a dock. (Figure 7–6) The carrier is suspended by pneumatic pressure caused by high-pressure air in the shaft. Once the carrier is loaded at dock level, the air in the shaft is heated, causing the elevator to rise and deliver the cargo to the ship's deck. When the carrier is emptied, the gas in the elevator is cooled, allowing the carrier to slowly drop back to dockside.

The gas in the elevator is doing the lifting and is referred to as the working medium. This gas is air, and from a thermodynamic point of view, the air in the shaft is the only important part of the machine. It is the air that is being heated, pressurized, cooled, and so on. The thermodynamic drawing of the

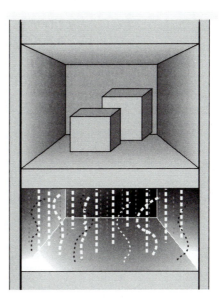

Figure 7–6. A gas elevator is suspended by air pressure.

lifting would strip away the boat, the dock, and the carrier [Figure 7–7(a)] and leave only the column of air. The two rectangles in Figure 7–7(b) show the shaft of air at its largest and smallest volumes. A Q has been drawn next to the large rectangle with an arrow pointing in. The thermodynamicist who drew this picture has done this to indicate that the most important occurrence in

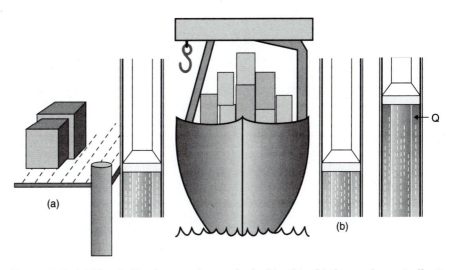

Figure 7–7. (a) Physically, the gas elevator looks like this; (b) thermodynamically, it looks like this.

going from the smaller rectangle to the larger was that some heat, Q, was added to the working gas.

As we investigate this gas elevator, let us consider that the elevator shaft continues 70 ft below the dock. The elevator car has a floor area of 10 sq ft and must carry one ton of cargo. The ship deck is located 70 ft above the dock.

With this information, it is easy to make some calculations to see not only how the elevator works, but also how well it works. First we can find the pressure required to suspend a one-ton cargo load at dockside.

$$P_b = \frac{W}{A} = \frac{2000 \ lb_f}{10 \ ft^2} = 200 \ psfg$$

The subscript "b" refers to the fact that these conditions are calculated for the elevator at the bottom of its run.

Next, let us determine how much 70°F air is in the elevator shaft to maintain this amount of pressure. The state of the working gas must satisfy

$$P_b = \rho R T_b$$

where

$$P_b = 200 \ psfg = 2316 \ psfa$$

$$R = \frac{53.3 \ ft-lb_f}{(°R-lb_m)}$$

$$T_b = 70°F + 460°F = 530°R$$

Therefore, the elevator shaft must be filled to a density of

$$\rho = \frac{2316 \dfrac{lb_m}{ft^2}}{53.3 \dfrac{ft}{°R}} \times 530°R = .082 \frac{lb_m}{ft^3}$$

Accordingly, the amount (lb_m) of air captured in the shaft is

$$m_b = \rho V_b$$

where V_b is the volume of the shaft

$$V_b = 10 \ sq \ ft \times 70 \ ft = 700 \ ft^3$$

$$m = .082 \times 700 \ ft^3 = 57 \ lb_m$$

When in operation, the elevator working gas is heated with large burners located in the bottom of the elevator shaft. As it is heated, the elevator begins to rise. What is the temperature of the working gas when the elevator and cargo reaches the top? The easiest way to realize this result is to recognize that while the working gas is being heated, the elevator is rising to maintain a constant pressure in the gas. The operation of the gas elevator is a constant pressure process. Therefore:

$$\frac{T_t}{T_b} = \frac{V_t}{V_b}$$

Subscript "t" implies that conditions are those at the top of the elevator run. The volume occupied by the gas at the top is twice the volume occupied at the bottom [see Figure 7–7 (b)]; $V_t/V_b = 2$. Then:

$$T_t = \left(\frac{V_t}{V_b}\right)T_b = 2 \times 530° \, \text{R} = 1060° \, \text{R} = 600° \, \text{F}$$

The final bit of information needed to complete the thermodynamic description of the elevator is the amount of heat that must be added to the elevator gas to raise it from dock to deck. This can be calculated from the energy equations for constant pressure processes, of which we now have two. The simplest one predicts:

$$Q = c_p m(T_t - T_b)$$

$$= .24 \frac{\text{Btu}}{\text{lb}} - °\text{F} \times 57 \, \text{lb} \times (1060° \, \text{R} - 530° \, \text{R})$$

$$= 7250 \, \text{Btu}$$

(The value for the c_p of air was found in Table 7–1). The other energy equation is longer but should predict the same thing:

$$Q = c_v m(T_t - T_b) + P\frac{(V_t - V_b)}{J}$$

$$= .17 \times 57 \times (1060 - 530) + \frac{2316(1400 - 700)}{778} = 5135 \, \text{Btu} + 2084 \, \text{Btu} = 7219 \, \text{Btu}$$

Within the accuracy of our calculations, these two results agree. From this longer equation, we can see that WORK = 2084 Btu.

7.4
HEATING
UNCONFINED AIR

A much different situation from the gas elevator is the heating of atmospheric air, which is also a constant pressure process. If a hot plate is turned on but nothing is placed on top of it, the air on top of the burner is heated and rises. A continuous stream of heated air expands and rises, confined only by the cold air around it. Since the pressure of the cold air is atmospheric, and this pressure is applied against the warm air stream, the pressure of the warm air is a constant atmospheric pressure. The warm air expands as if against a floating piston that confines it to 14.7 psia, but the floating piston is actually the atmosphere around it.

**SAMPLE
PROBLEM 7–8**

Although the air temperature around the Dallas airport is only 21°C, the black asphalt of the runway is much hotter because it is absorbing solar radiation. At the surface of the runway, the air temperature is 30°C. There is an optical distortion when passengers look back at the passenger terminal caused by the different density of the hot air lying on the runway. What is the density of the air on the runway? How much heat does each kg of air absorb? How much WORK does one kg of air do as it expands from the heat of the tarmac? What is this WORK done against?

SOLUTION:

The density of air at STP is

$$\rho = \frac{P}{RT} = \frac{100 \text{ kPa}}{\left(.287\dfrac{\text{kPa}-\text{m}^3}{\text{kg}-\text{K}}\times 294\right)} = 1.18\frac{\text{kg}}{\text{m}^3}$$

and since the heating is done at constant pressure,

$$\rho_2 = \rho_1 \times \frac{T_1}{T_2} = 1.18\frac{\text{kg}}{\text{m}^3}\times\left(\frac{294}{303}\right) = 1.146\frac{\text{kg}}{\text{m}^3}$$

$$Q = c_p m\,(T_2 - T_1) = 1.00 \times 1 \times (303 - 294) = 9 \text{ kJ (see Table 7–1)}$$

$$WORK = Q - c_v m\,(T_2 - T_1) = 9 - .71 \times 1\,(303 - 294) = 2.61 \text{ kJ} = 2610n - m$$

This WORK is done against the atmosphere in order to push cold air out of the way and to make room for the less dense air.

**TECHNICIANS
IN THE
FIELD**

INDUSTRIAL FURNACE DESIGNER

Furnaces that process steel to put it in its many useful forms are individually designed and erected to match their specific application. Engineering firms that perform this design and supervise the construction are located throughout the midwestern states of the United States, right in the steel belt. Atmospheric Furnace, Inc. (AFI) of Livonia, Michigan, is one of the most successful of these design-and-build firms. Most of the employees of AFI are engineers and technicians, but it is the technicians who are in charge of the detailed design of individual projects and who supervise the construction.

One of AFI's current jobs is to produce a carbonizing (carbon-hardening) furnace for a Metarie, Louisiana, plant. The design technician knows that his first task is to size the combustion burners for the job. The specifications of the furnace require that it heat 4000 lb_m per hour of strip steel ($c = .11$ Btu/lb-°R) from 70°F to 1500°F. Since this is a constant volume heating, the required amount of heat is $Q = cm(T_2 - T_1) = 849,200$ Btu/hr. The designer has a table (Table 7–2) that he uses often, and from this table he finds that if the fuel is natural gas, the amount of heat necessary from the burners is 1,451,624 Btu/hr. Why is the heat necessary from the burner not the same as the heat necessary for the steel? The designer knows.

Figure 7–8 shows the elements of an industrial gas furnace. Air is fed into the combustion chamber to participate in the combustion, and this air is heated to high temperatures during the process. These combustion gases rise until they contact the metal product, heating it. The metal product temperature rises and the combustion gas temperature falls until both are the same, then the gas

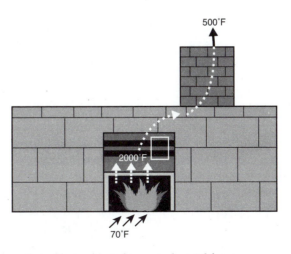

Figure 7–8. Air and steel being heated in an industrial furnace.

continues up and out the chimney. The flame never actually touches the metal but instead is transferred to the product by the gas.

Figure 7–8 traces the path of one lb$_m$ of air through the furnace, illustrating that before it enters the furnace, the air is a fairly compact volume at 70°F. As it participates in the combustion, it expands due to the addition of heat, then contracts as it cools over the product. The air parcel is unconfined and, therefore, the pressure of the air sample remains constant as it expands.

Sample Problem 7–9 illustrates how combustion technicians use the constant pressure process to size industrial furnace burners.

SAMPLE PROBLEM 7–9

Consider that one pound of air is fed to a flame at 70°F and is heated to 2000°F. It then rises and heats a metal product from 70°F to 500°F, then exhausts out the chimney.

I.   a.   How much heat is added to one pound of air as it goes through the flame?
     b.   How much heat is extracted from the lb$_m$ of air as it travels across the steel product? (Assume that the combustion gas temperature drops to 500°F.) What percent of the heat in part a does this represent?
     c.   How much of the original heat is "lost" with the combustion gases as they go up the chimney?
     d.   To combustion technicians, the term *available heat* refers to the heat that actually reaches the product (part b). What percent of the heat in part a is "available"?

II. Repeat all the calculations for Part I above but with the product temperature changed to 1000°F.

III. Repeat all the calculations for Part I above but with the product temperature changed to 1500°F.

SOLUTION:

I.   a.   Since the air is heated by a constant pressure process ($P = 14.7$ psia):
$$Q_{air} = c_p m(T_2 - T_1) = .24 \times 1 \times (2000 - 70) = 463 \text{ Btu}$$
     b.   The air is cooled over the steel to 500°F.
$$Q_{air} = c_p m(T_3 - T_2) = .24 \times 1 \, (500 - 2000) = -360 \text{ Btu}$$
$$\% = \frac{360}{463} \times 100 = 78\%$$
     c. $Q_{lost} = Q_{added} - Q_{extracted} = 463 - 360 = 103$ Btu
     d.   78% is available

II. a. $Q_{air} = .24 \times 1 \times (2000 - 500) = 463$ Btu
     b. $Q_{air} = c_p m(T_3 - T_2) = .24 \times 1 \times (1000 - 2000) = -240$ Btu
$$\% = \frac{240}{463} = 52\%$$
     c. $Q_{lost} = Q_{added} - Q_{extracted} = 463 - 240 = 223$ Btu
     d.   52% is available

III. a. $Q_{air} = .24 \times 1 \,(2000 - 500) = 463$ Btu

b. $Q_{air} = c_p m(T_3 - T_2) = .24 \times 1 \times (1500 - 2000) = -120$ Btu

$\% = \dfrac{120}{463} = 26\%$

c. $Q_{lost} = Q_{added} - Q_{extracted} = 463 - 120 = 243$ Btu

d. 26% is available

The sample problem demonstrates that the higher the temperature of the product being heated, the less the efficiency of the furnace will be, that is, the less the percent of the fuel that is actually used to heat the product will be, and the greater will be the heat loss up the chimney. Industrial furnace technicians and designers state this differently: the greater the product temperature must be, the less will be the percent of the heat generated from that fire that actually heats the product. This is termed *available heat*. Table 7–2 is an available heat table used by combustion technicians.

❖ **KEY TERM:**   **Available Heat:** The difference between the heat generated in combustion and the heat loss in the flue gas, given as a percentage of the total heat generated.

Table 7–2

	AVAILABLE HEATS FOR INDUSTRIAL BURNERS (% OF FUEL HEATING VALUE)				
Temp °F	Coke Oven 490 Btu/ft³	Natural Gas 1045 Btu/ft³	Propane 2500 Btu/ft³	Fuel Oil 140,000 Btu/gal	Anthracite Coal 13,700 Btu/lb
400	82	84	86	92	85
500	77	79	82	88	79
800	73	75	77	83	74
1000	69	70	73	78	67
1200	65	66	68	73	63
1400	60	61	64	68	58
1600	56	56	59	64	52
1800	52	52	54	58	47
2000	47	47	50	53	42
2200	44	43	45	47	35
2400	39	38	40	41	29

An industrial furnace is to be built that will carbon harden steel, which requires 4000 lb_m of steel per hour to be brought up to a temperature of 1500°F. What should be the capacity of the burners, in Btu/hr, in this furnace if the fuel is natural gas?

SOLUTION:

As the AFI technician found, Q steel = 849,200 Btu/hr. The available heat for 1500°F product is 58.5% (.585 as extrapolated from the table). This implies that the heat delivered by the combustion must be greater than that required by the product because some must be lost.

$$Q_{burner} = \frac{Q_{steel}}{\text{Avg Heat}\%} = \frac{849,200}{.585} = 1,451,624 \frac{\text{Btu}}{\text{hr}}$$

CHAPTER **7**

PROBLEMS

1. How much heat is required to raise a 4000-lb cargo seventy feet to the deck of a ship if the elevator floor area is 20 ft^2? Original height of the elevator shaft is 70 ft, and the deck is another 70 ft above that. How much heat would be required if the floor area were 40 ft^2?

2. One pound of air is heated under constant pressure conditions from an initial temperature of 500°R to a final temperature of 800°R. Find the WORK done during the process and the pressure at which the process occurs if the initial specific volume is 3.86 ft^3/lb.

3. A piston and cylinder device initially contains 10 lb of acetylene at 70°F. The piston is free to move to maintain 1000 psia on the gas. How much WORK is done by the gas as it is heated to 300°F?

4. A cylinder with a moveable piston holds .20 lb of hydrogen initially in a volume of 1 cu ft at a temperature of 1300°F. Then the volume is expanded to 2.5 cu ft while the gas pressure remains constant.
    a. What is the initial gas pressure?
    b. What is the final gas pressure?
    c. Find the final gas temperature.
    d. How much heat has been transferred to the gas?
    e. What mechanical WORK has been done on the piston by the gas?
    f. What has been the change in internal energy of the gas?

5. In a constant pressure process where a piston expands from 6.5 in^3 to twice that volume, the initial state of the air is P_1 = 826 psig and the temperature is T_1 = 1256°F. Find:
    a. T_2
    b. P_2
    c. WORK done
    d. Heat added

6. Find the WORK done in each of the processes described below:
    a. V_1 = .5 cu ft, V_2 = .5 cu ft, P_1 = 20 psia P_2 = 40 psia
        WORK done_____
    b. V_2 = .5 cu ft, V_3 = 1.5 cu ft, P_2= 40 psia P_3= 40 psia
        WORK done_____

   c. V_3 = 1.5 cu ft, V_4 = 1.5 cu ft, P_3 = 40 psia, P_4= 20 psia
      WORK done_____

   d. V_4 = 1.5 cu ft, V_5 = .5 cu ft, P_4 = 20 psia, P_5 = 20 psia
      WORK done_____

   e. Draw four cylinders on a sheet of paper and close their ends on top. Then place in the pistons in their location within the cylinder at the end of the process. Show whether the piston is fixed (constant volume) or floating. Show whether heat is rejected or added.

   f. Draw on a P-V diagram the line that represents each process above. Put values for volume and pressure on the axis. Make it according to scale. Put all four processes on one graph.

7. Methane gas is confined inside a loose bag (constant pressure) and undergoes a process in which 30 Btu are rejected and the volume changes from 20 ft³ to 5 ft³ with a pressure of 20 psia. For this process, what will be the change in temperature (increase or decrease and magnitude)?

8. A hot air balloon rises because the air inside is expanded by heat so that it weighs less than the air around it. (Figure 7–9) When this happens, it becomes "buoyant." To determine the density of the hot air that is necessary to raise a mass M, use the equation:

$$M_{cargo} + m_{air} = \rho_{cold\ air}V_{inflated}$$

   a. If the balloon initially has a volume of 1500 ft³ and the atmospheric conditions are T = 70°F and P = 14.7 psia, find the $\rho_{hot\ air}$ to raise a 200 lbm cargo, assuming the heating process to be constant pressure. (Hint: find $V_{inflated}$ and calculate $\rho_{hot} = m_{air}/V_{inflated}$.)

   b. Find the temperature of the heated balloon air.

   c. Determine the amount of heat that must be added to lift this weight if the heating process is taken to be constant pressure.

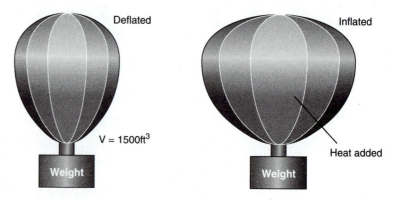

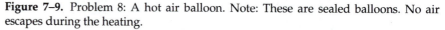

Figure 7–9. Problem 8: A hot air balloon. Note: These are sealed balloons. No air escapes during the heating.

9. The roof of the Pontiac Silverdome is made of an elastic fabric. (Figure 7–10) On a cool day, such as the Friday before Superbowl '83, the temperature of the air inside the dome gets as low as 30°F ($P = 14.7$ psia). At this temperature the roof is supported flat (no rise) by the air in the stadium. It encloses a volume of 30,000,000 ft³. To get the roof to raise for the big game, the air is heated for days with all doors shut. As the air expands under constant pressure, the roof rises.
   a. How much heat is needed to make the roof rise to a perfect half sphere with a radius of 100 ft?
   b. What is the temperature inside the dome at this point?

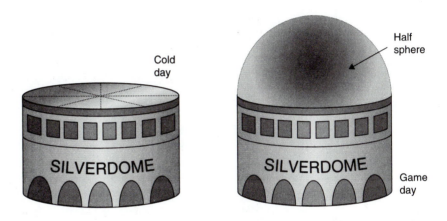

Figure 7–10. Problem 9: The roof of the Pontiac Silverdome.

10. An unknown gas is originally at a state $P_1 = 60$ psia and $V_1 = 3$ ft³ and is expanded to a pressure $P_2 = 40$ psia and $V_2 = 7.5$ ft³. If the internal energy of the gas decreases during the process by 35 Btu, find R. Furthermore, if $c_v = .35$, find c_p.

11. The tires on Indy cars are extremely important for success. Too low a pressure causes excessive friction and low speeds. Too high a pressure puts excessive wear on the tires and requires extra pit stops. Ideally, Indy car tires are inflated to 30 psig. Unfortunately as the race begins, the tires heat up from 70°F to a temperature as high as 300°F.
   a. If the tires are assumed not to stretch, what will be the pressure in the tire at 300°F? Assume constant volume and that air is a perfect gas.
   b. One ingenious performance team installs a pressure relief valve on their tires to maintain a constant pressure of 30 psig. What percent of the air will be released to maintain constant pressure when the tire reaches 300°F?

12. An elevator that uses the pressure from a gas as it is heated to raise itself to different floors utilizes which fundamental process?
    a. constant V
    b. constant P
    c. constant T

13. A chamber test that encloses a gas volume but has a free-floating piston demonstrates which fundamental process?
    a   constant volume
    b. constant pressure
    c. constant temperature
    d. constant density

14. Which specific heat is higher?
    a. c_V
    b. c_{IE}
    c. c_p
    d. R

15. A solid block of copper at 260°F is placed in a 10-ft³ insulated box filled with air (standard temperature and pressure). A lid is placed on the box and moves up and down to maintain a constant pressure inside. After a short time, the air and copper come to an equilibrium at 120°F. What is the weight of the copper placed in the box?

16. A cylinder is fitted with a piston and a set of stops is filled with air. (Figure 7–11) The piston cross-section is .8 ft² and the air inside is initially at 20 psig and 800°F. The air is then cooled at constant pressure as a result of heat transfer to the surroundings.
    a. What is the temperature of the air inside the cylinder when the piston reaches the stops? How much work is done by the gas during the cooling process? How much heat is rejected?
    b. The cooling continues after the piston reaches the stops until the temperature reaches 70°F. What is the pressure at this state? How much WORK is done during this process? How much heat is rejected?

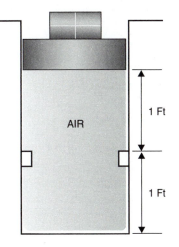

Figure 7–11. Problem 16.

17. Use the available heat table to determine the burner capacity (Btu/hr) required to add 4,000,000 Btu/hr to a metal product to get it up to 1000°F if the fuel is coke gas?

Problems in SI Units

18. A piston and cylinder device initially contains 10 kg of acetylene at 25°C. The piston is free to move to maintain 150 kPa on the gas. How much WORK is done by the gas as it is heated to 125°C?

19. Methane gas is used as a fuel to power buses in Cambodia. When filling such a bag, the volume typically increases from .01 m^3 to 2m^3. If the initial temperature in the bag is 22°C and the process is constant pressure (100 kPa), determine:
    a. The final temperature in the bag (°C)
    b. The WORK done to fill the bag (N-m)
    c. The heat (rejected or added) (J)

CHAPTER
8

Fundamental Processes: Constant Temperature

PREVIEW: Is it possible to heat an object without raising its temperature? Not only is it possible, but the process is an important one that occurs often in both nature and industry.

OBJECTIVES:
- ❑ Describe a typical vessel that supports a constant temperature process.
- ❑ Identify processes that occur in nature and in machines that maintain constant temperature.
- ❑ Calculate WORK done during a constant temperature process.
- ❑ Describe the perfect gas law as an "equation of state" and extend it to liquids and solids.

8.1 A MORE EFFICIENT GAS ELEVATOR

The gas elevator of Chapter 7 was designed to do useful WORK—that is, move cargo. The WORK was accomplished by a thermodynamic process: heating. This opens up a new world for thermodynamics. Not only will processes like these produce heat to cook food, warm houses, and melt iron ore, but they also may be used to move material and, with further ingenuity, to perform many other tasks.

The gas elevator considered in the previous chapter used 7219 Btu's of heat to produce 2084 equivalent Btu's of WORK (see section 7–3). The remaining 5135 Btu's went to increasing the internal energy of the gas that pushes the elevator. If you were in the business of moving cargo, you would say that the amount of energy necessary to increase internal energy did not move any cargo and, therefore, was "wasted." Only the 2084 Btu's that actually did WORK was useful. This process uses only 28% of the Btu's to actually do WORK. We say, then, it has an efficiency of 28%.

Is there any way we could move this elevator with 100% efficiency? Can we devise a process that will convert all the heat directly into WORK without any loss or storage of internal energy?

Figure 8–1 shows the consequences of such a process as pointed out by the energy equation, and compares the process against the constant pressure elevator. Mathematically, the only way to eliminate stored energy is to require:

$$c_v m(T_2 - T_1) = 0 \quad \text{or} \quad T_2 = T_1$$

To get 100% efficiency, the gas that moves the elevator may not increase in temperature throughout the heating!

ENERGY EQUATION FOR 100% EFFICIENT ELEVATOR

General Equation	Q	$=$	$c_V m (T_2 - T_1)$	$+$	WORK	$+$	$m_L L$
Constant pressure	7541 Btu		5354		2097		0
Ideal elevator	7541 Btu		0		7541		0

Figure 8–1. Energy comparisons for two elevators.

We might well imagine that the first people to experiment to find a 100% efficient elevator might have been creative university engineering technology students. They would have noticed that in a conventional constant pressure elevator, every time the elevator got to the top, the supporting gas was hotter than when it started rising. Couldn't there be more WORK drained from this elevator? Even though the elevator had exhausted its ability to move all the cargo, is it possible that it could move a partial load of cargo further? Testing this, the students threw off some cargo and the elevator moved up, without the addition of more heat. (Figure 8–2) After the elevator had moved up with reduced cargo, the temperature of the supporting gas had decreased! Each time the elevator moved up, the temperature of the gas dropped again. Finally, the gas temperature dropped down to its original value. No more WORK could be done by the heat that was originally put into the gas elevator.

Figure 8–2. Throwing cargo off makes the elevator go higher and reduces the gas temperature.

The result was an elevator that worked without overall changing the temperature of the working gas: a constant temperature process. Not only had the students discovered a unique new way of moving cargo, but they also had discovered a new way to heat a gas—*without increasing the temperature!*

In an effort to determine exactly where the heat goes during a constant temperature process, consider the general energy equation in Figure 8–1. If it is required that $\Delta T = 0$, then there is no internal energy buildup in the gas. Further, as in the constant pressure elevator, latent heat has not been employed. Therefore, the new energy equation for a constant temperature process can be written:

$$Q = WORK$$

All of the heat added goes to WORK!

To calculate the WORK done, we must rely on the equation for WORK:

$$WORK = F \times d = P \times A \times d$$

The pressure of the gas causes the force, and it is equal to the weight of the cargo plus the atmosphere pushing down. Unfortunately the weight on the elevator changes throughout the ride. During the first few feet the elevator weighs the most, and a large amount of WORK is used to move it. Later the elevator weighs much less, and the supporting gas pressure is correspondingly reduced. (Figure 8–3) Which pressure do we use to calculate the WORK

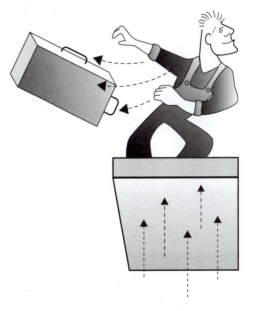

Figure 8–3. The gas pressure is decreased as cargo is thrown off.

done from the equation above? In fact, there is no single pressure that is appropriate. It may be necessary to calculate an average pressure. This could be very difficult because it depends on how fast the cargo is being thrown off.

We now introduce another "tool" of analysis that will eventually help calculate the WORK done in any process. It is a visual description in the form of a graph; a graph of the gas pressure during a process versus the volume occupied by the gas. Figure 8–4 shows this graph for a constant volume process. Since the graph is a vertical straight line, it indicates that pressure changes greatly but volume stays constant. Similarly, for a constant pressure process, Figure 8–5 demonstrates that pressure remains constant as the piston expands from V_1 to V_2, or compresses from V_2 to V_1. A rectangle is drawn on this graph, and if we define an area associated with this rectangle, it would be **Area** = *base* × *width*, where the base dimension is "$V_2 - V_1$" and the height dimension is "P." Therefore, the area under the P-V diagram is $P(V_2 - V_1)$ which is the equation for WORK in a constant pressure process!

Could it be that the area underneath the P-V diagram on any process is equal to the WORK done during the process? Figure 8–4 shows that during a constant volume process, the base of the rectangle has zero length; therefore, the volume under the P-V diagram is zero, which matches the WORK done in a constant volume process.

Figure 8–5 shows the P-V diagram for a constant temperature process. Notice that as the gas volume increases (indicating the raising of the elevator), the pressure decreases, since cargo is being thrown off.

The curve of Figure 8–6 looks like part of a hyperbola and can be described by a mathematical equation. The equation that describes this curve is readily found by checking the perfect gas law. It states that:

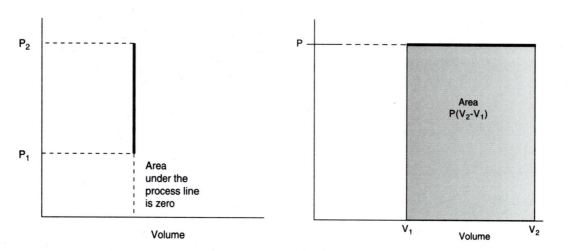

Figure 8–4. P-V diagram for constant volume process.

Figure 8–5. P-V diagram for constant pressure process.

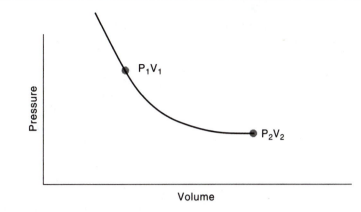

Figure 8–6. P-V diagram for constant temperature process.

$$PV = mRT$$

But if T does not change (constant temperature) and m does not change, then

$$PV = constant = \text{the equation of a hyperbola (see Sample Problem 1–10)}$$

SAMPLE PROBLEM 8-1   Verify that the equation

$$PV = 10 \text{ from } P_1 = .1 \text{ to } P_2 = 10$$

represents a curve similar to that in Figure 8–6.

SOLUTION:

We first make a table of points to plot on a graph

P	*V*
.1	100
1	10
2	5
5	2
10	1

These points are plotted on the graph in Figure 8–7 and a smooth curve is drawn through them.

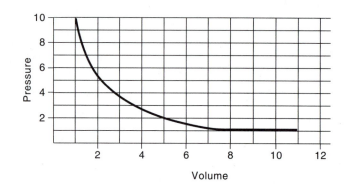

Figure 8–7. Sample Problem 8–1.

The area under the curve is equal to the WORK done, but the area is not as simple a calculation as it was for the constant pressure process. The area under this hyperbola from V_1 to V_2 is:

$$Area = WORK = P_1 V_1 \ln \left(\frac{P_1}{P_2} \right)$$

where *ln* is the natural logarithm of the ratio P_1/P_2 (see Section 1–11).

SAMPLE PROBLEM 8–2

For the conditions of Sample Problem 8–1, calculate the WORK done.

SOLUTION:

$$WORK = P_1 V_1 \ln \left(\frac{P_1}{P_2} \right)$$

$$= .1 \times 100 \ln \left(\frac{.1}{10} \right)$$

$$= 10 \ln (.01)$$

$$= -46.05$$

P_1/P_2 is dimensionless as long as P_1 and P_2 are measured in the same units and therefore $\ln (P_1/P_2)$ is dimensionless. Further, it should be mentioned that whatever unit is used for P_1 and P_2, it must be an *absolute pressure unit;* no gauge pressures are used here.

Checking the units of this equation for WORK:

$$WORK = \frac{lb_f}{ft^2} \times ft^3 = ft - lb_f$$

To write WORK in units of Btu, employ the conversion J:

$$WORK \text{ (Btu)} = \frac{P_1 V_1 \ln\left(\dfrac{P_1}{P_2}\right)}{J}$$

The energy equation becomes:

$$Q = \frac{P_1 V_1}{J} \ln\left(\frac{P_1}{P_2}\right)$$

Equation 8–1

for a constant temperature process.

8.2 THE CONSTANT TEMPERATURE PROCESS

In Section 8–1 the technology students provided an interesting introduction to constant temperature processes but yielded a clumsy machine that required a great number of personnel to operate and precise coordination to make it functional. It wasn't a true constant temperature process either, since the temperature increased initially when the burners were turned on, then decreased to the original value with the rise in the cargo. Constant temperature processes may occur in very simple devices.

Figure 8–8 is a drawing of a typical device that will support a constant temperature process. The device consists of a gas that is confined in a vessel and fitted at the top with a moveable piston. The gas can be compressed or rarified by the piston. The cylinder vessel itself is made of a very good conductor of heat and is surrounded by an outer container of ice water or hot water or any other material that will provide heat or cooling to the confined

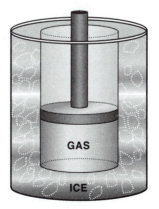

Figure 8–8. A constant-temperature vessel.

gas. Ideally, this outer vessel can be called a constant temperature bath. It imposes its temperature on the confined gas inside, no matter what happens. If for some reason the temperature of the confined gas attempts to decrease, heat will flow naturally from the constant temperature bath into the gas and instantly warm it to its original temperature. Similarly, if the temperature of the gas tries to increase, the external bath absorbs some heat by natural transfer and thus maintains the original temperature.

If the temperature of the gas cannot change, then how can a thermodynamic process take place inside the vessel? One way is by simply pushing the piston down and compressing the gas. The perfect gas law illustrates that if the gas is compressed, the temperature of the gas tends to increase. (See Section 3–6.) But the constant gas temperature vessel stifles this increase by absorbing sufficient heat to maintain a constant gas temperature. This is constant temperature compression of a gas.

SAMPLE PROBLEM 8–3

A constant temperature vessel uses a crushed ice bath to maintain 32°F (0°C). Originally air is confined in the chamber under 20 psig (239 kPa absolute) pressure. Answer the following questions using calculations in both ft-lb units and SI units.

a. What is the temperature of the air in the chamber?
b. If the gas is compressed to one-half of its original volume, what will the final pressure be?
c. If the gas originally occupied 3 ft^3 (.085 m^3) of volume, how much WORK is required to perform the compression?
d. How much heat is absorbed by the ice bath?
e. Draw the P-V diagram for the process.
f. What does it mean that the WORK is negative and that the heat transferred is negative?

SOLUTION:

a. $T_1 = 32°F = 492°R$

b. $P_1 = (20 + 14.7)\dfrac{\text{lb}}{\text{in}^2} \times 144\dfrac{\text{in}^2}{\text{ft}^2} = 4996.8 \text{ psfa}$

$\dfrac{P_2}{P_1} = \dfrac{v_1}{v_2} = \dfrac{V_1}{V_2}$ (assuming no weight change and T = constant)

$P_2 = \left(\dfrac{V_1}{V_2}\right) P_1 = \dfrac{2}{1} \times 4996.8 = 9993.6 \text{ psfa}$

c. $WORK = P_1 V_1 \ln\left(\dfrac{P_1}{P_2}\right)$

$= (4996.8 \times 3)\ln\left(\dfrac{4996.8}{9993.6}\right)$

$= -10388.34 \text{ ft-lb}_f$

d. $Q = WORK \text{ (Btu)} = \dfrac{-10388.34}{J} = -13.35 \text{ Btu}$

e.   (See Figure 8–9.)

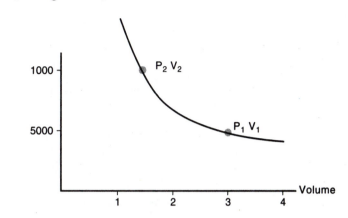

Figure 8–9. Sample Problem 8–3.

f.   The WORK is not done *by* the gas but rather *on* the gas. The gas is compressed. Heat is not put into the gas (absorbed), but rather it is taken out of the gas (rejected).

Solution in SI Units:

a. $T_1 = 0°C = 273K$

b. $P_1 = 239 \text{ kPa}$

   $P_2 = \dfrac{V_1}{V_2} P_1 = \dfrac{2}{1} \times 239 = 478.5 \text{ kPa}$

c. $WORK = P_1 V_1 \ln\left(\dfrac{P_1}{P_2}\right)$

   $= 239 \times .085 \times \ln\left(\dfrac{239}{478.5}\right)$

   $= -14.096 \text{ kJ}$

d. $Q = WORK = -14.096 \text{ kJ}$

SAMPLE PROBLEM 8–4

Consider the same apparatus as in Sample Problem 8–3, but now instead of compressing the gas, allow the pressure inside the vessel to push the piston up. Imagine that your hand is holding the piston in place against the pressure of the gas. Then let your hand, and the piston, move back slowly. As this is done, the gas is doing WORK on the piston.

a.   What will the temperature of the confined air be before you move your hand?

b.   If you allow the piston to move out to double its original volume, what is the final pressure?

c. If the gas originally occupied 3 ft^3 of volume, how much WORK was done?
d. Calculate the heat exchanged to the constant temperature bath.
e. Compare the sign of the values for Q and WORK for this case against the previous Sample Problem.

SOLUTION:

a. $T_1 = 32°F = 492°R$

b. $P_1 = 4996.8$ psfa

$$P_2 = P_1\left(\frac{V_1}{V_2}\right) = 2458.4 \text{ psaf}$$

c. $WORK = P_1 V_1 \ln\left(\frac{P_1}{P_2}\right) = 10632 \text{ ft–lb}_f$

d. $Q = WORK \text{ (Btu)} = \dfrac{10632}{J} = 13.66 \text{ Btu}$

e. Heat is added to the gas in this case (to keep the temperature up), and WORK is done *by* the gas.

Notice that since $P_1V_1 = P_2V_2$, WORK can also be written WORK $= P_2V_2$ ln (P_1/P_2) or $= P_1V_1 \ln (V_2/V_1)$.

8.3
THE "PERFECT GAS LAW" FOR SOLIDS, LIQUIDS, AND REAL GASES

If we consider a glass of water sitting on a table in the kitchen and wonder what the density of water in the glass is, we might attempt to apply the perfect gas law to it, $\rho = P/RT$. It is easy to estimate the pressure on the water as atmospheric, $P = 2116$ psf, and the temperature as room temperature, $T = 530°R$. The gas constant for water is 85.6 ft-lb$_f$/lb$_m$ – °R (this measurement for water vapor is taken from Table 7–1 and our assumption is that steam is water, no matter what phase it is in). The resulting calculation gives a value of ρ that is totally incorrect. The reason, of course, is that water is a liquid, and the perfect gas law works only for perfect or ideal gases; liquid water is not a gas.

The perfect gas law is one of many equations that relate state properties. Such equations are called *equations of state*. There is a corresponding equation of state for liquids. That equation is $\rho =$ constant, or equivalently $v =$ constant. This says that the density or specific volume of a liquid is constant, no matter what its temperature and no matter what pressure is exerted on it. Table 8–1 shows that the density of water is 62.4 lb$_m$/ft^3. We might say that the state equation for "perfect liquids" is $\rho =$ constant. Because the density and specific volume of a liquid is relatively the same for all temperatures and pressures, these values can be listed on tables such as Table 8–1. No calculations are necessary when finding the density or specific volume of a liquid.

A material such as a liquid is called incompressible: its specific volume does not change with pressure. Most solids are incompressible, too. Their equation of state is also $v =$ constant, although as Table 8–1 illustrates, the

Table 8–1

DENSITY AND SPECIFIC VOLUME OF SOLIDS AND LIQUIDS

	ρ lb_m/ft^3	v ft^3/lb_m
Solids		
Nonmetallic		
Glass	156.25	.0064
Ice	57.80	.0173
Hard rubber	74.62	.0134
Butter	60.50	.0165
Wood	40.00	.0250
Metallic		
Aluminum	169.49	.0059
Copper	555.55	.0018
Steel and iron	487.80	.0020
Liquids		
Ammonia	37.73	.0265
Oil (automobile)	55.55	.0180
Ethyl glycol	69.93	.0143
Water	62.4	.0160

specific volume or density of the solid may not be the same as that of the material as a liquid. Ice floats on water; the density of the solid is less than that of the liquid. Solid gold sinks in liquid gold; the solid density is greater than the liquid density.

Figure 8–9 illustrates the equation of state for a gas (air) plotted for one temperature (32°F). Figure 8–10 shows the equation of state for steam plotted for many temperatures. The equation of state for liquid steam (water) and for solid steam (ice) is v = constant, a vertical line on this graph. The specific volume of water is virtually constant under all conditions (pressure and temperature) and the same is true for ice.

This discussion makes our investigation of equations of state for all materials more complete, except for gases that do not act ideal. The perfect gas law, and the experiments of Chapter 2, depend on the ideal gas model, which assumes that molecules are not attracted to each other but instead go in a straight path. If, in fact, they are attracted by intermolecular forces, then deviations from the perfect gas law occur. For these cases, it is better to use data collected in the lab to establish the state of the gas, rather than a formula such as the perfect gas law. In fact, the data presented in Figure 8–10 for the gas region of steam did not come from the perfect gas law but from laboratory data. It looks much like the perfect gas law, because steam acts like a perfect gas throughout most of the graph. For example, at a pressure of 2116 psfa and a temperature of 350°F (the situation of Sample Problem 4–5 where we applied the perfect gas law), the perfect gas law yields $v = 2116/85.6 \times 810°R = 32.7$

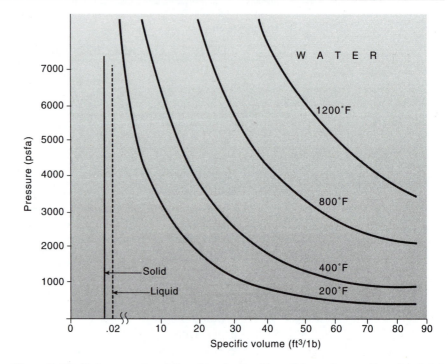

Figure 8–10. State property relationships for liquids, solids, and gases (from lab data).

ft³/lbm whereas the graph (interpolating) yields a value of $v = 32.1$, virtually identical. But for other values of P and T, there could be significant error (see Chapter 18).

SAMPLE PROBLEM 8–5

Compare the value of v for steam for the condition of $P = 1000$ psfa and $T = 400°F$ as determined from the perfect gas law and Figure 8–10.

SOLUTION:

$$v = \frac{85.6 \times 860}{1000} = 73.6$$

From the figure, $v = 75$

This represents a deviation of 3%.

A more detailed study of perfect gas inaccuracies is presented in Chapter 18, in which we will investigate processes that take a material from the gaseous state to the liquid state.

**APPLICATIONS
OF
CALCULUS**

Chapters 6, 7, and 8 all involved the energy equation; the only difference in the use of the energy equation between those chapters was the formulation of the WORK term. In each chapter, WORK was found from a different equation. Does this mean that there is no common definition of WORK? In the mathematics of algebra, WORK requires a different calculation for each process, but in the mathematics of calculus there is a common definition.

WORK has been described as $F \times distance$, or as $P \times (V_2 - V_1)$. If we were to consider a piston moving from the expansion of a gas from a differential viewpoint, we could say that the gas expands a small amount dV under the push of the pressure P, and the *differential* amount of WORK done is

$$dW = PdV$$

Since this is true no matter what the overall process is, we can write the differential form of the energy equation as

$$dQ = c_v mdT + \frac{P}{J}dV$$

To transform this equation into the overall process energy flows, we must integrate

$$Q = \int dQ = m\int c_v dT + \frac{1}{J}\int PdV$$

From this it is clear that

$$WORK = \int_{V_1}^{V_2} PdV \quad \text{(ft-lb}_f \text{ units)}$$

Chapter 6 involved constant volume processes, where $\int PdV$ has limits of V_1 to V_1 and therefore $= 0$, which agrees with our result from Chapter 6.

Chapter 7 involved constant pressure processes where

$$\int PdV = P\int dV = P(V_2 - V_1)$$

The integral form of WORK gives the proper algebraic result for constant pressure processes.

**SAMPLE
PROBLEM 8–6**

Show that the integral form for WORK yields the proper algebraic equation for the case of a constant temperature process.

SOLUTION:

For this case, the pressure changes during the expansion; therefore, the pressure is a function of the expanded volume, $P(V)$. To find this relationship, check the perfect gas law, $P = mRT/V$. Since T = constant, the complete numerator is a constant, and

$$WORK = \int \frac{mRT}{V} dV = mRT \int \frac{dV}{V} = mRT \left(\ln V_2 - \ln V_1 \right) = mRT \ln \left(\frac{V_2}{V_1} \right)$$

Since T is a constant,

$$mRT = P_1 V_1 = P_2 V_2 \text{, so } WORK = P_1 V_1 \ln \left(\frac{V_2}{V_1} \right) = P_2 V_2 \ln \left(\frac{V_2}{V_1} \right)$$

$$= P_1 V_1 \ln \left(\frac{P_1}{P_2} \right)$$

CHAPTER **8**

PROBLEMS

1. For the constant temperature processes indicated below, determine the missing property and tell whether the process involves compression or expansion:
   a. $P_1 = 10$ psia, $V_1 = 1$ cu ft, $P_2 = 15$ psia, $V_2 = ?$
   b. $P_1 = 10$ psia, $V_1 = 1$ cu ft, $P_2 = 5$ psia, $V_2 = ?$
   c. $V_1 = 5$ cu ft, $P_2 = 20$ psia, $V_2 = 7$ cu ft, $P_1 = ?$
   d. $P_1 = 14.5$ psia, $V_1 = 2$ cu ft, $V_2 = 4$ cu ft, $P_2 = ?$
   All processes occur without any escape of gases.

2. Calculate the WORK done for each of the situations in Problem 1.

3. During a constant temperature process, .05 lb_m of H_2 gas increases in pressure from 30 psig to 60 psig. Find the final volume if the gas density is initially .09 lb_m/ft^3.

4. Label each of the following thermodynamic situations as:
   a. constant pressure process
   b. constant temperature process
   c. constant volume process
       (1) Compressing air in a water-cooled compressor
       (2) Heating air in a closed container
       (3) Air-conditioning a room to maintain 72°F

5. The air in a cylinder is captured at 14.7 psia and $\rho = .02$ lb_m/ft^3.
   a. What will the specific volume of the air be as the cylinder compresses the air to 20 psia, 25 psia, 50 psia, and 100 psia? Assume that the air temperature did not change.
   b. What is the equation that relates P and v?
   c. On a graph, carefully plot P versus v as you determine them above.

6. A jar is sealed with 1 lb of atmospheric air (STP).
   a. Then it is heated to 500°F. What is the pressure in the jar?
   b. Air is now allowed to escape until the pressure falls back to 2116 psfa. How much air escapes? (Temperature remains constant)
   c. After equalizing the pressure, the jar is allowed to cool to its original temperature. What is the pressure inside after the temperature has reached 70°F?

7. A gas having an initial volume of 4 cu ft with a gauge pressure reading of 200 psi is cooled under constant temperature until its volume decreases to 1.75 ft^3.
   a. Determine the final pressure.
   b. How much WORK is done?

8. A round water barrel 3 ft high and 2 ft in diameter is half filled with water. (Figure 8–11) The remainder contains air at atmospheric pressure of 2116 psfa and temperature of 70°F. A faucet at the bottom is opened and water flows. Not all the water flows out, because a partial vacuum

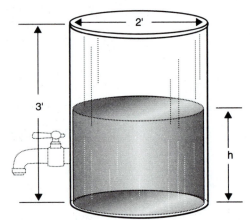

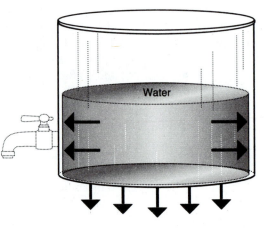

Figure 8–11. Problem 8: A water barrel.

is created over the water that holds some of the water in the barrel. If the pressure that holds water back in the barrel to height h (feet) is given by the equation:

$$P = 2116 - 62.4 \, \frac{lb_m}{ft^3} \times h$$

how many inches of water remain in the barrel if the air is expanded by a constant temperature process? (Hint: Find the pressure of the air cushion after it expands due to the falling water column, in terms of V_1 and V_2 (air volumes). Calculate V_1 and find V_2 in terms of h, then set this pressure equal to the hydrostatic formula given above.)

9. Lungs act like a human bellows or balloon; they inflate and deflate according to the action of the diaphragm. (Figure 8–12) The deflated capacity of an average adult lung system is 100 cu in (0 psig). When inflated, the lungs will pull in an additional 40 cu in of atmospheric air and pressurize it to about 1.5 psig. This pumping process is best described as a constant temperature process.
   a. Can you envision this thermodynamic process occurring with $m_1 = m_2$? If so, draw a picture of the volume of gas before and after the inhaling process.
   b. What is the volume of the inflated lungs?
   c. How much WORK (Btu's) is necessary to perform one inflation?
   d. If an adult breathes about 12 times per minute, how long will it take to use the energy from one order of McDonalds French Fries (120,000 calories)? (1 Btu = 252 calories)

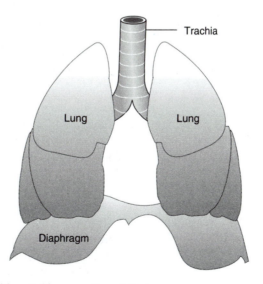

Figure 8–12. Problem 9: The operation of the lung.

10. During a process a gas undergoes a change of volume from 50 cu ft to 25 cu ft at a constant temperature of 200°F and an initial pressure of 20 psia.
    a. What is the curve representing the path on the P-V diagram?
    b. Is WORK done on or by the medium? How many foot-pounds of WORK are done?
    c. If 128.5 Btu leave the gas in the form of heat, what is the change in internal energy?

11. An air pump with a displacement of 144 in³ (Figure 8–13) is used to compress air into a 5-ft³ tank that is initially filled with atmospheric air at 70°F. Assume no temperature change during the pumping and do the following:
    a. Find the gauge pressure in the tank after 100 strokes of the piston.
    b. How much WORK (ft-lb) is required to fill the tank?
    c. Is any heat added or rejected from the gas during the pumping? If so, how much?

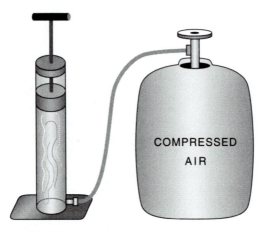

Figure 8–13. Problem 11: A hand pump.

12. When the water boils, small bubbles form at the bottom and get larger as they rise. (Figure 8–14) Bubbles get larger due to the decreasing pressure exerted on them by the water as they rise. The pressure on the bubble is predicted by the hydrostatic equation for gauge pressure:

$$P = 62.4 \ \frac{\text{lb}_m}{\text{ft}^3} \times h$$

where 62.4 lb$_m$/ft³ is the density of water and h is the depth of the bubble from the top water level. The expansion of the bubble as it is rising is a constant temperature process.

a. If water is boiling in a 10-ft deep vessel at 212° F, what is the volume occupied (ft³) of a bubble that contains .0001 lb of steam?
b. How big (volume) does the bubble get when it reaches the top? (Remember that the expansion is a constant temperature thermodynamic process)
c. How much WORK is done by the bubble as it expands?

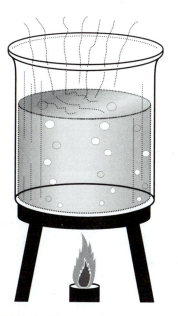

Figure 8–14. Problem 12: Water boiling.

13. The two-sided vessel in Figure 8–15 contains oxygen gas on one side, while the other side contains nothing at all. The sides are separated by a special device called a "WORK valve." When the valve is closed, it completely isolates tanks (A) and (B). When it is opened, the gas flows through a rotating wheel, causing a shaft to spin. The initial condition of tank A is $V_A = 1$ ft³, $P_A = 2000$ psia, $T_A = 70°F$; while in B, $P_B = 0$ psia and $V_B = 3$ ft³.

a. Find the pressure, volume, weight, and temperature in the vessel *after* the valve is opened if the resulting process occurs at a constant temperature.
b. According to the constant temperature energy equation, was any WORK done? If so, how much?
c. Something must move if WORK is done. What device moves?

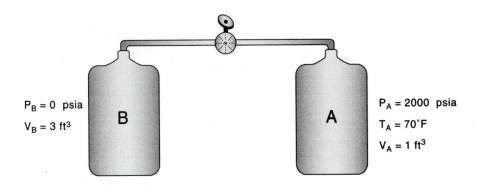

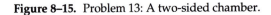

$P_B = 0$ psia

$V_B = 3$ ft³

B

A

$P_A = 2000$ psia

$T_A = 70°F$

$V_A = 1$ ft³

Figure 8–15. Problem 13: A two-sided chamber.

14. Which elevator is more efficient in turning heat into elevator height?
    a. constant P
    b. constant T

15. What is the equation for work in a constant T process?
    a. 0
    b. $P_1 V_1 \ln P_1/P_2$
    c. $P(V_2 - V_1)$

16. What process is governed by the equation $PV =$ constant?

17. WORK can be computed as the area under the pressure line on a
    a. P-V diagram
    b. P-T diagram
    c. P-ρ diagram
    d. P-Q diagram

Problems in SI Units

18. A gas having an initial volume of .5 m^3 and a pressure of 1000 kPa is cooled under constant temperature until its volume decreases to .3 m^3.
    a. Determine the final pressure.
    b. How much WORK was done?

19. An air pump (Figure 8–13) with a displacement of .01 m^3 is used to compress air into a .15 m^3 tank that is initially filled with atmospheric air at 21°C. Assume no temperature change during the pumping and do the following.
    a. Find the absolute pressure in the tank after 100 strokes of the piston.
    b. How much WORK (J) is required to fill the tank?
    c. Is any heat added or rejected from the gas during the pumping? If so, how much?

20. The two-sided vessel of Figure 8–15 has a volume B of .3 m^3 and V_A = .1 m^3, and the left side is initially evacuated while the right side is pressurized with 2000 kPa of oxygen at 20°C. Once the valve is opened, the gas flows through a rotating wheel, doing WORK.
    a. Find the pressure, volume, weight, and temperature in the vessel after the valve is opened if the resulting process occurs at constant temperature.
    b. According to the constant temperature energy equation, was any WORK done?
    c. Something must move if WORK is done. What device moves?

CHAPTER
9

Fundamental Processes: Adiabatic

PREVIEW: Now that we have studied processes that heat without increasing temperature, here is another that is equally strange: a thermodynamic process that involves no heating.

OBJECTIVES:
- ❑ Manipulate fractions with noninteger powers.
- ❑ Discover that the speed at which a process takes place is an important parameter in determining the type of process that occurs.
- ❑ Define *adiabatic process* and calculate properties of a gas after an adiabatic process.

9.1 CHARACTERISTICS OF THE ADIABATIC PROCESS

There is a fourth process, very distinct from the three previously studied, that may occur more often than any of the others—several billion times every second throughout the world. Yet, this fundamental process involves *no heat transfer* whatsoever to a gas! We have become accustomed to illustrating the other thermodynamic processes by adding heat to a gas and seeing how it reacts to the addition—whether it proceeds with a constant temperature, a constant pressure, or a constant volume. The energy equation was written specifically so we could tell where the heat went once it was added to the gas. Now we are going to take a look at a thermodynamic process that involves no heating.

First, let us begin by reviewing the energy equation as developed in Chapter 5.

Heat Added = Internal Energy + WORK + Latent Heat

In words, it states that all heat added to a substance is stored as internal energy, used to do WORK, or converted into latent heat. But what does the energy equation say if there is no heat added to a substance? Does the fact that the sum of internal energy and WORK and latent heat equals zero mean that each of the component terms must equal zero? The answer is, no. It is possible that a process can take place in which there is no heat added but some positive WORK is done and internal energy or latent heat is decreased to accomplish the WORK. In this way, the sum of the addends on the right side of Equation 9–1 are zero, without any of the individual terms being zero.

To clarify this concept, assume that the substance used in this process is a gas and remains a gas throughout the process. Then the latent heat is zero and the energy equation reads:

$$0 = \textit{Internal Energy} + \textit{WORK}$$

or

$$\textit{WORK} = - (\textit{Internal Energy})$$

For this situation, the energy equation states that even though $Q = 0$, a process can occur in which WORK is done at the expense of a decrease in internal energy. It is a comforting concept that sometimes we can take energy that has been stored and convert it into WORK.

SAMPLE PROBLEM 9–1

One pound of steam is confined in a cylinder. Although no heat is added or removed from the gas, the steam temperature decreases to indicate a 4-Btu loss of internal energy. If no steam was converted to water, how much WORK was done?

SOLUTION:

In this case, there is no latent heat involved, and the amount of heat added $Q = 0$.

$$0 = \textit{Internal Energy} + \textit{WORK}$$
$$0 = -4 + \textit{WORK}$$
$$= 4 \text{ Btu}$$

A process that involves no heat addition implies that there will be either a decrease in the stored energy of the gas and a corresponding positive WORK done *or* an increase in the stored energy of the gas with a negative amount of WORK being done. Negative WORK means that WORK is being done not *by* the gas, but rather *on* the gas. Such a process, in which no heat is added or removed, is called an *adiabatic* (ā-dee-uh-BAT-ic) *process*.

❖ KEY TERM:

Adiabatic Process: A process where no heat is added ($Q = 0$) even though the temperature of the substance increases or decreases.

Figure 9–1 illustrates one of the most typical sutuations in which an adiabatic process occurs. The device is simply a piston in a cylinder being mechanically held in place against the pressure and temperature of an encaptured gas. Once the piston is released, it will rise in the cylinder. Due to this rise, there is some

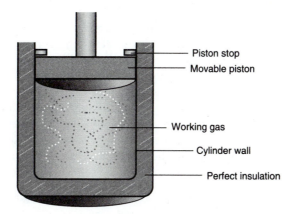

Figure 9–1. A device that sustains an adiabatic process.

WORK done by the gas. There is also a distinct change in pressure, density, and temperature. If no heat is added during the piston travel, the process is an adiabatic one. At least it appears that no heat is added. If, in fact, the cylinder is insulated from its surroundings so that no heat can flow through the walls of the cylinder, the process is guaranteed to be adiabatic. The process would begin with a gas containing a large amount of stored energy and end in a substantial amount of WORK being done on the piston.

To graphically illustrate the changes in temperature, pressure, and volume during this adiabatic process, consider the graphs in Figure 9–2. At the beginning of the process, the occupied volume is small, and the temperature is very high. As the process occurs and the piston starts to rise, the occupied volume becomes larger, and the temperature becomes somewhat smaller.

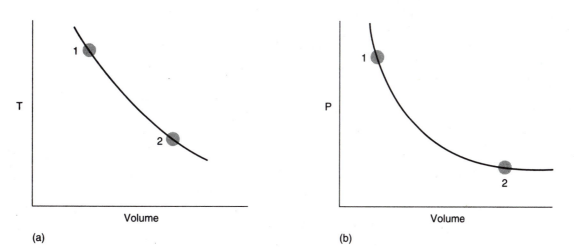

Figure 9–2. Graphs of an adiabatic process.

Figure 9–2 (a) shows that result. Similarly, Figure 9–2 (b) shows what happens to the pressure throughout the process: the pressure drops.

A pressure-volume diagram such as that in Figure 9–2 (b) was used in previous chapters to analyze other processes. In those cases, we were able to obtain an equation that represents the P-v curve. In this case, there is also an equation that describes that curve. However, the equation is not as easy to discover as it was with previous processes; that would require a great amount of mathematical work (see p. 202). The equation is Pv^k = constant; or in the special case where the mass of the gas also does not change, PV^k = constant. It is similar to the equation resulting from the constant temperature process, Pv = constant. The only difference is the constant exponent k.

SAMPLE PROBLEM 9–2

Plot the equation $PV^{1.5} = 25$ and show that it has an appearance similar to Figure 9–2 (b).

SOLUTION:

Rewrite the equation $P = 25/V^{1.5}$, and then substitute a variety of values for V. (See Figure 9–3.)

P	V
25.0	1
8.8	2
4.3	3
3.1	4
2.2	5

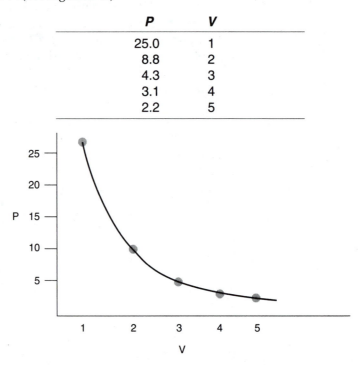

Figure 9–3. Sample Problem 9–2.

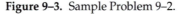

The exponent k is a constant that is different for each gas; k is equal to the ratio c_p/c_v where c_v and c_p are the two types of specific heats defined in Chapter 7. These two specific heats, which were defined with different processes in mind, now play an important role in adiabatic processes and are tabulated in Table 7–1.

The equation Pv^k = constant can be more descriptively written as:

$$P_1 v_1{}^k = P_2 v_2{}^k$$

Equation 9–1

but it has many forms, such as:

$$\frac{P_2}{P_1} = \left(\frac{v_1}{v_2}\right)^k$$

With some creative manipulation, Equation 9–1 has even more unique and helpful disguises:

$$\frac{v_1}{v_2} = \left(\frac{P_2}{P_1}\right)^{\frac{1}{k}}$$

Equation 9–2

$$\frac{v_1}{v_2} = \left(\frac{P_2}{P_1}\right)^{\frac{1}{k}} = \frac{\dfrac{RT_2}{V_2}}{\dfrac{RT_1}{V_1}} = \left(\frac{v_1 T_2}{v_2 T_1}\right)^{\frac{1}{k}}$$

or

$$\left(\frac{v_1}{v_2}\right)^k = \frac{v_1}{v_2}\frac{T_2}{T_1}$$

$$\left(\frac{v_1}{v_2}\right)^{k-1} = \frac{T_2}{T_1}$$

Equation 9–3

With a similar manipulation, the following relation also can be derived:

$$\frac{T_2}{T_1} = \left(\frac{P_2}{P_1}\right)^{\frac{k-1}{k}}$$

Equation 9–4

SAMPLE
PROBLEM 9–3   The cylinder in Figure 9–4 contains 3 ft^3 of steam at atmospheric pressure and 212°F. It is desired to adiabatically compress this gas to a pressure of 200 psig. To what volume must the gas be compressed, and what is the resulting gas temperature?

SOLUTION:

For steam $c_v = .335$ Btu/lb$_m$ – °F

$$c_p = .445 \frac{\text{Btu}}{\text{lb}_m - °F}$$

$$k = \frac{c_p}{c_v} = \frac{.445}{.335} = 1.33$$

The following conditions are known:

$T = 672°R, \qquad V_1 = 3$ cu ft, $\qquad m_1 = m_2$ (no gas lost)

$P_1 = 14.7 \times 144 = 2116$ psfa

$P_2 = 214.7 \times 144 = 30915$ psfa

Then

$$\frac{v_1}{v_2} = \left(\frac{P_2}{P_1}\right)^{\frac{1}{k}} = \left(\frac{30,916}{2116}\right)^{\frac{1}{1.33}}$$

$$= 14.6^{.75} = 10.98$$

But

$$\frac{v_1}{v_2} = \frac{V_1/m_1}{V_2/m_2} = \frac{V_1}{V_2}, \text{ since } m_1 = m_2$$

$$V_2 = \frac{V_1}{10.98} \qquad\qquad \frac{T_2}{T_1} = \left(\frac{V_1}{V_2}\right)^{k-1}$$

$$= \frac{3}{10.98} \qquad \text{AND} \qquad = \left(\frac{3}{.27}\right)^{33} = 2.21$$

$$= .27 \qquad\qquad\qquad T_2 = (1.83) \times 672°R = 1485°R$$

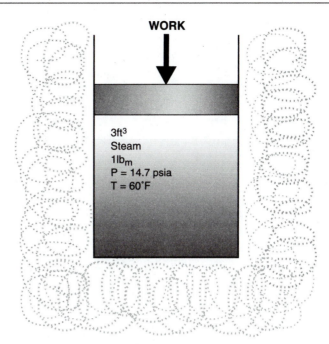

Figure 9–4. Sample Problem 9–3.

SAMPLE
PROBLEM 9–4

For an adiabatic process that occurs with no loss of gas ($m_1 = m_2$), the adiabatic relation PV^k = constant is true. For the process of Sample Problem 9–3, plot pressure versus volume and compare this graph to the same situation if it had occurred at a constant temperature. (See Figure 9–5.)

SOLUTION:

The equation we shall graph is:

$$PV^{1.33} = (2116)\,(3)^{1.33}$$

or

$$PV^{1.33} = 9122$$

Substituting values V = 1, 2, 3, 4, 5, the following table is generated:

V (ft³)	P (lb$_f$/ft²)
.5	22,932
.75	13,373
1	9122
2	3628
3	2116

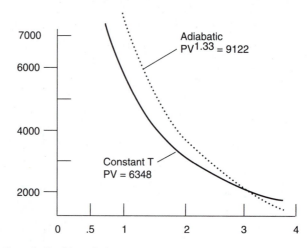

Figure 9–5. Sample Problem 9–4.

If this were a constant temperature process with the equation: $PV = $ constant $= P_1V_1$

V (ft³)	P (lb/ft²)
.5	12,696
.75	8464
1	6348
2	3174
3	2116

Comparison: Pressure rises faster for an adiabatic process in compression over a constant temperature process and falls faster in expansion.

9.2 WORK IN AN ADIABATIC PROCESS

The adiabatic process that we have investigated involves WORK. Sample Problem 9–3, in particular, dealt with compressing air through an adiabatic process by doing WORK on the gas. To answer the question of how much WORK is done during the process, we must turn again to the P-V diagram.

The WORK done during any process is simply the area under the corresponding P-V diagram. From Sample Problem 9–4, it appears that this area for an adiabatic process looks very similar to the area for a constant temperature process. Maybe this implies that the equations describing them are also similar. This is not the case. The WORK done in an adiabatic process is given by the simple relation:

$$WORK = \frac{(P_2V_2 - P_1V_1)}{(1-k)}$$

The energy equation for an adiabatic process is then:

$$0 = c_v m(T_2 - T_1) + \frac{(P_2 V_2 - P_1 V_1)}{J(1-k)}$$

SAMPLE PROBLEM 9–5

Calculate the WORK done in Sample Problem 9–3.

SOLUTION:

$P_2 = 30916$ psfa

$V_2 = .27$

$P_1 = 2116$ psfa

$V_1 = 3$ cu ft

$k = 1.33$

$$WORK = \frac{(30916 \times .27 - 2116 \times 3)}{-.33} = -6053 \text{ ft–lb}_f (-7.78 \text{ Btu})$$

9.3 APPLICATIONS: AIR COMPRESSORS

Adiabatic processes are very important to understand and analyze because they occur frequently in engineering analysis. Anytime a gas changes its state without heat being added, the process occurs adiabatically.

Figure 9–6 shows a typical situation in which an adiabatic process occurs—an air compressor. As the piston moves inward, the gas pressure gets higher, the temperature rises, and the gas gets more dense. The state of the gas is surely changing, but no heat is being added. Temperature rises even though no heat is added. This is an adiabatic compression.

SAMPLE PROBLEM 9–6

An air compressor (Figure 9–6) takes air into a .25-ft^3 (7×10^{-3} m^3) cylinder and compresses it to .025 ft^3 ($.7 \times 10^{-3}$ m^3). If the air is originally at standard temperature and pressure, and the compressor is adiabatic, at what temperature and pressure will the gas be exhausted? Assuming $k = 1.41$, how much WORK must be done on the gas to accomplish the compression? By how much is the internal energy of the captured gas changed?

SOLUTION:

Given: $P_1 = 14.7$ psia = 2116 psfa

$T_1 = 70°F = 530°R$

$V_1 = .25$ cu ft, $V_2 = .025$ cu ft

Figure 9–6. Adiabatic air compressor.

$$m = \frac{P_1 V_1}{R T_1} = \frac{(2116 \times .25)}{(53.3 \times 530)} = .0187 \text{ lb}$$

$$T_2 = T_1 \left(\frac{V_1}{V_2} \right)^{k-1} = 530 \left(\frac{.25}{.025} \right)^{.41} = 1331^\circ \text{R}$$

$$P_2 = P_1 \left(\frac{V_1}{V_2} \right)^{k} = 2116 \text{ psfa} \left(\frac{.25}{.025} \right)^{1.41} = 53{,}151 \text{ psfa}$$

$$WORK = \frac{(P_2 V_2 - P_1 V_1)}{(1-k)} = -1991 \text{ ft–lb}_f \left(-2.56 \text{ Btu} \right)$$

$$Internal\ Energy = c_v m (T_2 - T_1) = .17 \times .0187 \times (1331 - 530) = 2.56 \text{ Btu}$$

Solution in SI Units

$P_1 = 101.3 \text{ kPa}$

$T_1 = 21^\circ \text{C} = 294 \text{ K}$

$$m = 101.3 \times \frac{7 \times 10^{-3}}{(.287 \times 294)} = .0084 \text{ kg}$$

$$T_2 = 294 \left(\frac{7 \times 10^{-3}}{.7 \times 10^{-3}} \right)^{.41} = 738 \text{ K}$$

$$P_2 = 101.3 \times \left(\frac{7 \times 10^{-3}}{.7 \times 10^{-3}} \right)^{1.41} = 2544 \text{ kPa}$$

$$WORK = -2.68 \text{ kJ}$$
$$IE = .71 \times .0084 \times (738 - 294) = 2.68 \text{ kJ}$$

Notice that in both Sample Problems 9–3 and 9–6, the gas temperature initially was at standard temperature, which presumably was the same temperature as the air surrounding the vessel. This means that the gas inside the vessel was "in equilibrium" with the gas outside the vessel. But as the process goes on (either expansion or compression), the gas temperature in the vessel either increases or decreases. When this happens, heat starts to flow naturally from the warmer temperature to the colder temperature through the vessel walls. If this is allowed to happen, the process is not adiabatic. The assumption that the processes are adiabatic in Sample Problems 9–3 and 9–6 may not be a very good assumption. In fact, no situation is truly adiabatic, although many are very nearly so. If the vessel is surrounded with insulation, the heat transfer to the interior gas is cut down dramatically, and the process can be considered to be adiabatic. Similarly, if the process occurs very rapidly, heat has no time to transfer into or out of the gas. For an actual air compressor, the process can be assumed to be adiabatic because of the short time duration of each compression stroke, even though there is little insulation.

Whether a process is adiabatic or not may depend more on how long it takes than how it is done or what the apparatus looks like. Consider the inflation of an automobile tire. If the tire is filled with air very rapidly, the tire walls act as enough of an insulator so that the filling process may be considered to be adiabatic. If the process occurs very slowly, there is enough time for heat transfer through the tire wall to occur. Anytime the tire air temperature changes, heat will flow to bring the temperature back

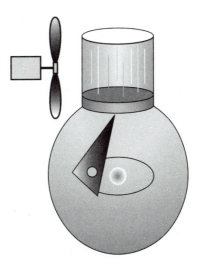

Figure 9–7. A constant temperature compressor.

Figure 9–8. An air-cooled refrigeration compressor. (*Courtesy Coppland Corporation.*)

to its initial point. In this situation, the process is best described as constant temperature.

A good rule of thumb by which to decide if a process is adiabatic or constant temperature is: If the process takes longer than one second, it is constant temperature. If it takes less than one second, it is adiabatic (unless special provisions have been made for constant volume or constant pressure).

Some air compressors are designed to improve the conditions for heat to escape the cylinder to ensure Q is not zero. They operate at slow speeds and have outside air blowing across the head to improve heat transfer, or extend cast-iron fins on the cylinder head to increase heat flow. Figure 9–7 shows a diagram of such a compressor, and Figure 9–8 is a photograph of an air-cooled compressor. If the heat transfer is one hundred percent perfect, the compression process can take place at a constant temperature.

SAMPLE PROBLEM 9–7

Consider a constant temperature compressor with the same specifications as the adiabatic compressor in Sample Problem 9–6. Calculate the temperature and pressure of the compressed air, the WORK done, and the change in internal energy. Then compare and evaluate this compressor against the adiabatic design.

SOLUTION:

$$P_1 = 14.7 \text{ psia} = 2116 \text{ psfa}$$

$$T_1 = 70°F = 530°R$$

$$V_1 = .25 \text{ cu ft}$$

$$V_2 = .025 \text{ cu ft}$$

$$m = .0187 \text{ lb}_m$$

$$T_2 = T_1 = 530°R$$

$$P_2 = P_1 \frac{V_1}{V_2} = 2115 \times \frac{(.25)}{.025} = 21{,}160 \text{ psfa}$$

$$WORK = P_1 V_1 \ln\left(\frac{V_2}{V_1}\right) = 2116 \times .25 \times \ln\left(\frac{.025}{.25}\right)$$

$$= -1213 \text{ ft–lb}_f (-1.56 \text{ Btu})$$

Internal Energy $= c_v m (T_2 - T_1) = 0$, since $T_2 = T_1$

Comparing this design with the adiabatic compressor shows that the constant temperature model requires far less WORK to perform the compression: 1213 ft-lb$_f$ compared with 1991 ft-lb$_f$.

9.4
WORK DONE BY/ON A GAS VERSUS WORK DONE BY/ON A PISTON

In the sample problems of this chapter, we have been calculating the WORK done on a gas to perform various compression and expansion processes. As the piston moves due to the action of the gas inside, is it true that all the WORK done will be transferred *to the piston itself*? If, in particular, the piston is surrounded by atmospheric pressure, and that pressure P_A is pushing down on the piston, doesn't some of the WORK done by the gas have to be used to push the piston against the atmospheric pressure?

Figure 9–9 illustrates that the atmospheric P_A is pushing against all faces of the piston and cylinder arrangement, but since the cylinder is rigid, the pressure there has no effect on the gas *through the cylinder*. But as the *piston* moves, the atmospheric pressure on the back side of the piston must be moved, which requires some of the WORK done by the gas. The remaining WORK is available at the piston.

During the expansion the WORK done on the atmosphere can be calculated by:

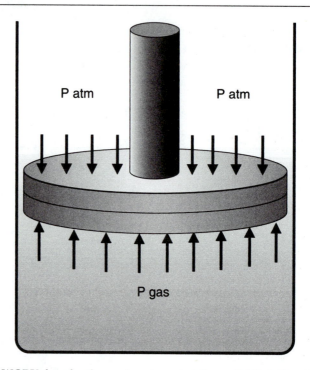

Figure 9–9. WORK done by the gas is not necessarily available to the piston because of the atmosphere.

$$WORK = F \times d = P_A \times A \times d$$

But $A \times d = V_2 - V_1$: that is, the change of the volume of the gas inside. So

$$WORK_A = P_A (V_2 - V_1)$$

This implies that some of the WORK done by the gas goes to push the atmosphere, while some of the WORK is delivered to the piston:

$$WORK_{gas} = W_{delivered} + W_A$$

Equation 9–5

where $WORK_{gas}$ is the work done by/on the gas as calculated by the thermo-dynamic formulas given in Table 9–1 (see end of chapter); $W_{delivered}$ is WORK delivered to/by the piston itself; and W_A is the WORK done by the atmosphere.

SAMPLE PROBLEM 9–8

How much of the WORK done on the gas during the compression described in Sample Problem 9–7 must be performed by the piston if the air compressor is surrounded by atmospheric air?

SOLUTION:

The WORK done on the gas is –1991 ft-lb$_f$. The WORK done by the the atmosphere is:

$$W_A = P_A \frac{(V_2 - V_1)}{J} = 2116\,(.025 - .25) = -474 \text{ ft-lb}_f\,(-.61 \text{ Btu})$$

Therefore:

$$W_{delivered} = W_{gas} - W_A = -1991 - (-474) = -1515 \text{ ft-lb}_f$$

❖ **KEY TERM:**

Work Delivered ($W_{delivered}$, or W_D): Different from WORK done by gas, because the gas must also push the surrounding environment: $W_D = WORK - W_A$

If we want to include the effect of atmospheric pressure in the energy equation, we would write:

$$Q = c_v m\,(T_2 - T_1) + W_D + P_A \frac{(V_2 - V_1)}{J}$$

Notice that this equation is good *for any process* and the equation to determine W_D is found in Table 9–1 (see end of chapter). Remember that the energy equation

$$Q = c_v\,m\,(T_2 - T_1) + WORK$$

is still valid, since *WORK* implies *WORK*$_{gas}$ and not W_D.

The idea that the atmosphere does some WORK also holds true if the process is compression. The atmosphere plays a part in constant temperature processes and constant pressure processes, not just adiabatic ones, as well.

MANUFACTURING TECHNICIAN

Automation has been the goal of manufacturing companies for the last twenty years, but no industry has come close to the automobile industry in the level of automation achieved. The Ford Motor Assembly Plant in Saline, Michigan, is fully automated, and you need not apply for employment there unless you have a degree in technology. There are no machine operators on the assembly line; there is no hand assembly on the assembly line. The few personnel actually on the floor of the plant are Line Coordinators. Each is responsible for monitoring the flow of parts through the section. They also are responsible for the ultimate quality of the product as it leaves the section; therefore, they are required to check tooling and to monitor the wear on the tooling. They schedule preventative maintenance for the machine tools in their sections, which often requires checking them for vibration with spectral vibratometers or checking them for excessive heat buildup with radiation spectrometers.

Each Line Coordinator is supported by three Line Specialists. Each has responsibility for one-third of the assembly line. It is their job to suggest and implement upgrades to the line, either new manufacturing equipment or modifications to existing equipment. For example, the specialist for the center of assembly line #1 has within her responsibility the massive paint spray booth, where a major modification has been going on for six months. The booth divides the line into separate tracts so that multiple cars can be painted simultaneously. Contractors have been building a third bay in the booth that is almost complete.

The final phase of construction is to increase the compressed air capacity to cover four additional robotic spray heads. The decision has been made to move the existing air compressor system to another portion of the line and replace it with a new system that can drive all twelve spray heads for all three booths. To specify the new compressor, the specialist must determine the *free air capacity* of the required air compressor—that is, how much air at standard temperature and pressure the compressor must process to get the required compressed air volume.

SAMPLE PROBLEM 9-9

The spray head manufacturer's catalog indicates that each of the spray heads requires 8 ft³/min (CFM) of compressed air at 120 psig pressure. Determine how much air at standard temperature and pressure the compressor must process to deliver the required compressed air capacity.

SOLUTION:

The total compressed air volume flow rate is 12×8 ft³/min = 96 ft³/min. The manufacturer indicates that the compression process is adiabatic, therefore:

$$\frac{V_1}{V_2} = \left(\frac{P_2}{P_1}\right)^{\frac{1}{k}}$$

where V_1 is the volume of free air ($P_1 = 14.7$ psia) and V_2 is the volume of compressed air. Solving for V_1 yields:

$$V_1 = 96 \times \left(\frac{(120+14.7)}{14.7}\right)^{\frac{1}{1.41}} = 462 \frac{\text{ft}^3}{\text{min}}$$

In the air compressor catalog, the specialist finds that a compressed air station is available for 450 CFM and the next model is for 500 CFM. Although tempted to pick the larger, she realizes that all spray heads never are firing at the same time. This is primarily because cars go through the booth in a staggered fashion, since they arrive at the head of the booth in a staggered fashion. At least one booth is waiting for a car for at least a portion of the time that the other two booths are firing. Therefore, the smaller air compressor station will be more than adequate.

APPLICATIONS OF CALCULUS

The mathematics of calculus can be used to derive the formula for WORK done in an adiabatic process (Section 9–2). For the adiabatic case, the pressure at any point of the process can be represented as $P = P_1 (V_1/V)^k$. Therefore,

$$WORK = \int P dV = P_1 \int \left(\frac{V_1}{V}\right)^k dV = P_1 V_1^k \int \frac{dV}{V^k}$$

Integrating this yields:

$$P_1 V_1^k \left(\frac{1}{(k-1)} V_2^{k-1} - \frac{1}{(k-1)} V_1^{k-1}\right) = \frac{P_1}{(1-k)} \left(\frac{V_1^k}{V_2^{k-1}} - \frac{V_1^k}{V_1^{k-1}}\right)$$

$$= \frac{\left(P_1 \dfrac{V_1^k}{V_2^{k-1}} - P_1 V_1\right)}{(1-k)}$$

But

$$P_2 = P_1 \left(\frac{V_1^k}{V_2^k}\right)$$

so

$$P_2 V_2 = \frac{P_1 V_1^k}{V_2^{k-1}}$$

Therefore,

$$WORK = \frac{(P_2 V_2 - P_1 V_1)}{(1-k)}$$

The mathematics of calculus also can help with the understanding of the adiabatic relation $P_1 V_1^k = P_2 V_2^k$. This chapter, using algebra, is silent on where these forumlas come from. They can be derived by using the differential form of the energy equation and creating a *differential form of the perfect gas law*.

If we differentiate the perfect gas law $PV = mRT$ and use the chain rule of differentiation $d(FG) = G\, dF + F\, dG$, we get

$$d\,(PV) = P\, dV + V\, dP = d\,(mRT) = mR\, dT$$

Now the differential form of the energy equation is:

$$dQ = 0 = m\ cv\ dT + \frac{P}{J}dV$$

or substituting for dT from the differential PGL:

$$0 = \frac{m\ c_v}{m\ R}(P\ dV + V\ dP) + \frac{P}{J}dV = P\left(\frac{c_v}{R} + \frac{1}{J}\right)dV + \frac{c_v}{R}V\ dP$$

but

$$\frac{c_v}{R} + \frac{1}{J} = \frac{c_p}{R} \quad \text{(Equation 7–6)}$$

Therefore,

$$c_p\ P\ dV = -c_v\ V\ dP$$

or

$$\frac{c_p}{c_v}\frac{dV}{V} = -\frac{dP}{P}$$

Now integrating both sides yields

$$k\left(\ln V_2 - \ln V_1\right) = k\ \ln\left(\frac{V_2}{V_1}\right) = -\ln\left(\frac{P_2}{P_1}\right)$$

or

$$\left(\frac{V_2}{V_1}\right)^k = \frac{P_1}{P_2}$$

Table 9–1

THERMODYNAMIC FORMULAS FOR FUNDAMENTAL PROCESSES (PERFECT GAS MODEL)				
Quantity	**Constant Volume**	**Constant Pressure**	**Constant Temperature**	**Adiabatic**	**Polytropic**
Pressure, volume, and temperature	$\dfrac{T_2}{T_1} = \dfrac{P_2}{P_1}$	$\dfrac{T_2}{T_1} = \dfrac{v_2}{v_1}$	$\dfrac{P_1}{v_1} = \dfrac{P_2}{v_2}$	$\dfrac{P_2}{P_1} = \left(\dfrac{v_1}{v_2}\right)^{k}$ $\dfrac{T_2}{T_1} = \left(\dfrac{v_1}{v_2}\right)^{k-1}$ $\dfrac{T_2}{T_1} = \left(\dfrac{P_2}{P_1}\right)^{\frac{(k-1)}{k}}$	$\dfrac{P_2}{P_1} = \left(\dfrac{v_1}{v_2}\right)^{n}$ $\dfrac{T_2}{T_1} = \left(\dfrac{v_1}{v_2}\right)^{n-1}$ $\dfrac{T_2}{T_1} = \left(\dfrac{P_2}{P_1}\right)^{\frac{(n-1)}{n}}$
WORK (gas)	0	$P(V_2 - V_1)$	$P_1 V_1 \ln\left(\dfrac{V_2}{V_1}\right)$	$\dfrac{(P_2 V_2 - P_1 V_1)}{1-k}$	$\dfrac{(P_2 V_2 - P_1 V_1)}{1-n}$
WD (work delivered)	0	$(P - P_a)(V_2 - V_1)$	$P_1 V_1 \ln\left(\dfrac{V_2}{V_1}\right)$ $-P_a(V_2 - V_1)$	$\dfrac{(P_2 V_2 - P_1 V_1)}{1-k}$ $-P_a(V_2 - V_1)$	$\dfrac{(P_2 V_2 - P_1 V_1)}{(1-n)}$ $-P_a(V_2 - V_1)$
Heat (first law)	$Q = c_v m(T_2 - T_1)$	$Q = c_v m(T_2 - T_1)$ $+P\dfrac{(V_2 - V_1)}{J}$ or $Q = c_p m(T_2 - T_1)$	$Q = \dfrac{P_1 V_1}{J} \ln\left(\dfrac{V_2}{V_1}\right)$	$Q = 0$ or $0 = c_v m(T_2 - T_1)$ $+$ $\dfrac{(P_2 V_2 - P_1 V_1)}{J(1-k)}$	$Q = c_v m(T_2 - T_1)$ $+\dfrac{(P_2 V_2 - P_1 V_1)}{J(1-n)}$

Note: For continuous processes, volume (V) is replaced by specific volume (v), $m = 1$, Q is replaced by q in terms of Btu/lb rather than Btu, and WD formulas are in Table 13–1.

CHAPTER **9**

PROBLEMS

1. One pound of air undergoes an adiabatic process during which the temperature increases from 500°R to 750°R.
   a. How much heat was transferred during the process?
   b. By how much was the internal energy increased?
   c. How much WORK was done? Was it done "on" or "by" the gas? Did the gas undergo an expansion or compression?
   d. By what percent did the specific volume decrease?
   e. By what percent did the pressure change?

2. The Mechanical Engineering Technology Department is showing high schools students through their labs on this day. To make the day interesting, the students are challenged with this problem. A two-chambered steel vessel (Figure 9–10) contains compressed air in the lower chamber, and the upper chamber is empty. Both chambers are of the same internal volume and have temperature probes sealed into them. The pressure in the lower chamber is 200 psia and the temperature is 70°F. The piston is held into position until the thermodynamic process is initiated. The students are asked to guess what the temperature of the gas will be

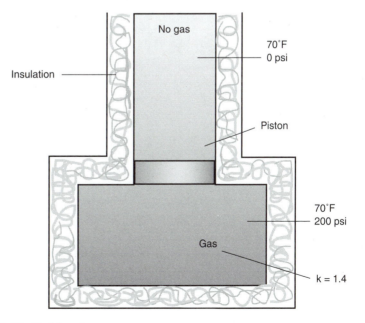

Figure 9–10. Problem 2.

immediately after the piston is released and reaches the top of the evacuated chamber.

3. Cybernetics, Inc., is in the business of thermodynamically freezing people at death in hopes of thawing them at a later date and giving them extended life. The chamber used is shown in Figure 9–11. It is originally 6 ft long and 3 ft high. It is filled with atmosphere air (70°F, 14.7 psia, c_v = .17, c_p = .24). How far will the piston have to be pulled out to reduce the temperature inside down to 100°R?

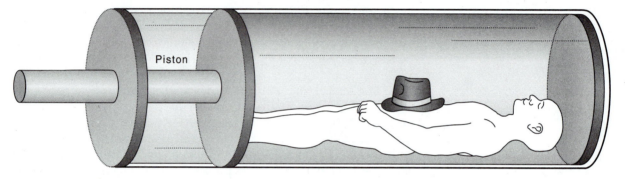

Figure 9–11. Problem 3: Cybernetic's personnel freezing machine.

4. Air held in a cylinder with a moveable piston has an initial pressure and temperature of 2000 psia and 0°F, respectively.
   a. What is the specific volume?
   b. If there is one cubic foot of gas in the cylinder, how much does the gas weigh?
   c. What will be the gas temperature if the gas is compressed adiabatically until its final volume is one-half of its initial volume?
   d. How much WORK is required for this compression?

5. How much mechanical WORK is done by the gas if 6 lb of nitrogen at 25 psig is allowed to expand adiabatically from 13.2 ft³/lb to 15.0 ft³/lb? Answer in ft-lb and Btu.

6. Six pounds of nitrogen are compressed from T=140°F with a volumetric compression ratio of V_1/V_2 = 5. If the process is considered to be adiabatic, find:
   a. the final temperature
   b. the pressurization ratio P_2/P_1
   c. the change in internal energy
   d. the WORK done (specify *on* or *by*)

7. Find the WORK done in the adiabatic compression of 5 lb of argon initally at a pressure of 20 psig and temperature of 80°F to a temperature of 300°F.

8. A gas (air) initially at P_1 = 30,000 psfa, V_1 = .05 ft³, and T = 4000°R can expand either by an isothermal process or by an adiabatic process to a final volume V_2 = .30 ft³. Which of the two methods of expansion will do the most WORK? (Make calculations)

9. Heating fuel can be generated on a farm by bacterial decay of livestock manure. The manure is put in a large barrel (digester) where bacterial action produces methane, which is stored at atmospheric pressure and temperature in a 125-cu-ft vessel. (Figure 9–12)
   a. If the heating value of the methane is 1005 Btu/ft³ (see Table 2–2), what is the total heat value of the methane stored in the vessel?
   b. This methane can be stored in a small container if it is pumped under pressure using a piston-cylinder pump. If the pump compresses the gas in the vessel to 100 psig adiabatically, what will the compressed volume be?
   c. How much WORK must be done to compress the gas? (ft-lb and Btu)
   d. What will be the resulting temperature?

Methane c_p = .59, c_v = .47,   R = 96.4   (Table 7–1)

Figure 9–12. Problem 9.

e. Sometimes the processing of natural fuel requires more energy than the heat energy will provide. Compare the amount of energy required to pump against the heat energy in the fuel.

10. Complete all four parts below:
   a. Air is compressed at constant temperature from standard temperature and pressure and volume = 258 in³ to a volume of 64.4 in³. Find:
      (1) final pressure
      (2) WORK done *on* or *by* the gas (specify)
      (3) amount of heat *added* or *rejected* from the gas (specify)
      (4) draw a rough P-V diagram of this process
   b. From part a, after the gas is compressed isothermally, it is further compressed adiabatically to a volume of 32.2 in³. Find:
      (1) final pressure
      (2) final temperature
      (3) WORK done *on* or *by* the gas (specify)
      (4) amount of heat added or rejected from the gas (specify)
      (5) on the same paper used for part a, draw a P-V diagram of this process
   c. From part b, the gas is next expanded at constant pressure back to its original volume (258 cu in). Find:
      (1) final temperature
      (2) WORK done *on* or *by* the gas (specify)
      (3) amount of heat *added* or *rejected* from the gas (specify)
      (4) draw a rough P-V diagram of the processes
   d. From part c, the gas is cooled at a constant volume until it reaches 70°F. Find:
      (1) final pressure
      (2) WORK done *on* or *by* gas (specify)
      (3) amount of heat *added* or *rejected* from the gas (specify)
      (4) draw a rough P-V diagram of the processes

11. Below are three descriptions of the energy equation of thermodynamics, more commonly called the *first law of thermodynamics*. Each description looks different from the energy equation in this chapter, but in fact, they are identical. Identify each term, or group of terms, by the common names we have used: *heat (Q)*, *internal energy*, and *WORK*.
   a. The first law of thermodynamics for a closed system is:

$$Q_2 - Q_1 = E_2 - E_1 + WORK$$

   b. The quantity of waste heat available from a specific process is usually determined by direct measurement or by calculating the amount of waste energy from an energy balance equation. An energy balance of one process or an entire factory is based on the first law of thermodynamics: "Energy cannot be created or destroyed." Mathematically, this law can be stated as: "net heat in, minus WORK out, equals change in internal energy:

$$Q_{added} - Q_{loss} - W = U$$

Therefore, this equation can be used to calculate the amount of heat lost, or waste heat from a closed system.

c. The first law of thermodynamics, one of the most important laws of nature, is the law of conservation of energy. Although the law has been stated in a variety of ways, all have essentially the same meaning. The following are typical statements: whenever energy is transformed from one form to another, energy is always conserved, being neither created nor destroyed; the sum total of all energy remains constant.

$$JQ - W = JE - (Je - pv)\, dw$$

12. In Problem 3, you found that it took a 387-ft stroke on the piston to reduce the temperature inside down to 100°R. If the gas inside were argon, how far would the piston have to be pulled?

13. Which one of the formulas below are correct manipulations of the adiabatic P-V-T relations of Table 9–1, and which are incorrect?

a. $T_2 = T_1 \left(\dfrac{P_2}{P_1}\right)^{\frac{(k-1)}{k}}$

b. $V_2 = V_1 \left(\dfrac{T_1}{T_2}\right)^{1-k}$

c. $\left(\dfrac{V_1}{V_2}\right)^{k-1} = \dfrac{T_2}{T_1}$

d. $\dfrac{T_2}{T_1} = \left(\dfrac{P_2}{P_1}\right)^{\frac{k}{k-1}}$

e. $V_1 = V_2 \left(\dfrac{T_1}{T_2}\right)^{\frac{1}{1-k}}$

f. $V_2 = V_1 \left(\dfrac{P_2}{P_1}\right)^{k}$

14. Figure 9–13 shows two concepts for an air compressor design. Both are reciprocating piston-in-cylinder approaches. The one on the left is a

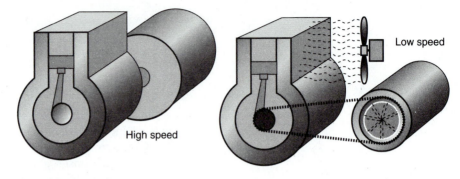

Figure 9–13. Problem 14: Comparison of two types of air compressors.

high-speed model, whereas the one on the right is a low-speed design outfitted with a fan for increased cooling.

a. What two fundamental processes are being described by the functions of these two designs above?

b. If both compressors have 100-cu-in displacement and a compression ratio V_1/V_2 of 20:1, how much WORK is required on the gas for each case to have one compression stroke (initial conditions are STP)?

c. Which design is more efficient (takes less WORK)? Also, by what percent is it more efficient than the other design?

15. a. Figure 9–14 is a diagram of a gas shock absorber. The piston supports a weight of 1000 lb in suspending a car. The enclosed air occupies 1 ft³ with a piston area of 1/4 ft². How much of a deflection is made in

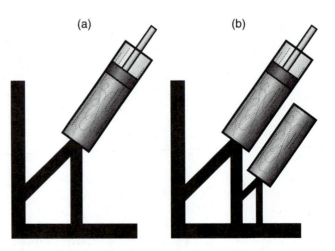

Figure 9–14. Problem 15: Shock absorbers.

the shock if the bump increases the suspended weight to 1500 lb? Be sure not to use gauge pressure.

b. An alternative design is to add a reservoir of air to double the enclosed volume without changing the piston area. How much deflection will be provided by the piston with a 1500-lb bump?

c. Which design shows the tightest suspension?

16. In Problem 3, determine the amount of WORK needed to "freeze" the corpse
   a. neglecting atmospheric pressure
   b. including atmospheric pressure

17. Hydrogen gas is compressed from an initial state of $P_1 = 20$ psia and 40°F to a final state of $P_2 = 60$ psia and 228°F. Is the process constant T, P, V, or adiabatic?

18. During an adiabatic change of the state of a gas, 15,000 ft-lb of WORK is done. What is the change in the internal energy of the gas? Will the temperature of the gas rise or fall as a result of the change?

19. Eight pounds of dichlorodifluoromethane gas (Table 7–1) at a pressure of 5000 psfa and 20°F is adiabatically compressed to 40,000 psfa.
   a. What is the final temperature of the gas?
   b. Find $P_2V_2 - P_1V_1$
   c. Calculate $\dfrac{(P_2V_2 - P_1V_1)}{J(1-k)}$
   d. Show that $c_v m(T_2 - T_1) + \dfrac{(P_2V_2 - P_1V_1)}{J(1-k)} = 0$

20. a. If WORK is defined as force times distance, how much WORK is required to raise a 2000 lb_m weight 70 ft into the air (ft-lb_f)?
   b. The elevator of Section 7–3 raised a 2000 lb_m weight 70 ft in the air. It was determined that this required 2084 Btu of WORK done by the gas. How many ft-lb_f is this?
   c. Why isn't the result of b the same as that of a?
   d. For the elevator of Section 7–3, find the WORK delivered (WD, Equation 9–5) by the gas to the elevator.
   e. Compare your result from d to a and then reconsider your answer to c.

Problems in SI Units

21. One kg of air undergoes an adiabatic process during which the temperature increases from 20°C to 300°C.
    a. How much was the internal energy increased (kJ)?
    b. How much WORK was done? Was it done "on" or "by" the gas?
    c. By what percent did the specific volume change?
    d. By what percent did the pressure change?

22. How much WORK is done by 6 lb_m of nitrogen gas at 200 kPa (absolute) if it is allowed to expand adiabatically from .3m³ to .4m³? Answer in kJ.

23. Four kg of chlorodifluoromethane at a pressure of 300 kPa and temperature of –5°C are adiabatically compressed to 2400 kPa (absolute).
    a. What is the final temperature of the gas?
    b. Find $P_2V_2 - P_1V_1$.
    c. Find $\dfrac{(P_2V_2 - P_1V_1)}{(1-k)}$.
    d. Show that $c_v m (T_2 - T_1) + \dfrac{(P_2V_2 - P_1V_1)}{(1-k)} = 0$.

CHAPTER
10

The Unified First Law: Polytropic Processes

PREVIEW: Although the title of this chapter is long and sounds complicated, the chapter is about simplification. Our thermodynamics could use some simplification, since we now have generated enough equations to fill a full page. (Table 9–1) This chapter reduces all these equations into just four equations, in the same way Chapter 2 combined diverse results into one simple equation by taking the results from three experiments and combining them into the perfect gas law.

OBJECTIVES:
❑ Expand the fundamental thermodynamic processes into a more general situation in which all properties change.
❑ Calculate the polytropic index for any thermodynamic process.

10.1 THE POLYTROPIC PROCESS

Chapters 6 through 9 describe simplified thermodynamic processes. From those we created enough equations to fill a whole page. (Table 9–1) Just a few of these are:

$$\frac{T_2}{T_1} = \frac{P_2}{P_1} \qquad \text{Constant } V$$

$$Q = c_v m(T_2 - T_1) + P\frac{(V_2 - V_1)}{J} \qquad \text{Constant } P$$

$$WORK = P_1 V_1 \ln\left(\frac{V_2}{V_1}\right) \qquad \text{Constant } T$$

$$\frac{T_1}{T_2} = \left(\frac{P_1}{P_2}\right)^{\frac{(k-1)}{k}} \qquad \text{Adiabatic}$$

Now we can attempt to compress this complete page of equations (Table 9–1) into one set of four equations that will work for all the simplified thermodynamic processes.

But more importantly, we will look for equations that expand our ability to solve thermodynamic problems beyond the simplified cases of the previous chapters. The idealized cases of Chapters 6 through 9 do not really exist. For example, the compressor in Sample Problem 9–7 was considered to be adiabatic; however, if you were to touch the compressor, it would feel hot. This means that heat is being transferred to your hand, which implies $Q \neq 0$. Some heat is escaping to the surroundings. The process is not precisely adiabatic.

Likewise, a constant temperature compressor seldom, if ever, holds the gas temperature constant as the gas is compressed. If the gas intakes at 70°F, it seldom exhausts, even with water cooling, at less than 140°F. The process is not precisely a constant temperature one.

In actual situations, the pressure, temperature, and volume change and some heat transfers into or out of the gas. These thermodynamic processes are truly more general, and we do not yet have the equations to make proper calculations.

Compare two of the equations that we know well:

$$P_1 V_1 = P_2 V_2$$

and

$$P_1 V_1^{k} = P_2 V_2^{k}$$

One of these equations is for a constant temperature process and one is for an adiabatic process, but notice the similarity. Both equations have the same form:

$$P_1 V_1^{n} = P_2 V_2^{n}$$

where $n = 1$ for constant temperature processes and $n = k$ for adiabatic processes. (Note on Table 7–1 that there is no gas for which $k = 1$, which would make an adiabatic process also a constant temperature process.)

Investigate further what would happen if all of the equations in the adiabatic column in Table 9–1 were replaced by equations involving a variable n. (Table 10–1) What would this variable n be? Is it just the same as k, which would mean we really haven't changed anything? The answer is no. For example, there is no such gas with $k = 0$, but consider what $n = 0$ implies.

If $n = 0$:

Equation 10–1 $\quad \dfrac{P_1}{P_2} = \left(\dfrac{V_2}{V_1}\right)^{0} = 1$ or $P_1 = P_2$ (constant pressure process)

Table 10–1

UNIFIED EQUATIONS OF THERMODYNAMICS

$$\frac{P_1}{P_2} = \left(\frac{V_2}{V_1}\right)^n \qquad\qquad \text{Equation 10–1}$$

$$\frac{T_1}{T_2} = \left(\frac{V_2}{V_1}\right)^{n-1} \qquad\qquad \text{Equation 10–2}$$

$$\frac{T_1}{T_2} = \left(\frac{P_1}{P_2}\right)^{\frac{n-1}{n}} \qquad\qquad \text{Equation 10–3}$$

$$Q = c_v m(T_2 - T_1) + \frac{(P_2 V_2 - P_1 V_1)}{J(1-n)} \qquad\qquad \text{Equation 10–4}$$

Similarly for $n = 0$

$$\text{Equation 10–2} \quad \frac{T_1}{T_2} = \left(\frac{V_2}{V_1}\right)^{-1} = \frac{V_1}{V_2}$$

$$\text{Equation 10–4} \quad Q = c_v m(T_2 - T_1) + \frac{(P_2 V_2 - P_1 V_1)}{J}$$

$$= c_v m(T_2 T_1) + P\frac{(V_2 - V_1)}{J}, \text{ since } P_1 = P_2 = P$$

Notice that every equation for $n = 0$ is correct for a constant pressure process. Therefore, the equations of Table 10–1 represent a constant pressure process if $n = 0$.

If $n = 1$:

$$\text{Equation 10–1} \quad \frac{P_1}{P_2} = \left(\frac{V_2}{V_1}\right)^1 = \frac{V_2}{V_1}$$

$$\text{Equation 10–2} \quad \frac{T_1}{T_2} = \left(\frac{V_2}{V_1}\right)^0 = 1 \text{ or } T_1 = T_2 \text{ (constant temperature process)}$$

$$\text{Equation 10–3} \quad Q = c_v m(T_2 - T_1) + \frac{(P_2 V_2 - P_1 V_1)}{J(0)}$$

$$= c_v m(T_2 - T_1) + \frac{P_1 V_1}{J}\ln\left(\frac{V_2}{V_1}\right)$$

These are the equations for a constant temperature process. (In Equation 10–3, $P_2V_2-P_1V_1=0$ and $1-n=0$, so the numerator and denominator are both zero. When this happens, differentiate (calculus) both numerator and denominator with respect to n to get this result.)

And if $n=+\infty$ or $-\infty$ (+ and – infinity):

$$\text{Equation 10–1}\quad \frac{P_1}{P_2}=\left(\frac{V_1}{V_2}\right)^n \ or \ \left(\frac{P_1}{P_2}\right)^{\frac{1}{n}}=\frac{V_1}{V_2}$$

but for $n=\pm\infty$:

$$\left(\frac{P_1}{P_2}\right)^{\frac{1}{n}}=\left(\frac{P_1}{P_2}\right)^{\frac{1}{\infty}}=\left(\frac{P_1}{P_2}\right)^{0}=1=\frac{V_1}{V_2}\quad\text{(constant volume process)}$$

$$\text{Equation 10–2}\quad \frac{T_1}{T_2}=\left(\frac{P_1}{P_2}\right)^{\frac{n-1}{n}}=\left(\frac{P_1}{P_2}\right)^{\frac{\infty-1}{\infty}}=\left(\frac{P_1}{P_2}\right)^{1}=\frac{P_1}{P_2}$$

$$\text{Equation 10–3}\quad Q=c_v m(T_2-T_1)+\left(\frac{P_2V_2-P_1V_1}{J(1-\infty)}\right)=c_v m(T_2-T_1)$$

These are the equations for a constant volume process.

Only if $n=k$ do the equations become correct for an adiabatic process:

$$\text{Equation 10–1}\quad \frac{P_1}{P_2}=\left(\frac{V_2}{V_1}\right)^{k}$$

$$\text{Equation 10–2}\quad \frac{T_1}{T_2}=\left(\frac{V_2}{V_1}\right)^{k-1}$$

$$\text{Equation 10–3}\quad \frac{T_1}{T_2}=\left(\frac{P_1}{P_2}\right)^{\frac{k-1}{k}}$$

What we have done is introduce a new thermodynamic property, n and a "simplified" set of equations (Table 10–1). The property n refers to the process with which the substance is involved. Figure 10–1 shows the range of n and the process to which each number refers.

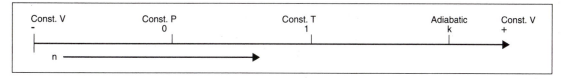

Figure 10–1. As the variable n changes numerically, the thermodynamic process changes nature.

10.2 PROCESSES OF A GENERAL NATURE

The line in Figure 10–1 seems to suggest that the value of n might be something other than 0, 1, k, or $\pm\infty$. What would it mean, though, if the value for n was .95? Or since $n = 1$ is a constant temperature process, then what kind of process is $n = .95$? Could it be that if $n = .95$, the process is "almost constant temperature"? The answer is a definite yes!

What is an "almost constant" process? Consider a large balloon made out of heavy rubber, like a satellite balloon. If the balloon is let go in the atmosphere, the gas inside will cool as the balloon rises because of the colder environment. As the gas cools, the balloon shrinks just a small amount, say less than 3%. What kind of thermodynamic cooling process is it? It is "almost" constant volume!

SAMPLE PROBLEM 10–1

A satellite balloon (Figure 10–2) is filled at the surface of the earth (70°F) to a pressure of 50 psig. It is sent aloft until at 30,000 feet, the surrounding (ambient) temperature reaches −10°F. If the balloon's volume decreases by 3% ($V_2/V_1 = .97$), find the value of n for the process. Find the final pressure and the amount of heat lost per pound by the balloon gas. Also make these calculations assuming that the process were truly a constant volume process.

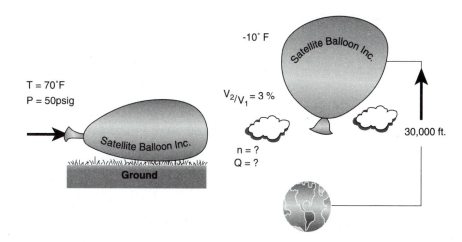

Figure 10–2. Sample Problem 10–1.

SOLUTION:

$$T_1 = 530°\text{R}, \qquad P_1 = 64.7 \text{ psia}$$

$$T_2 = 450°\text{R}$$

$$\frac{T_1}{T_2} = 1.177$$

$$\frac{V_2}{V_1} = .97 \ (V_2 \text{ is 3\% less than } V_1)$$

$$\frac{T_1}{T_2} = \left(\frac{V_2}{V_1}\right)^{n-1}$$

So

$1.177 = (.97)^{n-1}$ To solve this equation for n, take the ln of both sides.

$\ln 1.177 = \ln(.97) \times (n - 1)$ (see Section 1–9 for solving equations with logarithms)

$.163 = -.03(n - 1)$ $(\ln 1.177 = .163, \ln .97 = -.03)$

$-5.4 = n - 1$

$-4.4 = n$

Find P_2

$$P_2 = P_1\left(\frac{V_1}{V_2}\right)^{-4.4} = 65\left(\frac{1}{.97}\right)^{-4.4} = 56.8 \text{ psia}$$

If the process were assumed to be constant V ($n = \pm\infty$)

$$\frac{T_1}{T_2} = \frac{P_1}{P_2}$$

$$P_2 = P_1\left(\frac{T_2}{T_1}\right) = 65 \times \frac{450}{530}$$

$$P_2 = 55.19 \text{ psia}$$

Notice that there would be a small error if this process were assumed to be constant volume.

The thermodynamic process represented by Table 10–1 is the process we have been looking for all along. We didn't really want to be restricted to "constant" processes, because in real thermodynamic processes, all the properties change. We call the process that gives us this new freedom a polytropic process.

❖ **KEY TERM:**   **Polytropic Process:** The general thermodynamic process, the nature of which is described by the exponent n, which can take on any value and is called the polytropic index.

Imagine an "almost constant temperature" process.

SAMPLE
PROBLEM 10-2

A "constant temperature" vessel (Figure 10–3) uses a 70°F water heat sink for the compression of 3 lb of air inside. The air is originally at a volume of 1 cu ft and is compressed to .05 ft³, but the water "sink" temperature rises 10°F because there is not enough water to absorb the heat from the air without showing a change in temperature. Find:

a.   the polytropic index n
b.   how much WORK is done
c.   the amount of water in the sink
d.   contrast the WORK to the constant temperature WORK

SOLUTION:

$V_1 = 1$, $V_2 = .05$, $T_1 = 530°R$, $T_2 = 540°R$

a. $\dfrac{T_1}{T_2} = \left(\dfrac{V_2}{V_1}\right)^{n-1}$

or

$\dfrac{530}{540} = \left(\dfrac{.05}{1}\right)^{n-1}$ Simplifying:

$.981 = (.05)^{n-1}$ Now take the ln of both sides:

$\ln .981 = (n-1) \ln (.05)$

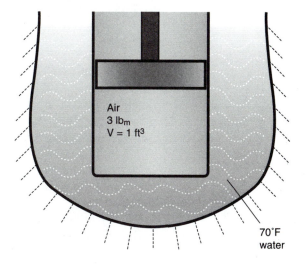

Air
3 lbₘ
V = 1 ft³

70°F
water

Figure 10–3. Sample Problem 10–2.

Solving:

$$\frac{\ln .981}{\ln .05} = n - 1 = .0064$$

$$n = 1.0064$$

b. $WORK = \dfrac{(P_2 V_2 - P_1 V_1)}{J(1-n)}$

but

$$P_1 = \frac{mRT_1}{V_1} = 3 \times 53 \times \frac{530}{1} = 84270 \text{ psfa}$$

$$P_2 = \frac{mRT_2}{V_2} = 3 \times 53 \times \frac{540}{.05} = 1{,}717{,}200 \text{ psfa}$$

$$WORK = \frac{(1{,}717{,}200 \times .05 - 84{,}270 \times 1)}{\left(778(1-1.0064)\right)} = -340.6 \text{ Btu}$$

c. $Q_{\text{heat rejected}} = c_v m(T_2 - T_1) + WORK = .17 \times 3 \times (10) - 340.6 = -335.5 \text{ Btu}$

$Q_{\text{absorbed by water}} = 335.5 = c_{\text{water}} \times m_{\text{water}} \times T$

$$= 1 \frac{\text{Btu}}{\text{lb}_m - {}^\circ \text{F}} \times m_{\text{water}} \times 10^\circ \text{F}$$

Solving

$$m_{\text{water}} = \frac{335.5}{10} = 33.55 \text{ lb}_m$$

d.   If there would have been sufficient water to keep the temperature precisely constant ($T_1 = T_2 = 530°F$), then the process would have proceeded in the following manner:

$$WORK = \frac{P_1 V_1}{J} \ln\left(\frac{V_2}{V_1}\right) = \frac{84270}{778} \times 1 \ln\left(\frac{.05}{1}\right) = -324.5$$

Comparing this with the first sections of this example, it appears it will cost an additional 15 Btu or 11,670 ft-lb$_f$ of WORK to perform this compression if there is insufficient cooling water to maintain the temperature constant.

Sample Problems 10–1 and 10–2 are examples of "almost" constant volume and temperature processes. Is there a process that is almost constant pressure?

A partially inflated garbage bag is "almost" a perfect constant pressure device. Its volume is easy to change. What happens when you blow into it? Does pressure build up? Yes, but very little. Mostly what happens is the bag gives way and expands for more space. When you blow into a garbage bag, you are mostly storing air in a constantly larger space, not increasing the pressure. Contrast this to what happens when you get hold of a "stiff" balloon; you blow and you blow but you can't get any air into it. If the balloon doesn't give, its volume is difficult to change. Table 10–2 illustrates that different types of 'balloons' represent processes that are 'almost' constant pressure to 'almost' constant volume.

Table 10–2

Balloon Type	Polytropic Index	Comment
Partially filled garbage bag	−.1	Holds only a little pressure
Party balloon	−.5	Stretch them before inflating
Commercial quality balloon (tire)	−1	Must inflate with an air compressor
Satellite balloon	−10	Thick-skinned, "almost" constant volume

SAMPLE PROBLEM 10-3

If a good quality party balloon with deflated internal volume of .05 ft^3 is blown up to 10 times its deflated volume, what is the pressure inside the balloon?

SOLUTION:

$n = -.5$, $V_2 = .5$ ft^3

$$P_2 = P_1 \left(\frac{V_1}{V_2} \right)^n = 2116 \left(\frac{.05}{.5} \right)^{-.5}$$

$$= 6691 \text{ psfa} = 50 \text{ psia}$$

The last kind of an "almost" process is "almost adiabatic." The following Sample Problem illustrates this one.

SAMPLE PROBLEM 10-4

In Sample Problem 9–6, an adiabatic compressor was considered. Actually some heat from the compressed gas is given off; the process is not quite adiabatic. If the polytropic index for this process was $n = 1.25$, find how much WORK was done on each compression, what the final temperature was, and how much heat was lost. Contrast these results to the fully adiabatic case.

SOLUTION:

$$T_2 = T_1 \left(\frac{V_1}{V_2} \right)^{n-1}$$

$$= 530\left(\frac{.25}{.025}\right)^{.25}$$

$= 954°R$ (compare to 1331°R, Sample Problem 9–6)

$$P_2 = P_1\left(\frac{V_1}{V_2}\right)^n$$

$$= 2116\left(\frac{.25}{.025}\right)^{1.25}$$

$= 37,628$ psfa (compare to 53,151 psfa)

$$Q = c_v m(T_2 - T_1) + \left(\frac{P_2V_2 - P_1V_1}{J(1-n)}\right)$$

$$= .17 \times .0187 \times 424 + \frac{(37600 \times .025 - 2116 \times .25)}{778 \times (1-1.25)}$$

$= 1.34 - 2.11 = -.77$ Btu (compare to $Q = 0$)

WORK $= -2.11$ Btu (compare to -2.56 Btu)

Solution in SI Units

$$T_2 = 294\left(\frac{7\times10^{-3}}{.7\times10^{-3}}\right)^{.25}$$

$= 522.8$ K (compared to 738 K)

$$P_2 = 101.3\left(\frac{7\times10^{-3}}{.7\times10^{-3}}\right)^{1.25}$$

$= 1801$ kPa (compared to 2544)

$$Q = .71 \times .0084 \times 228.8 + \frac{1801 \times .7 \times 10^{-3} - 101.3 \times 7 \times 10^{-3}}{1-1.25}$$

$= + 1.36 - 2.20 = -.84$ kJ (compare to $Q = 0$)

WORK $= -2.20$ kJ (compare to -2.68)

10.3 THE GENERAL THERMODYNAMIC PROCESS

The mathematics of the polytropic process combine all the previous special processes into one. By changing the value of n, the polytropic process can change greatly in its characteristics. The polytropic process can be constant volume, constant temperature, constant pressure, or adiabatic. It also can be a process that is very close to one of these.

But the polytropic process does not have to resemble any of the basic processes. All the fundamental thermodynamic variables can change, and they can change significantly. Still the mathematics of the polytropic process can readily find the resultant thermodynamic state.

Whereas the processes of Chapters 6 through 9 were special cases, the polytropic process is the general thermodynamic process. (Chapter 11 will expand on this.) The key to using polytropic mathematics is determining the value of n. Sometimes the value of n is totally unknown; then the process is assumed to be one of the basic processes in order to get approximate results. But for manufactured products, often the value of n is precisely determined in the laboratory, and indeed sometimes great research is expended to increase or decrease the value of n for a given process.

SAMPLE PROBLEM 10–5

A large manufacturer of air compressors is developing an air-cooled piston model with a diameter of five inches. The prototype was harnessed to a pressure gauge and the following readings taken during the compression stroke.

Pressure (psia)	Displacement (in^3)
14.70	120
21.15	90
34.91	60
138.60	20

Find the polytropic index of this process and calculate the WORK done.

SOLUTION:

$$P_1\left(\frac{V_1}{V_2}\right)^n = P_2$$

or

$$n = \frac{\ln(P_2/P_1)}{\ln(V_1/V_2)}$$

Using $P_1 = 14.7$ psia or 2116 psfa and $V_1 = 120$ in^3 or .069 ft^3 and conditions 2 being each of the other conditions of the table results in

$$n = \frac{\ln\left(\dfrac{21.15}{14.70}\right)}{\ln\left(\dfrac{120}{90}\right)} = 1.25$$

$$n = \frac{\ln\left(\dfrac{34.91}{14.70}\right)}{\ln\left(\dfrac{120}{60}\right)} = 1.25$$

$$n = \frac{\ln\left(\dfrac{138.60}{14.70}\right)}{\ln\left(\dfrac{120}{20}\right)} = 1.25$$

$$WORK = \frac{(P_2 V_2 - P_1 V_1)}{J(1-n)} = \frac{\left(138.6 \times 144 \times \dfrac{20}{1728} - 14.7 \times 144 \times \dfrac{120}{1728}\right)}{778(1-1.25)}$$

$$= -.44 \text{ Btu}$$

Notice that pressure has not been converted to psf in the ratio formulas, nor has volume been converted to ft³, but in the formula for WORK, these adjustments must be made.

SAMPLE PROBLEM 10–6

The same manufacturer is testing a water-cooled model compressor with the same cylinder diameter and gets the following results:

Pressure (psia)	Displacement (in³)
14.70	120
19.88	90
30.43	60
96.46	20

What is the value of n for the compression stroke, and how much WORK is done during one compression?

SOLUTION:

$$n = \frac{\ln\left(\dfrac{19.88}{14.70}\right)}{\ln\left(\dfrac{120}{90}\right)} = 1.05$$

$$= \frac{\ln\left(\dfrac{30.43}{14.70}\right)}{\ln\left(\dfrac{120}{60}\right)} = 1.05$$

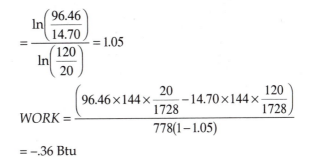

$$\frac{\ln\left(\dfrac{96.46}{14.70}\right)}{\ln\left(\dfrac{120}{20}\right)} = 1.05$$

$$WORK = \frac{\left(96.46 \times 144 \times \dfrac{20}{1728} - 14.70 \times 144 \times \dfrac{120}{1728}\right)}{778(1-1.05)}$$

$$= -.36 \text{ Btu}$$

SAMPLE
PROBLEM 10–7   Which one of the prototype designs will be more efficient?

SOLUTION:

Since each has the same displacement and will compress the same amount of air, the one with the least WORK required will be the most efficient: the water-cooled model.

PROBLEMS

1. If the gas in a thermodynamic process is acetelyne and the process is adiabatic, what is the value of n for the process? Repeat with the gas as helium.

2. A gas with original conditions of 16 psia pressure, 11 ft^3/lb specific volume, and 40°F temperature is compressed in a process that is described by a polytrophic index of $n = 1.3$. When the temperature during the compression reaches 80°F, what will the pressure be?

3. When a "commercial quality" balloon is inflated from atmospheric pressure and an internal volume of .05 cu ft to .5 cu ft, what is the pressure inside? (What is the pressure of a deflated balloon?)

4. One of the first satellites ever launched was called "Echo." It was a balloon satellite that was so big it was actually seen from earth with the naked eye. When slightly inflated to 30 psia, it showed a diameter of 950 ft. When it was totally inflated, the diameter was 1000 ft. What was the pressure required to inflate "Echo" (volume of a sphere = $4/3\pi r^3$)? Use Table 10–2.

5. A party balloon contains 2 lb_m of helium ($c_v = 1.24$ Btu/lb – °R, $R = 386$ ft-lb_m/lb_f –°R) at a pressure of 15 psia at a temperature of 75°F. When the balloon is sitting in the sun, the temperature rises to 100°F.
   a. What is the change in volume of the balloon?
   b. What is the change in internal energy of the balloon?

6. How much air must be pumped into a garbage bag to blow it up from .01 lb/ft^3 to .12 lb/ft^3? Assume initial air conditions are standard temperature and pressure. Repeat the exercise for a party balloon. Why is there a difference?

7. Air is pulled into a compressor at a rate of 300 ft^3/min from the atmosphere and compressed to 100 psia. If the compression is considered to be adiabatic, do the following:
   a. Sketch the process on a P-v diagram and indicate the area that represents the WORK done by or on the gas during the compression
   b. Find how many ft^3/min of air is exhausted from the compressor discharge
   c. Find the final air temperature

    d. Find the change in internal energy

    e. Find the heat transferred to the air

    f. Find the WORK per minute needed to perform the compression

8. Repeat the above problem in its entirety if the process were constant temperature (isothermal) rather than adiabatic.

9. If a compressor is fitted with a water jacket, the thermodynamic process of the compression might be best described as polytropic. Which of the following values of the polytropic index n best describes the resulting process:

    a. −4

    b. .2

    c. 1.15

    d. 2

Using this index, recalculate the results of problem 7.

10. The initial conditions for a thermodynamic expansion are: $T_1 = 500°F$, $P_1 = 350$ psig, $V_1 = 9$ ft^3. The final conditions are: $T_2 = 800°F$ and $P_2 = 460$ psig. Find the polytropic index for the process. Use this value of n to determine the final volume.

11. A constant temperature vessel is filled with glycol coolant that is supposed to maintain the compression of carbon dioxide at a constant 150°F. However, during the process of $V_1/V_2 = 6$, the temperature of the coolant actually rose to 170°F. What is the polytropic index?

12. A certain amount of air is initially at 30 psia and 100°F. If the final state after a thermodynamic process is 18 psia and 40°F, describe the kind of thermodynamic process that took place.

13. A vessel of pure oxygen at temperature of 100°F undergoes a polytropic process from a pressure of 30 psia and a volume of 15 ft^3 to a final pressure of 4 psia and a volume of 5 ft^3. Find the value of n for the process and the amount of heat transferred (added or reduced). Calculate the WORK done by the gas.

14. Hydrogen gas is compressed from an initial state of $P_1 = 20$ psia and $T_1 = 40°F$ to a final state of $P_2 = 60$ psia and $T_2 = 90°F$. Was the process constant T, P, V, or adiabatic?

Problems in SI Units

15. A gas with original conditions of 150 kPa (absolute) and 5°C is com-pressed in a process that is described by a polytropic index $n = 1.3$. When the temperature reaches 25°C, what will be the pressure?

16. Air is pulled into a compressor at a rate of 8 m³/min from the atmosphere and compressed to 500 kPa (absolute). If the compressioin is polytropic with $n = 1.3$, do the following:
    a. Determine how many m³/min of air is exhausted from the compres-sor discharge.
    b. Find the compressed air temperature.
    c. Find the change in internal energy.
    d. Find the heat transferred to the air.
    e. Find the WORK/min needed to perform the compression.

17. A certain quantity of air is initially at 200 kPa (absolute) and 35°C. If the final state after a thermodynamic process is 120 kPa and 10°C, find the polytropic index of the process and describe what type of process this index represents.

CHAPTER
11

Free Expansions and Irreversible Processes

PREVIEW: With the introduction of polytropic processes, it seems that we can find solutions for all problems of thermodynamics. There is a whole class of processes, however, that is not governed by these equations. When they are introduced, adaptations of these polytropic equations will have to be made.

OBJECTIVES:
- ❏ Distinguish between reversible and irreversible processes.
- ❏ Recognize irreversible processes.
- ❏ Solve problems that are partially or completely irreversible.

11.1 ADIABATIC FREE EXPANSION

Heat energy can be transformed into WORK energy. Pistons move. Elevators move. Spinning wheels move. They all move by the action of heat. Figure 11–1 illustrates this with a two-chambered tank fitted with a piston in which thermal energy is used to generate WORK. One chamber is filled with pressurized air and the other side evacuated. Once the piston is released, the compressed air fills the second chamber while pushing the piston, allowing WORK to be done.

SAMPLE PROBLEM 11–1

In Figure 11–1, both chambers are 3 ft³, and the air in the lower chamber is stored at 70°F and 20 psia. Releasing the piston creates an adiabatic process. What is the final pressure and final temperature in the tank, and how much WORK is done against the piston?

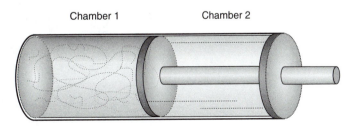

Figure 11–1. An adiabatic chamber.

SOLUTION:

$V_1 = 3, V_2 = 6$

$$P_2 = P_1\left(\frac{V_1}{V_2}\right)^k = 20(.05)^{1.41} = 7.6 \text{ psia} = 1091 \text{ psfa}$$

$$T_2 = T_1\left(\frac{V_1}{V_2}\right)^{k-1} = 530\left(\frac{3}{6}\right)^{1.41-1} = 403$$

$$WORK = \frac{(P_2V_2 - P_1V_1)}{J(1-k)} = \frac{(1091\times6 - 2880\times3)}{778(-.41)}$$

$$= 6.72 \text{ Btu} = 5230 \text{ ft-lb}$$

WORK is an important part of the adiabatic process; therefore, consider how different the process would be if there were no piston! Would the process be different if no WORK was done because there was nothing to do WORK on?

Figure 11–2 illustrates this case. It is a two-sided vessel, similar to that in Figure 11–1, where one side is filled with compressed air and the other side evacuated. Between the two sections is a diaphram, a membrane of material that isolates the two sides. The tank is properly insulated so that whatever happens inside is an adiabatic process.

When the diaphram is broken, compressed air rushes into the open space. The pressure falls. When the process is over, the gas occupies twice its original volume. How much WORK was done? None! There was *no piston* available to move. By breaking the diaphram without a piston, we lost the chance to do WORK.

This "oversight" of not installing a piston has serious consequences on the thermodynamic process. For an adiabatic process:

$$Q = 0 = c_v m(T_2 - T_1) + WORK$$

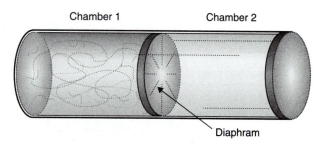

Chamber 1   Chamber 2

Diaphram

Figure 11–2. An adiabatic chamber with no piston.

Now if *WORK* = 0, then:

$$0 = c_v m(T_2 - T_1)$$

or

$$T_2 = T_1$$

This process occurs at constant temperature! Such a result suggests that this adiabatic process is also a constant temperature one. This new process is called a *free expansion*. It has much different consequences than the adiabatic process of Sample Problem 11–1.

❖ **KEY TERM:**   **Free Expansion:** Any thermodynamic process in which the gas volume changes but no WORK is done.

This new process can be added to our list of thermodynamic processes only after we answer a final, puzzling question: how do we calculate the pressure of the air after the process is over? If the process is adiabatic:

$$\frac{P_2}{P_1} = \left(\frac{V_1}{V_2}\right)^k$$

This indicates that if the diaphram is broken in a vessel like that of Sample Problem 11–1, the final pressure would be P_2 = 1091 psfa. But the process is also constant temperature, and the constant temperature perfect gas law for a changing situation states:

$$\frac{P_2}{P_1} = \left(\frac{V_1}{V_2}\right)$$

For this situation:

$$P_2 = 20\left(\frac{3}{6}\right) = 10 \text{ psfa} = 1440 \text{ psfa}$$

There is a lot of disagreement here. Which is the way it will happen? Before this can be answered, we must talk about equilibrium.

11.2 EQUILIBRIUM AND IRREVERSIBILITY

The bursting of the diaphram in the free expansion is a catastrophic occurrence. It is more like a rifle shot than a thermodynamic process. The noise from the rupture and the rapid release of gas can be deafening. The gas expands down the cylinder in a very turbulent way. The pressure in front of the expansion is low, while behind the expansion it is high. Temperatures at different points in the gas may not be the same. (Figure 11–3) Only after the rupture and after the expansion will all the gas properties settle and the gas reach a uniform state: a state of equilibrium.

❖ **KEY TERM:**   **Equilibrium:** When the properties of a gas are the same throughout the gas.

The perfect gas law is valid for a gas only if it is in equilibrium. For the short period of time during free expansion, the perfect gas law does not apply! This suggests deep trouble, since at the end of Section 11–1, we were attempting to determine which of two perfect gas laws for a changing situation was the correct one, and now we see that the perfect gas law does not even apply *during* this process.

After the process is over, however, the pressures, temperatures, and specific volumes reach an equilibrium—a new equilibrium. The perfect gas law applies to the final state as it did to the initial state. Since the perfect gas laws we are considering link the final and initial states, we again have hope that one of them (adiabatic or constant temperature) is correct.

The fact that free expansion creates "nonequilibrium" states has drastic consequences for the P-V diagram. This diagram is most helpful because the area underneath the P-V diagram for a process represents the WORK done during the process. But how can the process line be drawn if the perfect gas law is not valid during the process? Adiabatic free expansions do not follow the adiabatic process line. Constant temperature free expansions do not follow the constant temperature process line. (Figure 11–4)

This requires a word of caution for our well-understood processes studied in Chapters 6 through 10. While those processes occur, the gas must remain in equilibrium. The process cannot proceed so fast that the state does not remain the same everywhere during the process, or else the P-V lines will not be well defined and the WORK done will not be as we predicted.

Look again to the free expansion. If one distinct process line could be defined (say, the average state), what good would it do us? The area under this curve is not the WORK done during the process. The WORK in a free expansion is zero! The area underneath the process line is not equal to the WORK done in a nonequilibrium process.

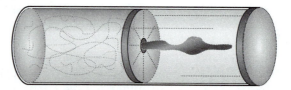

Figure 11–3. Properties of this gas are not the same everywhere.

Consider one more distinction between equilibrium processes and free expansions. Recall the elevator carrying the cargo in the constant pressure process (Section 7–3). When heat is added to the gas, the elevator goes up, the temperature of the gas rises, pressure remains the same, specific weight of the gas drops, and WORK is done by the gas. Then when the cargo goes back down, heat must be taken out of the gas, the temperature drops, the pressure stays the same, the specific weight rises, WORK is done on the gas by the cargo, and when the process is complete, the state of the gas returns to its original state. The amount of heat put in on the upward travel equals that taken out on the downward travel. Likewise, the WORK done by the gas on the way up is equal (but opposite in sign) to the WORK done on the gas to get the elevator down. The constant pressure process we have described can be reversed: it is reversible.

❖ **KEY TERM:**   **Reversible Process:** A process that may be made to occur in the forward then reverse direction with no overall net change in state, and no net heat or WORK needed to accomplish it.

The adiabatic processes studied in Chapter 9 were reversible. The WORK put into a gas to compress it adiabatically was the same amount of WORK that could be gotten out of the process by allowing it to expand adiabatically back to its original state. The constant volume and constant temperature processes that we studied also are reversible. Consider any problem given in Chapters 6 through 10 and note that the process can be reversed back to its original state.

Now consider the free expansion described above. Can you imagine the ruptured diaphram repairing itself? Can you imagine all the compressed air going back into the tank without any help? Certainly not. That doesn't mean that we can't get the compressed air back into the tank by one means or another. For example, we could place a piston in the cylinder and compress

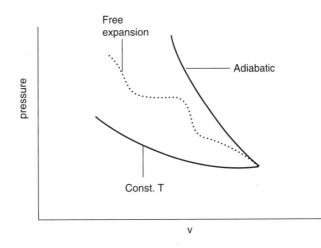

Figure 11–4. Process lines for a free expansion are difficult to draw.

the gas back into its original volume. To do this takes WORK, even though no WORK was done during the expansion. Even then, we have no confidence that the state of the compressed air will be the same as its original state.

The free expansion process is an irreversible one, making it even more distinct from the previous processes studied.

❖ **KEY TERM:**   **Irreversible Processes:** Thermodynamic processes that do not maintain equilibrium conditions while they occur and cannot be returned to their original state with the same amount of WORK they generated.

Irreversible processes are quite common, even in areas outside thermodynamics. If you sprint a half mile around a one-mile track, you end up out of wind, exhausted, and on the opposite side of the track. You do not imagine that by turning around and sprinting back you will arrive at the original position and have all your wind and energy back. When you cut your finger with a knife, would you imagine running the knife backward through your finger and sealing the cut? The cutting is an irreversible process. Not that you couldn't go to the hospital and find the best plastic surgeon in town to sew the cut and return your finger to its original "state" with no scar. But the healing process requires far more WORK than the cutting did—both on the part of the surgeon and on your part in paying the bill.

One of the greatest folklore stories in the English language focuses around irreversibility. Remember "Humpty Dumpty" where "All the king's horses and all the king's men, couldn't put Humpty Dumpty together again"? The author seems to indicate that breaking an egg is a totally irreversible process.

This section has introduced you to the world of irreversible processes. Here are two important conclusions that can be drawn:

1. Irreversible processes have no distinguishable process lines on the P-V diagram, and
2. The area under the "indistinguishable" process line cannot be interpreted as the WORK done during an irreversible process; therefore, the equations for WORK in Table 9–1 are not valid for an irreversible process.

Now, finally, we may attack the question posed at the end of the previous section: "What perfect gas law predicts the pressure at the end of a free expansion?" From the description of the adiabatic free expansion, here are the things we know to be true:

$$T_1 = T_2$$

$$m_1 = m_2$$

Also, the state of the gas before and after the process is governed by the perfect gas law, since at these two conditions, the gas is in equilibrium:

$$P_1v_1 = RT_1$$

$$P_2v_2 = RT_2$$

It follows by algebra that:

$$R = \frac{P_1v_1}{T_1} = \frac{P_2v_2}{T_2}$$

But since $T_1 = T_2$ and $m_1 = m_2$:

$$P_1V_1 = P_2V_2$$

This is the constant temperature perfect gas law for a changing situation, and it is also valid for the adiabatic free expansion. Consider another of the adiabatic equations:

$$\frac{v_2}{v_1} = \left(\frac{T_1}{T_2}\right)^{\frac{1}{(k-1)}}$$

Since we know that for an adiabatic free expansion, $T_1 = T_2$, the right side of the equation equals 1. This predicts that $v_2 = v_1$ or $V_2 = V_1$, which definitely is not true. Therefore, the adiabatic perfect gas laws are not valid for the free expansion, but the constant temperature equations are valid.

SAMPLE PROBLEM 11-2

What is the pressure of the gas in the vessel in Figure 11–2 after the free expansion? Contrast this result with the solution for the reversible adiabatic process of Sample Problem 11–1. Also summarize the final pressure, temperature, WORK done, and heat input of the reversible expansion (Sample Problem 11–1) in comparison with the free expansion solution.

SOLUTION:

$$P_2 = \frac{P_1V_1}{V_1} = 20 \text{ psia} \times \frac{1}{2} = 10 \text{ psia} = 1440 \text{ psfa}$$

This is higher than the pressure for the reversible process, indicating that in this process, the pressure was not used effectively to generate the maximum amount of WORK. Following is a comparison of reversible and irreversible states and properties after the process:

	Reversible	Irreversible
$T_2(°F)$	−57	70
P_2(psf)	1091	1440
WORK	5230	0
Q	0	0

SAMPLE
PROBLEM 11-3

A propane torch used for brazing contains 16 oz (.45 kg) of gas under a pressure of 200 psig (1480 kPa) at room temperature. (Figure 11–5) As the gas comes out of the bottle, it expands to atmospheric pressure and is immediately ignited. If the expansion process is adiabatic, how many brazes will one cylinder make if the average braze connection requires .5 cu ft (.014 m³) of expanded gas?

$$R = \frac{34.8\,\text{ft}-\text{lb}_m}{\text{lb}_f\,°\text{R}} \left(.189\frac{\text{kPa}-\text{m}^3}{\text{kg}-\text{K}}\right)$$

SOLUTION:

Possibly the most difficult aspect of this problem is visualizing it in a framework that is understandable from the thermodynamics we have studied. The expansion of the gas through the burner port is a free expansion. There is no WORK done. Consider the propane cylinder to be the bottom chamber of the vessel. (Figure 11–6) The diaphram in this case is the burner valve, and the upper chamber represents the volume that the expanded propane will occupy.

To solve the problem, find the volume occupied by the expanded gas—that is, the volume of the upper chamber—and consider each

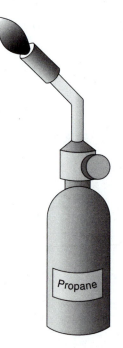

Figure 11–5. Sample Problem 11–3.

.5 ft^3 to be one solder connection. Then:

P_1 = 200 psig = 30,916 psfa

T_1 = 70°F = 530°F

P_2 = 0 psig = 2116 psfa

$$V_1 = \frac{mRT_1}{P_1} = 1 \text{ lb} \times 34.8 \times \frac{530}{30,916} = .596 \text{ ft}^3$$

$$V_2 = V_1\left(\frac{P_1}{P_2}\right) = .596 \text{ ft}^3 \times \frac{30,916}{2116} = 8.707 \text{ ft}^3 \text{ (this is an adiabatic } \textit{free}$$

expansion equation)

The volume created by the expansion is:

$$V_2 - V_1 = 8.111 \text{ ft}^3$$

or

$$\frac{8.11 \text{ ft}^3}{\left(\dfrac{.5 \text{ ft}^3}{\text{braze}}\right)} = 16.22 \text{ brazes}$$

Solution in SI Units

P_1 = 1480 kPa

T_1 = 21°C = 294 K

P_2 = 101.3 kPa

$$V_1 = .45 \times .189 \times \frac{294}{1480} = .017 \text{m}^3$$

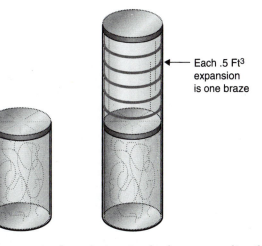

Figure 11–6. The propane torch can be imagined to be an expanding thermodynamic vessel.

Each .5 Ft^3 expansion is one braze

$$V_2 = .0169 \times \left(\frac{1480}{101.3}\right) = .247\text{m}^3$$

$$V_2 - V_1 = .247 - .017 = .230 \text{ m}^3$$

$$\frac{.230}{.014} = 16.22 \text{ brazes}$$

11.3 CONSTANT TEMPERATURE FREE EXPANSIONS

In this section we will investigate whether there is such a thing as constant temperature free expansion. Remember that the model for a reversible constant temperature process occurred in a chamber that was immersed in a constant temperature bath. Consider the same situation, but now the piston is replaced with a diaphram. (Figure 11–7) On one side of the diaphram is a compressed gas in thermal equilibrium with the constant temperature bath. On the other side of the diaphram is a volume with no gas in it. The thermodynamic free expansion takes place when the diaphram is ruptured.

What happens to the thermodynamic state of the gas? The energy equation for a constant temperature process states that:

$$Q = WORK$$

but $WORK = 0$ for a free expansion. Therefore:

$$Q = 0$$

This indicates that no heat will be absorbed into or rejected from the constant temperature bath. During the expansion, the gas has no tendency to drop its

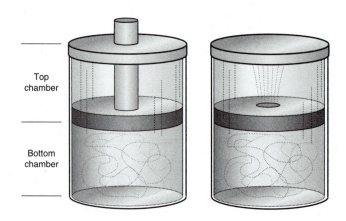

Figure 11–7. Reversible and irreversible constant temperature process.

temperature; therefore, no heat is required to keep the temperature up. This process is identical to the adiabatic free expansion process.

**11.4
CONSTANT
PRESSURE FREE
EXPANSION**

In both the adiabatic and constant temperature free expansions, the pressure of the gas drops. (Table 11–1) In a sudden free expansion in which the pressure does not drop, heat will have to be added in order to keep the initial and final pressures the same. Figure 11–7 shows the familiar two-chambered vessel with diaphram that is our model vessel for free expansions. In this case, the bottom chamber is not only filled with compressed gas but also a fuel burner to allow for a rapid and sudden addition of heat to keep the pressure up.

The energy equation for free expansions is

$$Q = c_v m(T_2 - T_1)$$

since $WORK = 0$. Also, the state of the gas before and after the process is consistent with the perfect gas laws:

$$P_1 v_1 = RT_1$$

and

$$P_2 v_2 = RT_2$$

even though the perfect gas law is not valid during the process. But since

$$P_1 = P_2$$

then

$$\frac{v_1}{v_2} = \frac{T_1}{T_2}$$

or since

$$m_1 = m_2$$

then

$$\frac{V_1}{V_2} = \frac{T_1}{T_2}$$

is the proper equation to predict the gas temperature after the process occurs.

Table 11–1

THERMODYNAMIC FORMULAS FOR IRREVERSIBLE FREE EXPANSIONS

Quantity	Constant T	Constant P	Adiabatic
Temperature - Volume	$T_1 = T_2$	$\dfrac{T_1}{V_1} = \dfrac{T_2}{V_2}$	$T_1 = T_2$
Pressure - Volume	$P_1V_1 = P_2V_2$	$P_1 = P_2$	$P_1V_1 = P_2V_2$
WORK (gas)	0	0	0
WORK (delivered)	0	0	0
Heat	0	$c_vm(T_2 - T_1)$	0

SAMPLE PROBLEM 11–4

For the situation of Sample Problem 11–1, find the final temperature of the gas if the process is a constant pressure free expansion, and determine the heat added per pound of gas. Contrast this with the results of a reversible constant pressure expansion in the same device.

SOLUTION:

Free Expansion:

$$P_1 = P_2 = 20 \text{ psia} = 2880 \text{ psfa}$$

$$T_2 = T_1\frac{V_2}{V_1} = 530°\text{R} \times \frac{2}{1} = 1060°\text{R}$$

Heat added per pound

$$\frac{Q}{m} = c_v(T_2 - T_1) = .17(1060 - 530) = 90.1\frac{\text{Btu}}{\text{lb}}$$

Reversible Expansion:

$$P_1 = P_2 = 2880 \text{ psfa}$$

$$T_2 = T_1\frac{V_2}{V_1} = 1060°\text{R}$$

Heat added per pound

$$\frac{Q}{m} = c_p(T_2 - T_1) = .24(1060 - 530) = 127.2\frac{\text{Btu}}{\text{lb}}$$

11.5 CONSTANT VOLUME FREE EXPANSIONS

Free expansion requires a volumetric change of a gas without doing any WORK. Constant volume processes do not allow for an expansion; therefore, there is no such thing as a free expansion, constant volume process. It is impossible, in Figure 11–2, to have the diaphram to rupture and the gas remain in the original volume.

Another process that does not exist is a "free *compression*." There is no compression that occurs spontaneously and without WORK.

11.6 SEMI-FREE EXPANSIONS

We have discussed reversible processes in which pistons move and WORK is done and free expansion processes in which no pistons are present and no WORK is done. There are times when, realistically, the process that occurs is somewhere between these two extremes.

For example, in the case of the reversible adiabatic chamber (Figure 11–1), we have assumed that the piston is attached to something that can use the WORK that is supplied. But what if that is not the case? What if, for example, the piston is connected to nothing? Doesn't it then just act as a diaphram and the process does no WORK? Essentially yes, except for one difference. If the piston itself has weight, then some WORK is done to move the piston, whereas the diaphram has no weight, nor does it move. Nevertheless, the WORK required to move the piston is less than the reversible WORK the gas can do.

Consider the adiabatic chamber (Figure 11–2), but with a free-floating piston just above the diaphram. When the rupture occurs, the piston is driven upward until the volume of the gas doubles. Assume that the weight of the piston is 500 lb and that it moves up 3 ft during the doubling of the volume. The WORK done on the piston equals the weight of the piston times the distance that it travels (assuming no atmoshperic pressure):

$$WORK = F \times d = 500 \text{ lb}_f \times 3 \text{ ft} = 1500 \text{ ft-lb}_f$$

The amount of WORK that the gas can do against this piston was calculated in Sample Problem 11–1 to be 5230 ft-lb$_f$. This can be called the reversible WORK, the WORK done if the process was reversible. Since the piston is not able to have all this WORK done on it (it is not heavy enough), the process is not reversible. But it is not a free expansion, either, since *some* WORK was done. This process is somewhere between a reversible and a free expansion process. Such a process definitely will be irreversible, since it will take more WORK to reset the piston than we got out of the process in the first place. To be specific, $1500/5230 = .286$ is the fraction of the maximum WORK that this process accomplished. We might classify this process as having an *irreversibility index* of

$$\beta = \frac{(5230 - 1500)}{5230} = .714 \text{ or } 71.4\%$$

of the possible WORK from the situation was not realized.

❖ **KEY TERM:**   **Irreversibility Index of a Process** (β): The fraction of the reversible WORK that is lost to a partially free expansion.

SAMPLE PROBLEM 11-5

What is the final temperature and pressure of the irreversible adiabatic process that takes place under the conditions of Sample Problem 11–2 (P_1 = 20 psia, $V_1/V_2 = 1/2$, $T_1 = 70°F$, m = .3 lb$_m$) and has a free expansion index of $\beta = .714$?

SOLUTION:

$$Q = 0 = c_v m(T_2 - T_1) + (1 - \beta) \times 5230$$
$$0 = .17 \times .3 \times (T_2 - 530) + 1500$$

Solving for

$$T_2 = -38 + 530 = 492°R = 32°F$$

Then

$$P_2 = \frac{mRT_2}{V_2} = .3 \times 53.36 \times \frac{492}{6} = 1312 \text{ psfa}$$

Table 11–2 summarizes the results of Sample Problems 9–6, 11–2 and 11–5, which are the reversible case, the free expansion case, and the general irreversible case.

The irreversible index allows us to classify thermodynamic processes. If a process has a free expansion index $\beta = 0$, then the process is reversible. If $\beta = 1$, the process is a free expansion. If β is less than one but greater than zero, the process is a general irreversible process. Notice that b can never be less than zero because the reversible WORK is the maximum amount of WORK a gas can do.

The definition of the irreversibility index allows us to solve problems concerning irreversible processes of constant temperature and constant pressure.

SAMPLE PROBLEM 11-6

Ten pounds of compressed air are contained in 7 cu ft of a constant temperature (70°F) vessel. The gas expands to a volume of 21 cu ft against a piston

Table 11–2

Sample Problem	Heat Added	WORK Done (ft-lb)	Increase in Volume	Final Pressure	Final Temperature
9–6 (reversible)	0	5230	Twice	1091	–57°F
11–2 (free expansion)	0	0	Twice	1440	70°F
11–5 (general irreversible)	0	1500	Twice	1312	32°F

which is coupled to a water pump that does not use all of the WORK of the gas, $\beta = .40$.

Find how much WORK is done during the process, how much heat is absorbed from the constant temperature source, and the final pressure of the air behind the piston.

SOLUTION:

Solve the problem first as if it were reversible:

$$T_1 = 530°R, \ V_1 = 7 \text{ ft}^3, \ m = 101 \text{ lb}$$

$$P_1 = \frac{mTR_1}{V_1} = 10 \times 53.36 \times \frac{530}{7} = 40{,}401 \text{ psfa}$$

$$T_2 = 530°R, \ V_2 = 21 \text{ ft}^3$$

$$P_2 = \frac{P_1 V_1}{V_2} = 40{,}401 \times \frac{7}{21} = 13{,}453 \text{ psfa}$$

The reversible work can be calculated from

$$WORK = P_1 V_1 \ln \left(\frac{V_2}{V_1} \right) = 310{,}695 \text{ ft-lb}_f$$

and to be specific, we shall call this $WORK_{rev}$.

Since $T_2 - T_1 = 0$:

$$Q_{rev} = WORK_{rev} = \frac{310{,}695}{778} = 400 \text{ Btu}$$

Now for the actual situation that is irreversible:

$$T_2 = 530°R, \ V_2 = 21 \text{ ft}^3, \ P_2 = 13453 \text{ psfa}$$

$$WORK_{irrev} = (1 - \beta) \times WORK_{rev} = 186{,}417 \text{ ft-lb}$$

$$Q_{irrev} = WORK_{irrev} = \frac{186{,}417}{778} = 240 \text{ Btu}$$

again, since $T_2 - T_1 = 0$.

SAMPLE PROBLEM 11–7

In Section 7–3, a constant pressure elevator was described that would move a 200-lb cargo from the dock to the deck of a ship. For that case, 7250 Btu's of heat were added to the "working" gas under the elevator to perform the 2097 Btu's of WORK, and the resulting gas temperature was raised 530°F. If the elevator were allowed to rise to the deck without any cargo (assume the elevator itself has no weight and that there is no atmosphere pushing down on the carrier bed), how much heat would be needed and what would be the final temperature?

SOLUTION:

For that situation:

$$T_b = 70°F, \ m = 57 \ lb_m, \ P_b = 200 \ psfg, \ V_t = 1400 \ ft^3, \ V_b = 700 \ ft^3$$

(see Section 7–3, b is for the bottom position of the elevator)

If the empty and weightless elevator carrier is allowed to go up to the deck, the expansion process of the gas does no WORK and the process is a free expansion, with the result that the final temperature is the same as the original temperature: $T_t = 70°F$. Since ΔT and WORK = 0, then $Q = 0$, and this is an adiabatic, constant temperature free expansion:

$$P_t = P_b\left(\frac{V_b}{V_t}\right) = 2316 \times \frac{700}{1400} = 1156 \ psfa$$

This is a very realistic situation where the elevator is held in place at the dock, and then let go empty to rise up to the deck using the pressure that was available in the initial compressed gas.

The results of Sample Problem 11–6 would be very useful if the problem included the effect of the atmosphere resisting the motion of the carrier with a pressure of 2116 psia.

SAMPLE PROBLEM 11-8

Recalculate the WORK done, the heat added, the final temperature, and the final pressure of the gas for the situation of Sample Problem 11–6, but include the effects of the atmospheric pressure pushing against the carrier.

SOLUTION:

The WORK done by the atmosphere against the carrier during the rise is

$$WORK = F \times d = P_{atm} \times A \times d$$

where A is the area of the carrier, $A = 10 \ ft^2$

$$WORK = 2116 \ \frac{lb_f}{ft^2} \times 10 \ ft^2 \times \frac{70 \ ft}{778 \ \frac{ft-lb_f}{Btu}} = 1903 \ Btu$$

The reversible work from Section 7–4 is 2097 Btu; therefore, the process of this sample problem is irreversible, and the irreversibility index is $\beta = (2097–1903)/2097 = .092$.

Analyzing the situation, the pressure of the gas must remain greater than 2116 psfa during the process in order to keep the carrier moving. Without adding any heat, the pressure must fall below this value, as illustrated in Sample Problem 11–6. Therefore, some heat will be added so that the pressure remains high enough to push, and this implies that the gas pressure will fall to 2116 just at the moment the carrier reaches the top:

$$P_t = 2116 \text{ psfa}$$

which allows us to calculate the final temperature from the perfect gas law:

$$T_t = \frac{P_t V_t}{mR}$$

$$= \frac{2116 \text{ lb}_f}{\text{ft}^2} \times \frac{1400 \text{ ft}^3}{57 \text{ lb}_m \times \frac{53.3 \text{ ft} - \text{lb}_f}{°R - \text{lb}_m}} = 975°R$$

Then

$$Q_{\text{irrev}} = c_v m (T_t - T_b) + WORK_{\text{irrev}}$$

$$= .17 \times 57 \times (975 - 530) + 1903$$

$$= 6215 \text{ Btu}$$

11.7 APPLICATION: FILLING PROCESSES

One of the situations to which an engineer or technician often has to apply irreversible thermodynamics to find proper results is the problem of transporting compressed gases. Gases such as oxygen, hydrogen, and acetylene, to name a few, are stored at different points throughout a plant and transferred to tanks where they are used for various processes. The operators and designers of such facilities must understand the thermodynamics of the transfer, in particular the temperatures and pressures that will result, so they can evaluate the need for cooling, insulation, pressure relief, and so on.

Figure 11–8(a) provides a simplified picture of the transfer of a gas from one vessel (A) to another vessel (B). Figure 11–8(b) shows the thermodynamicist's vision of the process, one that is familiar to us. Each of the vessels are filled with a gas to conditions T_{A1} and P_{A1} and P_{B1} and T_{B1}. We are interested in what happens to these properties once the valve is opened (or sliding diaphram is released), either momentarily or completely.

The perfect gas law applies to each side individually, and the first law applies both to each side individually *and* to the combined set of vessels as a whole. For example, if we insulate both vessels, then the opening of the valves may cause heat to transfer from one side to the other, but no heat flows into or out of the combined set. Therefore, for the combined set:

$$Q = Q_A + Q_B = c_v m_A \, \Delta T_A + WORK_A + c_v \, m_B \, \Delta T_B + WORK_B = 0$$

However, since there is no WORK-generating device that takes WORK out of the system, the only WORK done by the gases, whether reversible or irreversible, is done by one gas against the other. Therefore, $WORK_A = -WORK_B$ and

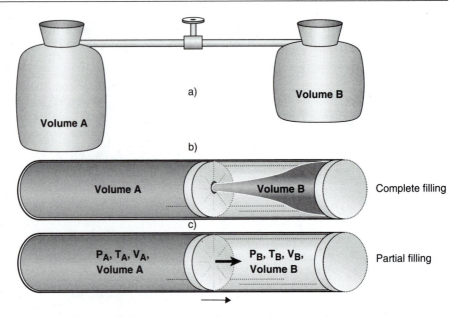

Figure 11–8. Staged filling of compressed-gas tanks.

$$Q = 0 = c_V \left(m_A \, \Delta T_A + m_B \, \Delta T_B \right)$$

or

$$\Delta T_A = \frac{-m_B}{m_A} \, \Delta T_B$$

Equation 11–1

SAMPLE PROBLEM 11–9

Acetylene is contained in a 10-ft³ storage vessel in a 50°F storage room at 200 psia. A portable acetylene welding tank (1 ft³) that has been discharged to 15 psia is brought in to be filled. Its temperature is 80°F. A hose is connected between both valves, and the tank valves are opened until there is pressure equalization between both tanks, and the valves remain open long enough for gases to mix until there is temperature equalization.

a. What is the common final temperature of the two vessels? (Assume complete mixing.)

b. What is the common final pressure of the two vessels?
c. How much (lb$_m$) acetylene was in the portable cylinder initially, and how much did it contain after it was filled?

SOLUTION:

a. m_A (storage cylinder) $= \dfrac{P_A V_A}{R T_A} = 9.5$ lb$_m$

m_B (portable cylinder) $= \dfrac{P_B V_B}{R T_B} = .067$ lb$_m$

Since $T_{A2} = T_{B2} = T_2$:

$T_2 - T_{A1} = \dfrac{-m_B}{m_A}\,(T_2 - T_{B1})$ (using Equation 11–1)

$T_2 - 50 = \dfrac{.067}{9.5}\,(T_2 - 540)$

$T_2 = 506°$R

b. $P_{A2} = P_{B2} = P_2$

$P_2 = \dfrac{(m_A + m_B)R T_2}{(V_A + V_B)} = 183.18$ psia

c. $m_{Bfilled} = \dfrac{P_2 V_B}{R T_2} = .795$ lb$_m$

Equation 11–1 provides for some interesting results concerning one special case that was discovered over one hundred and fifty years ago by several thermodynamists, including Mayer and Joule, but which now has become known as Joule's Experiment. If the two sides are *initially at the same temperature*, $T_{A1} = T_{B1}$, and *if the two gases are allowed to mix completely* after opening the valve, ($T_{A2} = T_{B2}$) then $\Delta T_A = \Delta T_B$. Equation 11–1 states that this is impossible unless $\Delta T_A = \Delta T_B = 0$. Therefore, when the gases are mixed under the Joule Experiment (JE) conditions, the temperature in the two vessels does not change. This implies that this adiabatic process is also a constant temperature one, and the reverse also must be true—a constant temperature JE process is also adiabatic. This implies that there is no need to insulate these two vessels, since the gas exchange will naturally occur adiabatically and at constant temperature. Joule and others proved this by immersing the vessels in water and carefully measuring the water temperature to confirm that there was no rise in temperature.

The JE was actually done with side B evacuated, but according to our analysis above, this was not necessary. However, with side B evacuated, the situation is identical to that illustrated in Figure 11–2, which was an adiabatic free expansion with the result that the temperature of the gas remained the same (in the current case, $T_{A1} = T_{A2}$), which is the same result that we have analyzed here ($\Delta T_A = 0$). Opening the valve is equivalent to bursting the diaphram.

SAMPLE
PROBLEM 11-10

Acetylene is contained in a 10-ft^3 storage vessel in a 80°F storage room at 200 psia. A portable welding tank (1 ft^3) containing 15 psia of acetylene at 80°F is to be filled from the storage tank. Once the gas is exchanged:

a.   What is the common final temperature of the two vessels?
b.   What is the common final pressure of the two vessels?
c.   How much (lb$_m$) acetylene was in the portable cylinder initially, and how much did it contain after it was filled?

SOLUTION:

a.   This situation has the same conditions as the JE and, therefore, $T_{A2} = T_{A1} = T_{B2} = T_{B1} = 80°F$

b. $m_A = \dfrac{P_A V_A}{RT_A} = 8.97 \text{ lb}_m$

$m_B = .075 \text{ lb}_m$ (see Sample Problem 11–9)

$P_{A2} = P_{B2} = \dfrac{(m_A + m_B)RT_{A2}}{(V_A + V_B)} = 29{,}375 \text{ psfa} = 183 \text{ psia}$

c. $m_{B\text{filled}} = \dfrac{P_{B2} V_B}{RT_{B2}} = .82 \text{ lb}_m$

Now consider what happens if the filling process of Sample Problem 11–9 is interrupted before the pressures have equalized. For example, let's say that the portable cylinder is filled until the storage pressure drops to 180 psia, then the valve is shut off and the two tanks are isolated. Equation 11–1 is still valid, but it now contains two unknowns, since $T_{A2} \neq T_{B2}$. Another equation is needed to complete the solution (two equations, two unknowns).

The second equation can be found by investigating the energy equation for each side (A and B) individually. Also consider that what is meant by partial filling (Figure 11–8(c)) is that the membrane between the two gases does not burst, giving almost immediate pressure equalization, but instead the membrane is moved by the pressure of A into tank B, compressing the gas of B and expanding the gas of A. (Figure 11–9) The gas in B is compressed reversibly, since the diaphram piston is pushing it, but the gas in A is expanding irreversibly, since the WORK necessary to compress B is less than the reversible WORK that A can do (once the gas expands from A, it will take more WORK to put it back than it took to move gas B out of the way). Still $WORK_A = -WORK_B$ and $WORK_B = m_B R (T_{B2} - T_{B1})/J(1-k)$, since it is reversible. Also, $Q_A = Q_B = 0$, since no heat is exchanged with the surroundings or between the two gases. Therefore:

$$Q_A = 0 = c_v m_A (T_{A2} - T_{A1}) - \frac{m_B R (T_{B2} - T_{B1})}{J(1-k)}$$

Equation 11–2

Since side B was compressed reversibly:

$$T_{B2} = T_{B1}\left(\frac{P_{B2}}{P_{B1}}\right)^{\frac{(k-1)}{k}}$$

If the portable cylinder of Sample Problem 11–9 is partially filled until the adiabatic compressed pressure P_{B2} is 50 psi, find the pressure and temperature of the storage cylinder.

SOLUTION:

First find the adiabatic compressed temperature T_{B2}.

$$T_{B2} = T_{B1}\left(\frac{P_{B2}}{P_{B1}}\right)^{\frac{(k-1)}{k}} = (80 + 460)\left(\frac{50}{15}\right)^{\frac{.41}{1.41}} = 766.4°\,R$$

Then Equation 11–1 results in

$$T_{A2} = T_{A1} - \frac{m_B}{m_A}\left(T_{B2} - T_{B1}\right) = 510 - \frac{.067}{9.5}(766.4 - 540) = 508.42°\,R$$

The adiabatically compressed volume V_{B2} is:

$$V_{B2} = V_{B1}\left(\frac{T_{B1}}{T_{B2}}\right)^{\frac{1}{(1-k)}} = 1 \times \left(\frac{540}{766.4}\right)^{\frac{1}{.41}} = .426\ \text{ft}^3$$

Therefore, the volume occupied by A is

$$V_{A2} = 10 + (1 - .426) = 10.574\ \text{ft}^3$$

From the perfect gas law:

$$P_{A2} = \frac{m_A R T_{A2}}{V_{A2}} = 27{,}144 = 188.5\ \text{psia}$$

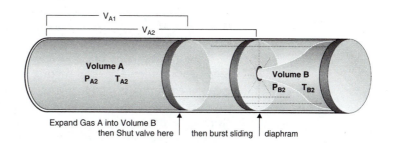

Figure 11–9. The left side expands irreversibly; the right side compresses reversibly.

Notice that once the valve is shut, the pressure and temperature in the storage cylinder is determined, but the portable cylinder contains some high pressure and some low pressure gas. (Figure 11–9) We can now achieve the mixing of these gases by bursting the diaphram to find the resultant pressure and temperature in the portable cylinder. This calculation is similar to Sample Problem 11–9.

SAMPLE PROBLEM 11–12

What is the final pressure and temperature in the portable cylinder of Sample Problem 11–11?

SOLUTION:

In Figure 11–9, this problem involves the mixing of volumes $V_{A2} - V_{A1}$ and V_{B2}.

$$m_A = \frac{P_{A2}(1 - .426)}{RT_{A2}} = 27,144 \times \frac{.574}{59.4} \times 508.4 = .516 \text{ lb}_m$$

After mixing

$$T_f - T_{A2} = \frac{-m_B}{m_A}(T_f - T_{B2}) = \frac{-.067}{.516}(T_f - T_{B1})$$

$$T_f = 538.16$$

Then from the perfect gas law:

$$P_f = \frac{(m_A + m_B)RT_f}{V_B} = 129.5 \text{ psia}$$

Overall, the portable cylinder was filled with .516 lb$_m$ of additional acetylene and the vessel pressure increased from 15 to 129 psia. The portable cylinder temperature decreased to 538 while the storage cylinder pressure dropped to 188 psia and its temperature dropped to 508.4.

The mathematics of this partial filling process are quite lengthy, but the solution is a useful one for some design considerations. Now let us look at one final type of filling process that also often occurs.

Consider in Figure 11-8 that V_B is very large, say the whole atmosphere. If we open the valve, we are actually exhausting the gas to the atmosphere. What will the temperature be in vessel A once all the gas is exhausted?

There are actually two answers to this question. The first one is for the case in which we allow the total mixing between A and B. If we consider that the gases were initially at the same temperature and are finally at the same temperature, then we have the conditions of the JE and the temperature in the storage cylinder will not change.

If, however, we consider that the valve is shut immediately after achieving atmospheric pressure in cylinder A, then the temperature of the remaining gas in the cylinder is not the same as the atmosphere. In this case, the solution follows the solution in Sample Problem 11–11 precisely except the compression in the right-side vessel is constant pressure rather than adiabatic. Therefore, Equation 11–2 is modified to

$$Q_A = 0 = c_v m_A (T_{A2} - T_{A1}) - \frac{P_B(V_{B2} - V_{B1})}{J}$$

This time the equations are quite difficult using algebra, but calculus (see the Applications of Calculus section in this chapter) gives a solution of

$$\frac{T_{A2}}{T_{A1}} = \left(\frac{P_{A2}}{P_{A1}}\right) \frac{(P_{A1} + (k-1)P_B)}{(P_{A2} + (k-1)P_B)}$$

Equation 11–3

The calculus section also shows this equation to be valid for the situation where P_B is not atmospheric but any other pressure, such as when a cylinder feeds a process pipe that carries the gas away at a constant pressure P_B. (Figure 11–10)

SAMPLE PROBLEM 11–13

If the valve on the acetylene cylinder in Sample Problem 11–9 is opened and the contents exhausted to the atmosphere until the pressure drops to 15 psia (atmospheric), then the valve is shut, what will be the temperature in the supply cylinder?

SOLUTION:

$P_{A2} = 15$ psia
$P_{A1} = 200$ psia
$P_B = 15$ psia

$$\frac{T_{A2}}{T_{A1}} = \left(\frac{15}{200}\right) \frac{(200 + .41 \times 15)}{(15 + .41 \times 15)} = 15 \times \frac{206}{(200 \times 210 = .72)}$$

$T_{A2} = (50 + 460) \times .72 = 367°R = -93°F$

In summary, this section has investigated filling processes and has identified five different situations, all with different results:

1.   adiabatic with temperature and pressure equalization and $T_{A1} = T_{B1}$: Sample Problem 11–10;

2. adiabatic with temperature and pressure equalization but $T_{A1} = T_{B1}$: Sample Problem 11–9;

3. adiabatic partial filling (neither pressure nor temperature equalization): Sample Problem 11–11;

4. constant pressure exhausting with or without pressure equalization but with no temperature equalization: Equation 11–3; and

5. constant pressure exhausting with pressure equalization but no temperature equalization: Sample Problem 11–12.

TECHNICIANS IN THE FIELD

Compressed Gas Operator

Not all jobs in technology are exciting and require great powers of concentration and deduction. Some jobs are routine and basic. Usually these positions are filled by technicians seeking their first job, their first experience. Welders Supply in Torrance, California, hires a part-time technician to fill portable cylinders with compressed gas from larger supply cylinders. The gases are typically oxygen, nitrogen, or acetylene. Their technician is completing his coursework at a local college and works at this job to earn his tuition and to gain experience for when he graduates and seeks a career position.

The procedure for filling individual vessels is quite unorthodox. The size of the supply cylinders allows them to hold about ten times more gas than the portable cylinders, but the supply tank pressure is only slightly higher than the pressure to which the individual cylinders will be filled. Therefore, the individual cylinders cannot be filled directly from the supply tank because the supply tank pressure would quickly drop below that necessary to fill the individual vessels.

Instead, the individual vessels, for this reason, are filled from three supply tanks in stages. At the first stage, the portable vessel is filled from a supply tank that is almost empty and contains the lowest pressure. As the individual vessel moves to the second stage, it is filled from a supply tank that has a higher pressure than the first supply tank. Finally, in the third stage, the portable vessel is filled by a fresh supply tank to a pressure very close to the pressure of the supply tank. Once a supply tank in any stage falls below the pressure for that stage, then all cylinders are rotated to the next stage and the lowest pressure supply tank is retired. (Figure 11–8)

The technician realizes that each stage filling is similar to the solution of Sample Problems 11–11 and 11–12. Since he will have to perform these calculations for each stage, he decides to program them on spreadsheet.

COMPUTER LESSON

Use the computer lesson to find the temperatures and pressures in the portable cylinder after each stage of filling:

Stage 1   Supply Tank $T = 540°R$, $P = 60$ psia
Stage 2   Supply Tank $T = 540°R$, $P = 120$ psia
Stage 3   Supply Tank $T = 540°R$, $P = 200$ psia

SOLUTION:

The computer lesson requires that the following information be input (see Sample Problem 11–11):

$T_{A1} = 540°R$

$$P_{A1} = 60 \text{ psia}$$
$$T_{B1} = 540°\text{R}$$
$$P_{B1} = .01 \text{ psia (computer cannot accept } P_{B1} = 0 \text{ here)}$$

It also requires a value of P_{B2}, the adiabatic pressure of the partial gases after the filling but before the mixing. The technician does not know this pressure, but he does know that this pressure must be equal to P_{A2}; therefore, he guesses a value for P_{B2}, checks the resulting value of P_{A2}, then adjusts P_{B2} until $P_{B2} = P_{A2}$.

After the calculation for each stage is complete, the technician starts the next stage by inputting T_{A1} and P_{A1} from the given data, then setting T_{B1} and P_{B1} equal to the final mixed temperature and pressure from the previous stage. Here is the result from the spreadsheet:

	P_{A1}	T_{A1}	P_{B1}	T_{B1}	P_{B2}	T_{B2}	m_A	m_B	T_{A2}	V_{B2}	P_{A2}	m_A	T_f	P_f
Stage 1	60	540	.01*	540	54.5	6589	2.69	0	540	0	54.5	.24	540	54.5
Stage 2	120	540	54.5	540	114	669	5.38	.24	534	.59	114	.21	606	114
Stage 3	200	540	114	606	192	705	8.97	.45	535	.69	192	.27	642	192

* Computer will not take '0' here.

From the results, the technician can show that what he has sensed during the staged filling can actually be predicted by thermodynamic calculation. The temperature of the portable cylinder during first-stage filling does not change, whereas during the second stage, it shows a dramatic increase, and during the third-stage filling, a smaller increase.

The technician spent many weeks considering this problem before he arrived at this complex solution. He was never able to put the results to work for him on the job, but it gave him great comfort to satisfy his curiosity, and he felt much more comfortable with irreversible processes.

**APPLICATIONS
OF
CALCULUS**

The equation of state of a filling process for a gas that flows from a high-pressure vessel to a pipe that has a constant low pressure can be derived using calculus and the differential form of the energy equation combined with the differential form of the perfect gas law (see the Applications of Calculus section in Chapter 9). Figure 11–10 illustrates the expansion of gas A into volume B under the opposition of the constant pressure P_B. For a differential expansion of gas A:

$$dQ = 0 = c_v m dT + P_B/J\, dV = c_v/R\,(VdP + PdV) + P_B/J\, dV$$

(see Applications of Calculus, Chapter 9)
Then

$$\frac{c_v}{R} VdP = -\left(\frac{c_v}{R} P + \frac{P_B}{J}\right)dV$$

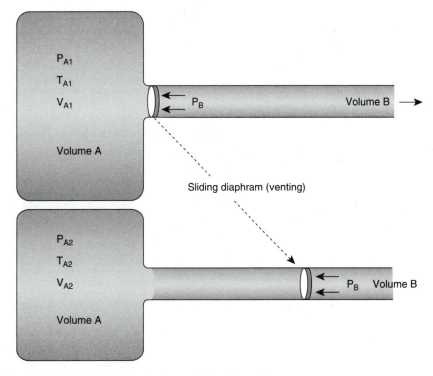

Figure 11-10. A compressed gas emptying into a pipe.

or

$$\frac{dP}{\left(P+(k-1)P_B\right)} = \frac{-dV}{V}$$

using $R/c_v J = 1-k$ (see Equation 7–3).
Integrating between P_{A1} and P_{A2} yields:

$$\frac{\left(P_{A2}+(k-1)P_B\right)}{\left(P_{A1}+(k-1)P_B\right)} = \frac{V_{A1}}{V_{A2}}$$

This is the equation of state for the situation described above. Notice that this solution has the interesting result that if $P_{A2} = P_{A1}$ (the valve is not opened), then $V_{A1} = V_{A2}$. Also if $P_B = 0$ (Joule's Experiment), then $P_{A2} V_{A2} = P_{A1} V_{A1}$, which is to say that $T_{A2} = T_{A1}$, which is the result stated by Joule. To write the equation of state in terms of pressure and temperature, use the perfect gas law:

$$\frac{V_{A1}}{V_{A2}} = \left(\frac{T_{A1}}{T_{A2}}\right)\left(\frac{P_{A2}}{P_{A1}}\right)$$

with result

$$\frac{P_{A1}}{P_{A2}} \frac{\left(P_{A2}+(k-1)P_B\right)}{\left(P_{A1}+(k-1)P_B\right)} = \frac{T_{A1}}{T_{A2}}$$

For the special case that Tank A is exhausted completely to P_B (or to determine the temperature of the exhausted gas), then $P_{A2} = P_B$, and the above result simplifies to:

$$\frac{T_{A2}}{T_{A1}} = \frac{1+(k-1)\dfrac{P_B}{P_{A1}}}{k}$$

CHAPTER **11**

PROBLEMS

1. If the vessel in Figure 11–2 is filled with hydrogen to a pressure of 20 psia and a temperature of 70°F, what is the pressure in the vessel after the diaphram breaks and the gas occupies twice its original volume?

2. A 2-ft diameter pipe is 10 ft long and sealed at both ends. A plastic diaphram isolates one half of its length from the other half. Air is pumped into one chambered section of the pipe at 100°F. When the pressure reaches 455 psig, the diaphram breaks. If the process is considered to be an adiabatic free expansion, what is the gas pressure in the pipe when equilibrium is achieved?

3. Two vessels, each of 3 ft^3 volume, sit side by side, connected by a pipe fitted with a closed isolation valve. One vessel is evacuated to 0 psia, and the other is filled with nitrogen to 50 psig and 70°F. If the temperature of the room is also 70°F, what will the resulting temperatures be in each of the vessels when the valve is opened? Also, what will the resulting pressures be? How much heat will be exchanged between the gas inside the vessel and the room?

4. The vessel in Figure 11–2 is filled with 20 psia of hydrogen at 70°F. When the diaphram is broken, heat is added to maintain a constant pressure. What is the final temperature and how much heat is added (per pound)?

5. A 2.5-ft^3 oxygen tank is filled with 20 lb of oxygen at 70°F. A scuba diver uses this pure oxygen to breathe. His lung capacity is 1 ft^3 and the process of filling his lungs is considered an adiabatic free expansion.
   a. What is the pressure of the oxygen in the vessel when it is full?
   b. What is the temperature of the oxygen when it expands into the lungs?
   c. How many breaths can he take from the tank?
   d. How much WORK could the compressed oxygen do if it were expanded reversibly and adiabatically against a piston down to atmospheric pressure?
   e. What is the free expansion index for the breathing process?

6. A 20-ft^3 oxygen supply cylinder at 2000 psia is used to fill a 2-ft^3 individual tank that initially contains 200 psia of residual pressure. What is the maximum pressure to which the small tank can be filled, how much (lb$_m$) oxygen is put into the individual tank, and what is the final temperature in the individual tank if the valve is shut immediately after filling?

CHAPTER
12
Putting Heat to Work

PREVIEW: Thermodynamic processes do WORK, but only for a short time, usually just long enough to move a piston to the end of its cylinder. For WORK to be done continuously, as in an automobile engine, a variety of processes must be put together back-to-back.

OBJECTIVES:

❑ Describe the operation of an internal-combustion engine in terms of thermodynamic processes that create a cycle.

❑ Calculate the efficiency (thermal efficiency) of an automobile engine.

❑ Evaluate the design of an internal-combustion engine based on specific-performance parameters.

12.1 THE AUTOMOBILE ENGINE REVEALED

The automobile engine is a machine familiar to most people. It runs using spark plugs, cylinders, a carburetor, and other familiar items. What may not be familiar is the fact that it operates using many of the thermodynamic processes we have studied. It can be said in all truthfulness that an automobile engine uses the perfect gas law, and it uses the energy equation, in order to create its power.

The operation of an automobile engine involves four strokes: intake, compression, power, and exhaust. Figure 12–1 shows a four-stroke internal-combustion engine. The strokes of an engine refer to the position of the piston in the cylinder. The intake stroke occurs when the piston is on its way down and the intake valve at the top is open to draw in a fresh supply of gasoline and air. The compression stroke occurs as the piston begins to rise in the cylinder and the intake valve closes. The air-gasoline mixture is compressed to a pressure about eight times what it was initially. The power stroke actually includes two parts: first a spark plug ignites the gasoline, creating an explosion while the cylinder is at the very top of its travel, then the piston moves down the cylinder, generating the power of the engine. The last stroke, the exhaust stroke, begins when the piston reaches the bottom of the cylinder. The exhaust valve opens and the piston pushes the spent air-gasoline mixture out into the atmosphere.

This description of the internal-combustion engine is very accurate and can be used to build such an engine. Using it, we could mass-produce the engine, or take it to the Indianapolis 500, or improve and modify it. But one thing the description will not do is to analyze the engine from the point of view of how efficient it is—how efficiently it converts fuel energy (gasoline) into WORK. It won't determine the maximum power you can expect from this engine or what kinds of things can be changed in order to improve the engine's efficiency or power.

If we look at the engine as a series of thermodynamic processes acting on the gas in the cylinder, we can determine the efficiency of the engine and identify the factors that contribute to it. The idea of looking at a engine as a thermodynamic process was first introduced by Sadi Carnot. The engine he studied was not the automobile engine, but rather the steam engine (see Chapter 14). By the time Carnot got to it, the steam engine was an operating machine used primarily in the coal mines of England. Carnot was not involved with the invention of the engine. James Watt, before Carnot's time, is credited with that. Carnot's contribution was to identify what could be done to the steam engine in order to improve its power and increase its efficiency.

Nikolaus Otto (1876) applied Carnot's ideas to the automobile engine by analyzing the four strokes of the automobile engine and breaking them down into the proper thermodynamic processes: constant volume, constant temperature, constant pressure, or adiabatic.

Take first the intake stroke. What happens to the thermodynamic state of the air-gas mixture during the intake? As air is taken into the cylinder chamber, the pressure of the air does not change; it remains atmospheric pressure. The temperature of the air does not change; it is the temperature of the outside air. As the chamber fills with more and more gas, the density of that gas does not change. No heat is added to the air or taken away. During the intake stroke,

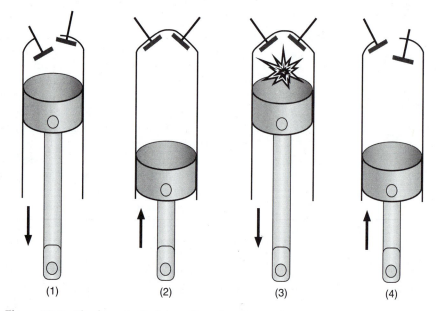

Figure 12–1. The four-stroke internal-combustion engine.

therefore, absolutely nothing happens to the air to change the thermodynamic state. (Figure 12–2)

The compression stroke is something different. While the air-gasoline mixture is being compressed, the pressure of the gas is certainly rising. The density is also increasing, and temperature is going up due to compression. However, no heat is being added to the gas during this compression. Because there is no hot plate or heat source around, there is no mechanism for adding heat or taking heat away.

The temperature of the gas is rising even without heat being added or taken away. You may wish to argue that heat is taken away by being transferred through the cylinder walls as the air gets hot. This is true. In fact, in a water-cooled engine, the cylinder walls are being cooled by radiator water. However, the piston rises in the compression stroke thousands of times per minute. The time allowed for the piston to raise once is very short, yet heat takes a long time to transfer through the wall. Therefore, during one stroke, not much heat is lost. We can safely say that $Q = 0$ during the compression stroke. The compression stroke is adiabatic. (Figure 12–3)

The power stroke actually consists of two processes, as previously described. Initially, the gasoline-air mixture is exploded by the spark plug. During the explosion, the pressure becomes fantastically high, as does the temperature of the gas. However, the specific volume of the gas does not change during the explosion because the piston remains at top dead center of the cylinder and no air is added or taken out. Large amounts of heat are added

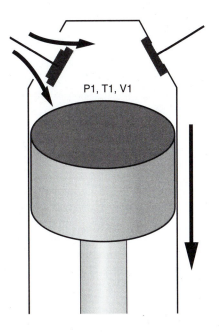

Figure 12–2. The intake stroke is no thermodynamic process at all.

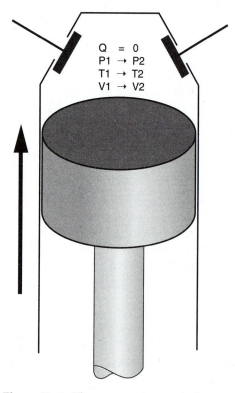

Figure 12–3. The compression stroke is adiabatic.

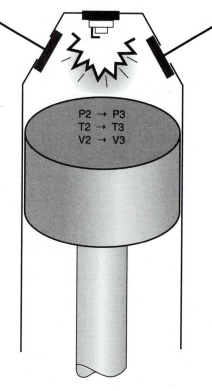

Figure 12–4. The first part of the power stroke is a constant volume explosion.

during the explosion. The explosion process, then, is a constant volume process. (Figure 12–4)

The second part of the power stroke is the downstroke. As the piston moves down, the pressure is dropping. The specific volume is also changing, as the air is getting less dense. Temperature is dropping as the gas expands against the piston. But the process happens so fast that no appreciable amount of heat is added or taken away. This is an adiabatic process. (Figure 12–5)

The exhaust stroke takes this hot, high-pressure, burned-gasoline-and-air mixture and forces it out of the cylinder. While this happens, the pressure is being relieved on the gas. The exhaust and intake strokes combined, however, cause no net change in the volume occupied by the gas. The hot gas is replaced by cool gas, and the pressure is reduced. This is equivalent to a constant volume heat rejection. Since the mass of the gas also has no net change, the process is also constant specific volume. (Figure 12–6)

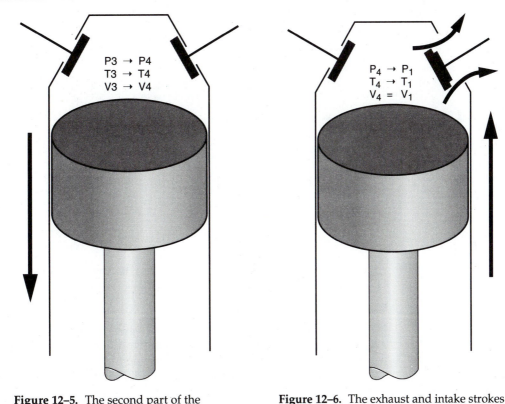

Figure 12–5. The second part of the power stroke is an adiabatic expansion.

Figure 12–6. The exhaust and intake strokes combine for a constant volume process.

12.2 THERMODYNAMIC ANALYSIS OF THE AUTOMOBILE ENGINE

We have analyzed the automobile engine from the point of view of thermodynamic processes and found that there are four processes involved with the thermodynamics of the air-gasoline mixture in the engine. Plotting the processes on a pressure-volume diagram gives tremendous insight into the operation of an internal combustion engine. The first process is the adiabatic compression of the gas. Process 1–2 of Figure 12–7 shows that during the adiabatic compression, the initial pressure P_1 increases to P_2. The initial volume V_1 of the gas decreases to V_2, and the initial process occurs along an adiabatic line on the pressure-volume chart. The second process is the constant volume explosion. During this process, pressure rises from P_2 to P_3. Volume does not change, so the final volume is again V_2, and the final temperature T_3 is increased considerably from the initial temperature T_2. During the downstroke (the adiabatic expansion), the pressure decreases from P_3 to P_4, the volume increases from V_2 to V_4, and the temperature decreases slightly from

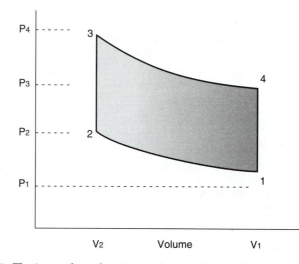

Figure 12–7. The internal-combustion engine as a thermodynamic cycle.

T_3 to T_4. During the exhaust phase, the gas is let out of the cylinder and the pressure in the cylinder is relieved from P_4 back down to the original P_1. The specific volume of the gas does not change during this process, and the temperature T_4 is reduced drastically down to the intake temperature T_1.

Figure 12–7 shows that the internal combustion engine is really four thermodynamic processes that take the air-gasoline mixture from a certain state and eventually bring the air back to the original state. This is called a *thermodynamic cycle.*

❖ **KEY TERM:**   **Cycle:** A series of thermodynamic processes on a gas that begin at a state and return to the same state.

In the language of thermodynamics, the automobile engine cycle consists of the following four processes:

1. Adiabatic compression
2. Constant volume heat addition
3. Adiabatic expansion
4. Constant volume heat rejection

Because Nikolaus Otto did the first analysis of this cycle, it has been named the Otto cycle.

An automobile engine produces power. How do we find from Figure 12–8 how much WORK it does? Recall that the WORK done during a process is the area under the P-V diagram. Therefore, during the adiabatic compression (process 1), it is obvious that WORK is being done to the gas. That WORK is the area underneath the process 1 curve, shown by the right sloping lines. WORK is also done during process 3. This time the gas does WORK on the

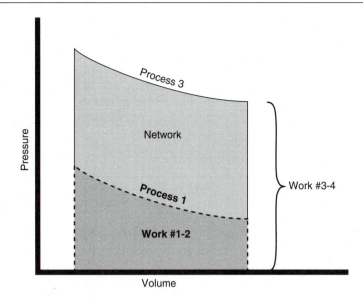

Figure 12–8. The WORK diagram.

piston instead of having WORK done to it, accounting for the WORK of the power stroke. That WORK is shown by the left sloping lines on the diagram. The gas both does WORK and has WORK done on it. What is the net amount of WORK that the gas does? Subtract the left sloping area from the right sloping area and what remains is the area inside the P-V diagram. This area represents the WORK done by the cycle.

Since there is a net WORK from the cycle, you might expect that we are getting something free. Is there a cost of running this cycle? The cost is the amount of heat that is added in order to keep the cycle going. Process 2 is the constant volume explosion of fuel, which adds a tremendous amount of heat. This heat comes from gasoline, which costs money. How much heat is added in this cycle?

During process 2, $Q_2 = c_v m(T_3 - T_2)$. During process 4, heat is also involved. This time, however, heat is being taken out of the air and removed from the cylinder with the spent gases. Unfortunately, this heat cannot be put to use and, therefore, is wasted heat. The net cost of operating the engine is the heat that is added during process 2 of the cycle. (Figure 12–9)

Can we analyze the automobile engine for efficiency using these thermodynamic processes? One thing we would like to know is the efficiency of the four-stroke internal-combustion engine.

❖ **KEY TERM:**   **Efficiency (η):** The ratio of the "useful output" of the device divided by the "costly input," applicable to any device, whether thermal, mechanical, and so on.

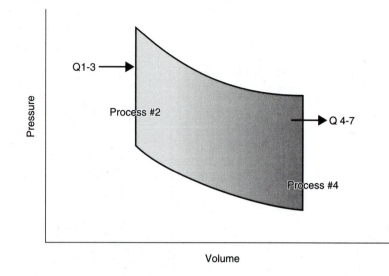

Figure 12–9. Heat added and rejected in the Otto cycle.

For an automobile, the useful output is the distance it has traveled or the load it has moved. In thermal terms, this is "net WORK." The costly input, as every car driver knows, is the fuel it burns, Q_{2-3}.

$$Efficiency = \frac{Useful\ Output}{Costly\ Input} = \frac{Net\ WORK}{Q_{2-3}}$$

Now we are prepared to do a complete thermodynamic analysis on an automobile engine.

SAMPLE PROBLEM 12–1

A 1999 Ford Taurus has the following specifications: 6 cylinders, 258 CID (cubic-inch displacement), 8/1 compression ratio, and 17 mpg mileage rated at 55 mph and 3000 RPM. What is the thermodynamic efficiency of this cycle? (Figure 12–10)

SOLUTION:

A rough pressure-volume diagram of the processes involved is shown in Figure 12–10. For this situation, the given data is as follows: $c_v = .17$, $c_p = .24$, $k = 1.41$, $V_1 = 258\ in^3 = .149\ ft^3$, $P_1 = 14.7\ psia = 2116\ psfa$, $T_1 = 70°F = 530°R$.

Process 1 Adiabatic Compression

$$m_{air} = \frac{P_1 V_1}{R T_1} = \frac{2116 \times .149}{53.3 \times 530} = .011\ lb$$

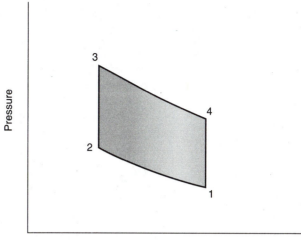

Figure 12–10. Sample Problem 12–1.

$$\frac{V_2}{V_1} = \frac{1}{8} = \frac{v_2}{v_1}$$

$$\frac{P_2}{P_1} = \left(\frac{v_1}{v_2}\right)^k = 8^{1.41} = 18.77$$

$$P_2 = 18.77 \times P_1 = 39,717 \text{ psfa}$$

$$\frac{T_2}{T_1} = \left(\frac{v_1}{v_2}\right)^{k-1} = \left(\frac{1}{8}\right)^{1.41-1} = 2.35$$

$$T_2 = 2.35 \times T_1 = 1245.5°R$$

$$V_2 = \frac{V_1}{8} = .019 \text{ ft}^3$$

$$WORK_{1-2} = \frac{P_2 V_2 - P_1 V_1}{J(1-k)} = \frac{39,717 \times .019 - 2116 \times .149}{778(1-1.41)}$$

$$= -1.38 \text{ Btu or } -1071 \text{ ft-lb done on the gas}$$

Internal Energy $= c_v m (T_2 - T_1) = 1.38$ Btu

Process 2 Constant Volume Heat Addition

$P_2 = 39,717$ psfa, $V_2 = .019$ ft^3, $T_2 = 1245.5°R$

The key to determining the state after this ignition and combustion process is to determine the amount of heat that is added to the engine on each cycle (Btu/cycle). This can be done by converting the fuel efficiency (mpg) into Btu/cycle with the following equation, created by making the units balance (dimensional analysis):

$$\frac{Btu}{cycle} = \frac{Btu}{gal} \times \frac{gal}{mile} \times \frac{mile}{hr} \times \frac{hr}{min} \times \frac{min}{rev} \times \frac{rev}{cycle}$$

Btu/gal is the heating value of gasoline, which can be found from Table 2–2 to be 150,000 Btu/gal. Gal/mile is the mathematical reciprocal of mpg; gal/mile = 1/17. Min/rev is the inverse of RPM, and rev/cycle is revolutions of the piston and crankshaft per cycle, which is two for a four-stroke engine (intake and compression are one revolution, power and exhaust are a second revolution).

$$\frac{Btu}{cycle} = 150,000 \times \frac{1}{17} \times 55 \times \frac{1}{60} \times \frac{1}{3000} \times 2 = 5.4\ Btu$$

so

$Q_{2-3} = 5.4\ Btu = c_v m(T_3 - T_2) = .17 \times .011 \times (T_3 - 1245.5)$

$T_3 = 4133°R$

$P_3 = P_2\left(\frac{T_3}{T_2}\right) = 131,811\ psfa$

$WORK_{2-3} = 0$

Process 3 Adiabatic Expansion

$P_3 = 131,811\ psfa, \quad T_3 = 4133.5 \quad V_3 = .019\ ft^3 \quad V_4 = V_1 = .149\ ft^3$

$P_4 = P_3\left(\frac{V_3}{V_4}\right)^k = 131,811 \times .0555 = 7250\ psfa$

$T_4 = T_3\left(\frac{V_3}{V_4}\right)^{k-1} = 4133.5 \times .43 = 1777°R$

$WORK_{3-4} = \frac{P_1 V_1 - P_3 V_3}{J(1-k)} = \frac{7250 \times .149 - 131,811 \times .019}{778(1-1.41)}$

$= 4.46\ Btu\ or\ 3473\ ft\text{-}lb$

Summarizing:

Net WORK = $WORK_{1-2} + WORK_{3-4} = -1.38 + 4.46 = 3.08\ Btu$

Costly Energy Input = 5.4 Btu

$Efficiency = \frac{Net\ WORK}{Costly\ Energy\ Input} = \frac{3.08}{5.4} = .57\ or\ 57\%$

Process 4 Constant Volume Heat Rejection

$Q_{4-1} = c_v m(T_1 - T_4) = .17 \times .011 \times (530 - 1777) = -2.32\ Btu$

$W_{4-1} = 0$

12.3 CALCULATING HORSEPOWER

The analysis of the automobile engine in the last section provides information that automobile engineers need: the temperature throughout the cycle, the pressures, and the overall efficiency. Probably the most important characteristic of an engine is the horsepower. Horsepower is not the same as the WORK put out by an engine. A small-horsepower lawn mower engine can do just as much WORK as a large-horsepower steam shovel engine; it just takes a much longer time to do it.

Horsepower is the measure of WORK done in a unit of time. Horsepower is equivalent to accumulating the amount of ft-lb$_f$ of WORK done by a cycle for a one-minute time span.

$$Power = \frac{Net\ WORK}{Time}$$

The time for one cycle is similar to the number of revolutions per minute but is the inverse of that, minutes per revolution. Since there are two revolutions for every cycle (four-stroke engine), the number of cycles per minute equals RPM/2. Thus, minutes per cycle equals 2/RPM.

The power of the engine equals:

$$Power = \frac{Net\ WORK}{\frac{2}{RPM}} = \frac{Net\ WORK \times RPM}{2}$$

To convert ft-lb$_f$/min to horsepower, use 1 horsepower = 33,000 ft-lb$_f$/min. Then:

$$Horsepower = \frac{Net\ WORK \times RPM}{2 \times 33,000}$$

SAMPLE PROBLEM 12–2

Calculate the horsepower (hp) of the engine specified in Sample Problem 12–1.

SOLUTION:

$$Net\ WORK = 3.08\ Btu \times 778\ \frac{ft-lb}{Btu} = 2396\ ft\text{-}lb_f$$

$$Hp = \frac{Net\ WORK \times RPM}{2 \times 33,000}$$

$$= \frac{2396 \times 3000}{2 \times 33,000} = 109\ hp$$

12.4 EFFICIENCY OF THE OTTO CYCLE

The results of Sample Problem 12–1 indicate that the efficiency of this automobile engine is very low, below 60%. This low efficiency could mean the engine needs a tune-up, but the condition of the engine did not play any role in the calculations of that problem. In fact, the efficiency calculated is the highest possible efficiency. Why is it so low?

This low efficiency comes from the nature of the cycle itself. The Otto cycle uses heat in order to create high pressures in the cylinder. These pressures force the piston to move and do a great deal of WORK. The cost of operating the engine is the cost of the heat that is put in to cause these high pressures. But all the heat is not utilized. Much of it is lost in the exhaust gases going out the tailpipe. In the sample problem, 2.32 Btu were exhausted from every cycle, yet 5.4 Btu were put into each cycle. This means that almost half of the heat is not used, thus causing the inefficiencies of the Otto cycle. It is the price we pay for using an engine that performs the task of changing heat into WORK.

Engines take heat in the form of gasoline and turn it into WORK in the form of miles traveled have been historically termed *heat engines*. The conversion is not one hundred percent efficient. The efficiency that has been calculated in our sample problem is called *thermodynamic efficiency* because it involves the basic cycle itself. There are other efficiencies that have not been considered: how well the engine is tuned, how completely combustion occurs in the engine, how completely the system transfers energy through the universal joint and transmission, and how hard the tires are inflated. These efficiencies are classified as *mechanical efficiency*. We might say that a certain engine has a mechanical efficiency of 80%. This means that after the power is generated by the heat engine, an additional 20% is lost by mechanical means. Total efficiency can be calculated as:

$$\eta_{total} = \eta_{thermal} \times \eta_{mechanical}$$

SAMPLE PROBLEM 12-3

The Ford Taurus in Sample Problem 12–1 has a transmission loss of 20% and a mechanical efficiency otherwise of 85%. Find its total efficiency.

SOLUTION:

Since there are two mechanical losses plus a thermal loss:

$$\eta_{total} = \eta_{thermal} \times \eta_{trans} \times \eta_{mech} = .57 \times .80 \times .85 = .39$$

Another way of looking at efficiency or performance that is of interest to automobile engineers is the mean effective pressure (MEP).

❖ KEY TERM:

Mean Effective Pressure (MEP): A measure of the amount of WORK an engine will generate for a specific amount of displacement (volume) of the engine.

There are two ways to get more WORK from an engine. The easiest way is to make the cylinder bigger (increasing the displacement). The second way is to be clever about the way we design the engine so that we get greater WORK from the same cylinder volume. The more creative engine will have the highest MEP.

$$MEP = \frac{Net\ WORK}{Engine\ Displacement}$$

On the P-V diagram for the Otto cycle, MEP is the average operating engine pressure if the net WORK area were allowed to slide down the vertical direction and settle at a constant pressure. (Figure 12–11)

SAMPLE PROBLEM 12–4

a. For the engine specified in Sample Problem 12–1, calculate the MEP.
b. Calculate the MEP for a more powerful engine in which the net WORK is increased to 3.93 Btu per cycle because the cubic-inch displacement is increased to 329 in^3.
c. If the engine WORK is increased to 3.93 Btu per cycle but the displacement is not increased, calculate the MEP.

SOLUTION:

a. $MEP = \dfrac{Net\ WORK}{Engine\ Displacement} = 3.08 \times \dfrac{778}{.149} = 16{,}082\ \dfrac{lb_f}{ft^2}$

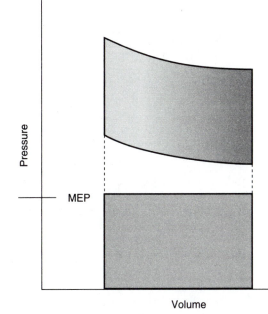

Figure 12–11. Calculating mean effective pressure.

b. $MEP = 3.93 \times \dfrac{778}{\left(\dfrac{329}{1728}\right)} = 16{,}059$ psf

c. $MEP = 3.93 \times \dfrac{778}{.149} = 20{,}520$ psf

While the MEP is a measure of the WORK per cubic-inch displacement, another measure of how creatively an engine is designed is the specific fuel consumption (SFC).

❖ KEY TERM:

Specific Fuel Consumption (SFC): The amount of fuel consumed per hour by an engine to generate each horsepower.

Specific fuel consumption is to the automotive engineer what gas consumption is to the automobile owner. The more efficiently the automobile uses fuel, the better will be its gas mileage and specific fuel consumption. To calculate SFC:

$$SFC = \frac{Gal/hr}{Hp}$$

SAMPLE PROBLEM 12–5

Calculate the SFC for the engine specified in Sample Problem 12–1.

SOLUTION:

To calculate the rate at which fuel is burned, multiply:

$$\frac{Gal}{Hr} = \frac{Gal}{Mi} \times \frac{Mi}{Hr}$$

$$= \frac{1}{17} \times 55 = 3.23$$

Then

$$SFC = \frac{Gal/hr}{Hp} = \frac{3.23}{109} = .0296 \frac{Gph}{Hp}$$

12.5 ANALYZING INTERNAL COMBUSTION ENGINES

The foundation for the thermodynamic analysis of the Otto cycle has been set and we can use it to determine how to increase the efficiency of such an engine. For instance, looking at the P-V diagram of the cycle, what would happen if we increased the compression ratio? Figure 12–12 shows exactly that situation. It a compares two Otto cycles, one with a low-compression ratio and one with

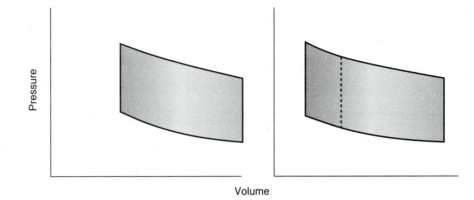

Figure 12–12. Otto cycles with different compression ratios.

a high-compression ratio. Notice that the WORK done per cycle is increased in the high-compression cycle; therefore, the horsepower of the engine will be increased. The same amount of fuel is used for both, which means that the same amount of heat will be added on each cycle. Since the net WORK is increased and the heat added is the same, the efficiency for the higher-compression ratio engine is increased. To show this mathematically, we can convert efficiency directly into compression ratio for the Otto cycle by the following steps:

$$\eta = \frac{Net\ WORK}{Q_{23}} = \frac{W_{12} + W_{34}}{Q_{23}} = \frac{P_2 V_2 - P_1 V_1 + P_4 V_4 - P_3 V_3}{c_v (T_3 - T_2)}$$

Realizing

$$\frac{T_4}{T_3} = \left(\frac{V_3}{V_4}\right)^{k-1} = \left(\frac{V_2}{V_1}\right)^{k-1} = \frac{T_1}{T_2}$$

Then, by careful manipulation, the following results:

$$\eta = 1 - \left(\frac{V_2}{V_1}\right)^{k-1}$$

Equation 12–1

SAMPLE PROBLEM 12–6

Compare the efficiency from Sample Problem 12–1 with Equation 12–1.

SOLUTION:

$$\eta = 1 - \left(\frac{1}{8}\right)^{(1.41-1)} = 1 - .43 = .57$$

This shortcut formula for efficiency of the Otto cycle clearly shows that higher compression ratios mean higher efficiencies.

What happens if we change the octane rating of our gasoline? More Btu/cycle will be added. Figure 12–13 shows the P-V diagram for the Otto cycle with increased octane. Notice that the net WORK done by the engine is increased; therefore, the power of the engine is high, and the horsepower of the engine is increased.

What about the efficiency of the cycle? More heat is added on each cycle with the higher-octane gasoline, and more WORK is the result. It is unclear whether we have increased the efficiency. It "costs" more to run the engine, but the WORK done is greater. The equation relating thermal efficiency to compression ratio (Equation 12–1) indicates that the octane rating of the gasoline plays no part in determining efficiency; therefore, efficiency will not increase. Calculating the MEP for the high-octane engine would result in a higher MEP for this engine, meaning that horsepower per displaced volume would be increased. The SFC would increase, indicating that we are using our fuel more efficiently. This is misleading, however, since actually the fuel is of better quality and is more costly.

**12.6
ANOTHER ENGINE,
ANOTHER CYCLE**

The Diesel engine is another cylinder-and-piston engine, but it differs somewhat from the Otto cycle. Figure 12–14 illustrates the thermodynamic processes of the Diesel cycle. This cycle is named for Rudolf Diesel, even though the

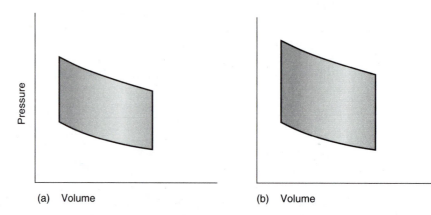

(a)   Volume   (b)   Volume

Figure 12–13. Otto cycles with different heat inputs.

principles were advanced by the French engineer Alphonse Beau de Rochas in 1862. As in the case of the Otto cycle, there are intake, compression, power, and exhaust strokes. However, the compression occurs without fuel being added, and during the compression stroke the compression ratio is much higher than that in the Otto cycle, so that the temperatures created are high enough to ignite the fuel spontaneously as it is injected in the chamber.

The processes of the Diesel cycle are:

a.   Adiabatic compression
b.   Constant pressure heat addition
c.   Adiabatic expansion power stroke
d.   Constant volume heat rejection

The cycle begins the intake stroke as the piston moves down and pulls air into the cylinder. Next the piston rises and compresses the air. During the compression stroke the temperature may reach a high of 900°R. When fuel is injected into the cylinder, it mixes with the hot air and explodes. (Figure 12–15) Gases heated by this combustion action push the piston down in the power stroke. The exhaust stroke completes the cycle as the piston moves up again and forces the burned gases out of the cylinder.

The Diesel engine differs from other internal-combustion engines mainly in its method of introducing fuel and ignition. In the Otto cycle constant volume heat addition process, fuel and air are mixed before entering the combustion chamber and are spark-ignited. In the Diesel engine, the fuel is slowly sprayed directly into the combustion chamber and ignited by the high temperature of the air in the combustion chamber. The fuel injection occurs at

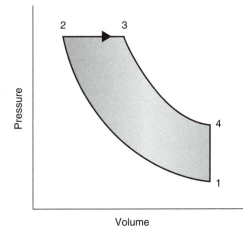

Figure 12–14. The Diesel cycle.

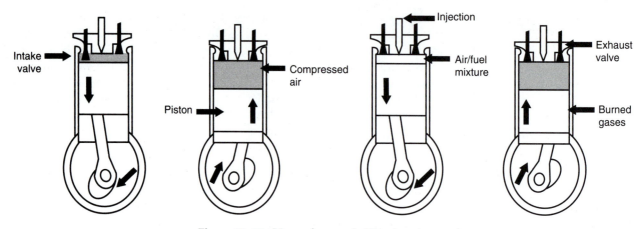

Figure 12–15. How a four-cycle Diesel engine works.

the top of the power stroke and maintains a continuous supply of fuel as the piston begins its power stroke expansion. When the piston reaches about one-sixth of the way down the cylinder, the gas injection is shut off.

The manner of fuel introduction is the key to satisfactory and efficient operation. A high-pressure fuel pump is used and injects fuel during precisely the proper period in the stroke, accurately metered in rate and total quantity and properly atomized. This necessitates relatively expensive fuel-injection components of extremely closely controlled manufacturing tolerance.

Because of its characteristic fuel-supply-and-combustion system and its use of heavy fuels, the Diesel engine has certain limitations. Smoking occurs in the exhaust if an attempt is made to use more than 60% to 70% of the air in the cylinder during the combustion process. Consequently, a Diesel engine producing the same power as an Otto engine is much bulkier and heavier.

Although relatively costly, the Diesel engine has high efficiency, long life, and reliability; it needs servicing infrequently. A large number of Diesel engines are used where these factors are important, as in motor vehicles where fuel is expensive, and in locomotives, trucks, and buses where high reliability and long life are important. The largest engines are to be found in facilities for generating electricity and for propelling ships. These engines are typically very heavy because weight is not a restrictive consideration. They produce up to 10,000 horsepower from their 12 cylinders at a speed of about 200 RPM.

SAMPLE PROBLEM 12-7

Using the specifications of the engine in Sample Problem 12–1, including the same amount of Btu/cycle and the same maximum temperature in the engine ($T_3 = 4133°R$), calculate the efficiency and horsepower of the cycle if it were a Diesel engine.

SOLUTION:

$$V_1 = 258 \text{ in}^3 = .149 \text{ ft}^3, \quad P_1 = 14.7 \text{ psia} = 2116 \text{ psfa}$$

$$T_1 = 70°F = 530°R, \quad m_{air} = \frac{P_1 V_1}{R T_1} = .011 \text{ lb}_m$$

$$c_v = .17 \frac{\text{Btu}}{\text{lb}_m - °F}, \quad c = .24 \frac{\text{Btu}}{\text{lb}_m - °F}, \quad k = 1.4$$

$$T_3 = 4133°R, \quad RPM = 3000$$

$$Q_{2-3} = 5.4 \text{ Btu} = .24 \times .011 \times (4133 - T_2)$$

$$= 10.9 - .00264 \, T_2$$

or

$$T_2 = 2083°R$$

$$V_2 = V_1 \left(\frac{T_1}{T_2}\right)^{\frac{1}{(k-1)}} = .149 \left(\frac{530}{2083}\right)^{2.5} = .0048 \text{ ft}^3$$

(Notice that the compression ratio for the Diesel cycle (V_2/V_1) is .149/.0048 = 31, a much higher value than that in the Otto cycle.)

$$P_2 = P_1 \left(\frac{T_2}{T_1}\right)^{\frac{k}{(k-1)}} = 2116 \left(\frac{2083}{530}\right)^{3.5} = 254,661 \text{ psfa}$$

$$WORK_{1-2} = \frac{P_2 V_2 - P_1 V_1}{J(1-k)} = -2.91 \text{ Btu}$$

$$V_3 = V_2 \left(\frac{T_3}{T_2}\right) = .0095 \text{ ft}^3$$

$$WORK_{2-3} = \frac{P(V_3 - V_2)}{J} = 1.53 \text{ Btu}$$

$$P_4 = P_3 \left(\frac{V_3}{V_4}\right)^k = 5399 \text{ psfa}$$

$$T_4 = T_3 \left(\frac{P_4}{P_3}\right)^{\frac{(k-1)}{k}} = 1377°R$$

$$WORK_{3-4} = \frac{P_1 V_1 - P_3 V_3}{J(1-k)} = 5.19 \text{ Btu}$$

$$Net\ WORK = W_{1-2} + W_{2-3} + W_{3-4} + W_{4-1} = -2.91 + 1.53 + 5.19 - 0$$
$$= 3.81 \text{ Btu} = 2964 \text{ ft-lb}_f$$

$$\eta = \frac{3.81}{5.4} = .70$$

$$Hp = \frac{2964 \times 3000}{2 \times 33,000} = 134.7$$

$$MEP = \frac{2964}{.149} = 19,892 \text{ psfa}$$

$$SFC = \frac{3.23}{134.7} = .00239 \frac{\text{Gph}}{\text{hr}} \quad \text{(see Sample Problem 12–5)}$$

AUTOMOBILE MECHANIC

Ms. Williams pulls into the dealership. She has been dreading this morning ever since she made the appointment two weeks ago. She doesn't like automobile repair garages. They are dirty, they are rude, and they are lucky if they can fix your car. She remembers what it was like in 1974 when she last had to take her car in for major service.

The sign says "Service Center," and an arrow points to a garage door that opens when Ms. Williams pulls up. Inside she drives right up to the service desk, and to two women who are standing there. It is spotless here, but Ms. Williams suspects the dirty floors begin right on the other side of the wall. One of the service technicians checks her in, saying yes, they do have her appointment on record. "Can you describe what is wrong with your car, in detail?" the technician asks. Ms. Williams tells her what she knows: that the vehicle has had no pep, that it labors on the expressway even to keep up with the speed limit. The technician types something into her computer and asks Ms. Williams whether she has noticed the problem being more severe when she first starts her trips or after the car has warmed up. Now that she thinks about it, Ms. Williams recalls that the problem doesn't start until about five minutes into her trip, that it isn't so bad on cold mornings, and that it's much worse on her way home. The technician types a little more into the computer, then asks whether the gas mileage is worse now than when she first bought the car. Ms. Williams knows well the answer to this question. She has been stopping at gas stations more often than before.

Now the two service technicians look at the computer screen together and make some comments back and forth. Finally the other technician asks, "Ms. Williams, when was the last time you had the spark plugs changed?" Within the last three months is the answer, as soon as she had noticed the loss of power. There is more computer typing, then the technicians are through with the questioning. They pull a corrugated hose over to the car and ask Ms. Williams to start it up. With the hose clamped to the tailpipe, they take measurements from several meters attached to the ends of the hose. It takes only a minute, and then they are ready to announce their findings. They tell Ms. Williams that she can pick up the car the following evening, and they give her a preliminary estimate of $350.

That afternoon the car is brought into bay #18. The technician there first looks at the repair analysis sheet and then starts the car, pulls a spark plug from a cylinder, and replaces it with a gauge. The technician is checking compression on the cylinder. Eventually he has checked all six cylinders and finds two with low compression; low enough to cause loss of power. This means the piston rings on the two cylinders will have to be replaced. Piston rings are an important part of a piston-cylinder arrangement because the spacing between the walls of the piston and cylinder includes a small gap. If the piston and cylinder were positioned metal-to-metal, there would be too much friction and too much wear of the cylinder wall. To close the gap and eliminate compressed gases that can "blow by" the gap, an expandable ring is placed in a groove on the piston and rides against the cylinder wall, closing the gap. If these rings wear down or break or bend, the compression on the cylinder is sacrificed.

After the technican orders the parts he will need, he puts the car on the lift and drops the oil pan from the engine, letting the oil drain while he begins analyzing the car that was just brought into bay #17.

The following morning the technician begins the delicate operation of replacing the piston rings. Usually this is a weeklong job, requiring that the engine be pulled out of the car, and the driveshaft pulled out of the engine to get the pistons to drop down. This technician will attempt to do the replacement in four hours by exposing the driveshaft, unbolting the piston rod from the driveshaft, then delicately rotating the driveshaft to a position in which the connecting rod will drop down, exposing the piston. Replacing the ring is a simple task, but positioning it back into the cylinder is not, since the ring is oversized and must be compressed to get it inside the cylinder wall. A special tool had to be built in order for the technician to be able to compress the ring on the piston while blindly fitting the piston into cylinder, and then release the tool to set the piston. This tool is the critical link in performing this new repair technique.

Ms. Williams picks up her car that evening. She has a good feeling about the job they've done; she had expected the bill to be higher. Still she is unaware of the new technology that was involved in her repair. The technicians that she spoke to at the service desk, and the technicians in the back, were all hired after finishing a two-year technology degree. Then they were certified through several weeks of training at the automobile manufacturer's training center in Dearborn, Michigan. At the service desk, the technicians were trying out a new troubleshooting program developed by the automobile manufacturer that used the elements of artificial intellegence. The questions they asked were prompted by the program. The technicians relayed the customer's responses back to the computer in the form of "key words." The technicians were so good that their preliminary finding indicated that the piston rings were bad. They performed the test on the exhaust of the car to detect whether raw gasoline or unburned oil was a component of the exhaust gases, indicators that there was blow-by past the rings.

The repair technician used a procedure that he had learned at the last recertification course. The success of the procedure relied on the special tool developed and provided by the manufacturer and on the skill of the repair technician. Ms. Williams had no idea of the technology behind her routine and satisfactory repair.

CHAPTER **12**

PROBLEMS

1. Redo Sample Problem 12–1, with the only change being the type of fuel used. The heating value of this fuel is 100,000 Btu/gal. For this fuel, find the cycle properties P, V, and T at each state point, the WORK done during each process, net WORK, efficiency, and horsepower.

2. A newly discovered cycle consists of:
   a. constant pressure compression $(V_2/V_1 = 5)$
   b. constant volume heat addition $(P_3/P_2 = 10)$
   c. constant pressure expansion
   d. constant volume heat rejection
   Draw a P-V diagram of the process. Calculate the net WORK done, heat added, heat rejected, and the efficiency of the cycle. Allow state 1 to be atmospheric conditions of

$$T_1 = 70°F, P_1 = 14.7 \text{ psia}, m_1 = 1 \text{ lb}_m, c_v = .17, \text{ and } c_p = .24$$

3. Determine the hp, efficiency, and the MEP for a 1999 Ford with the following specifications: 289 CID, 10.5:1 compression ratio, 15 mpg @ 3000 RPM. Use air standard:

$$c_v = .17, c_p = .24$$

4. Fill in the following data from your car:
   CID _____, Compression ratio _____
   Mileage _____, Mpg at _____ RPM and _____ mph

   Using this information, perform a cycle calculation similar to that in Sample Problem 12–1. Fill in the table below with the pertinent information from your calculation. Also complete the table by comparing your automobile characteristics with the results of Sample Problem 12–1.

Your Auto	Maximum Pressure	Maximum Temperature	Net WORK/Cycle	Hp	EFF	MEP	SFC
1991 Ford Taurus							
Your Car							

Write a short paragraph comparing the engines, using each performance criterion. Tell what each criterion measures. Turn in all calculations neatly done.

5. A Wankel engine (Figure 12–16) uses triangular rotors in a specially shaped combustion chamber to replace the traditional piston and cylinder. As the rotor turns, each of its three sides goes through a four-step cycle. During the intake stroke, fresh air is mixed with fuel as it is drawn into the engine. The rotor then seals off the fuel-air mixture and commences to compress it. In the power stroke, the spark plug ignites the mixture. The burning gases expand, adding momentum to the rotor. As the rotor completes its cycle, the burned gases leave the engine through the exhaust port, and the cycle begins over again.

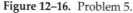

Intake   Compression   Power   Exhaust

Figure 12–16. Problem 5.

From the above description of a Wankel engine, identify the four thermodynamic process that drive this device. Can you identify the cycle (Otto, Diesel, Carnot, and so on)?

6. The 1985 Corvette engine is claimed to have thermodynamic efficiency η of .61. What is the compression ratio of the engine?

7. Pollutants are an important aspect of engine design, although understanding of them is more of a chemical nature than a thermodynamic one. In the case of nitric oxide (NO) in the exhaust, however, it is well understood that the higher the combustion temperatures, the more the NO pollution. To reduce these temperatures, less fuel must be added per stroke.
   a. In the case of Sample Problem 12–1, how high would the gas temperature get if the combustion was reduced to 3 Btu/cycle?
   b. How much combustion heat is allowable to keep the maximum temperatures below 2000°F?

8. Rudolf Diesel's first successful engine, described in Section 12–6, developed a maximum pressure of 1160 psia at 2000 RPM with a fuel con-

sumption of 8 Btu per cycle. What was its efficiency, horsepower, and mean effective pressure?

9. Specifications for an Otto cycle are given on the table below. Fill in all the quantities left blank in that table. ($c_v = .17$, $c_p = .24$)

STATE	P	V	T	WORK (+ or −)	Q (+ or −)
1	$P_1 = 2116$ psfa	$V_1 = .149$ ft^3	$T_1 = 530°R$		
2	$P_2 = 131{,}811$ psfa	$V_2 = $ _____	$T_2 = $ _____	$W_{12} = $ _____	$Q_{12} = $ _____
3	$P_3 = $ _____	$V_3 = .075$ ft	$T_3 = $ _____	$W_{23} = $ _____	$Q_{23} = $ _____
4	$P_4 = $ _____	$V_4 = $ _____	$T_4 = $ _____	$W_{34} = $ _____	$Q_{34} = $ _____
5	$P_5 = $ _____	$V_5 = $ _____	$T_5 = $ _____	$W_{45} = $ _____	$Q_{45} = $ _____

Now calculate:

$$\text{Thermal Efficiency} = \frac{\text{Useful Output}}{\text{Costly Input}}$$

Also calculate hp.

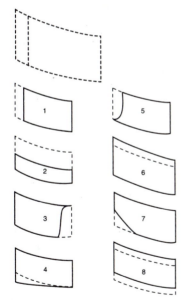

Figure 12–17. Problem 10.

10. The cycle shown at the top of Figure 12–17 is the simple Otto cycle. This cycle is reproduced in dashes eight more times in the remaining diagrams. The solid black lines represent similar cycles that each demonstrate one modification from the simple cycle. Match the modifications in the list below to the appropriate diagram. Write the letter of the modification inside the proper diagram.

a. fouled spark plug (slow ignition)
b. timing advanced (ignition after top dead center)
c. sticking exhaust valve
d. acceleration
e. turbocharged
f. blown engine (pistons don't move)
g. idle speeds
h. timing retarded (ignition before top dead center)
i. overheating
j. friction in cylinder (low oil pressure)
k. intake and/or exhaust valves leaking

11. From the information for thermodynamic processes given in parts a, b, c, and d below, find the quantities that have not been given.

   a. $V_1 = .15$ ft^3 $V_2 =$ _____

   $P_1 = 2116$ psfa $P_2 = 131,814$ psfa

   $T_1 = 530°R$ $T_2 =$ _____

   $m_1 =$ _____ $Q = 0$

   $WORK =$ _____

   b. $V_2 =$ as calculated in part a $V_3 = .075$ ft^3

   $P_2 = 131,811$ psfa $P_3 = 131,811$ psfa

   $T_2 =$ as calculated in part a $T_3 =$ _____

   $Q_{2-3} =$ _____ $WORK =$ _____

   c. $V_3 = .075$ ft^3 $V_4 = .15$ ft^3

   $P_3 = 131,811$ psfa $P_4 =$ _____

   $T_3 =$ as calculated in part b $T_4 =$ _____

   $Q_{3-4} = 0$ $WORK =$ _____

   d. $V_4 = .15$ ft^3 $V_5 = .15$ ft^3

   $P_4 =$ as calculated in part c $P_5 = 2116$ psfa

   $T_4 =$ as calculated in part c $T_5 =$ _____

   $Q_{4-1} =$ _____ $WORK =$ _____

12. a. The compression stroke of Sample Problem 12–1 is ideally adiabatic, but more realistically it would test out to be polytropic with a polytropic index of $n = 1.3$. Using the 8:1 compression ratio and a 258-cu-in displacement from that problem, find the following information using a polytropic compression:

   (1) P_2

   (2) T_2

   (3) V_2

   (4) $WORK$ (gas)

   (5) $WORK$ (delivered)

   (6) Q

   (7) Note the difference between the results of this calcuation and those of the sample problem. Describe why these results appear to more realistic.

   b. The constant volume combustion of gasoline in Sample Problem 12–1 is more realistically a polytropic process with $n = -3$. Using the state 2 conditions from part a, and $n = -3$ for this heat addition process with $Q = 5.4$ Btu, find:

   (1) P_3

   (2) T_3

   (3) V_3

(4) *WORK* (gas)
(5) *WORK* (delivered)
(6) Compare the results of this calculation with the solution in the sample problem and describe why this solution is more realistic.

c. The power stroke expansion in Sample Problem 12–1 is theoretically adiabatic, but realistically it is a polytropic process with $n = 1.5$. Using the state 3 condition from part b as a starting point, and expanding to a volume of 258 cu in, find:
(1) T_4
(2) P_4
(3) *WORK* (gas)
(4) *WORK* (delivered)
(5) Q
(6) Compare the results of this calculation with the solution to Sample Problem 12–1. Describe why this result is more realistic.
d. Find the net WORK for the cycle consisting of the processes of parts a, b, and c.
e. Calculate the efficiency of the cycle you have evaluated in a, b, c, and d above.

13. Using Figure 12–18 with the following conditions for air:
   $P_1 = 100$ psia, $P_2 = 400$ psia, $v_1 = 1$ ft^3/lb
   a. What is the temperature at B?
   b. What is v_2?
   c. What is the temperature at C?

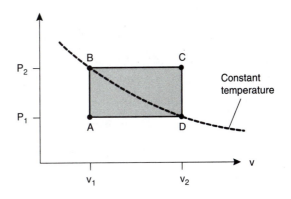

Figure 12–18. Problem 13.

14. Using Figure 12–19 with the following conditions for air:
$P_1 = 100$ psia, $P_2 = 400$ psia, $v_1 = 10$ ft^3/lb
   a. What is the temperature at B?
   b. What is v_2?
   c. What is the temperature at C?

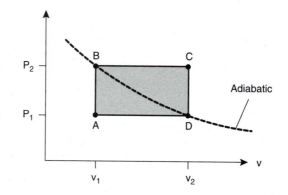

Figure 12–19. Problem 14.

15. Figure 12–20 is a P-V diagram of a simple piston cylinder cycle. Find the state properties at each point of the cycle and determine the heat transfer (Q) and the WORK done on/by the gas during each process. Use the specifications that condition 1 is STP (air), compression ratio is 6:1, and the engine displacement is 258 CID.

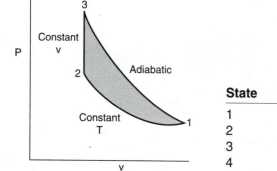

Figure 12–20. Problem 15.

State	T	P	V	WORK	Q
1					
2					
3					
4					

Calculate net WORK, thermal efficiency, MEP, and hp (@ 3000 RPM, 4-stroke).

CHAPTER

13

Open Processes, Cycles, and Systems

PREVIEW: Processes and cycles without pistons often pack great power in a small device.

OBJECTIVES:
- ❑ Put simple devices together to make a variety of cycles.
- ❑ Distinguish between the WORK done by a gas during a process and the WORK delivered from the process.
- ❑ Understand and analyze simple machines that operate using open thermodynamic processes.

13.1 OPEN PROCESSES

The thermodynamic cycles of Chapter 12 are the result of the development of complicated equipment to carry on these processes: cylinder-and-piston systems, perfectly timed valves, and others. Some devices that perform similar processes are simpler in design. Figure 13–1 shows one of these. It is a chamber or duct with gas flowing through it by the push of a fan. The gas comes to a point where it is compressed and confined to a smaller space before moving on. As the gas is compressed, the pressure goes up, the temperature increases with the compression, and specific volume decreases. This is surely a thermodynamic process. The process is accomplished by merely forcing the gas to go through this converging restriction in the duct line.

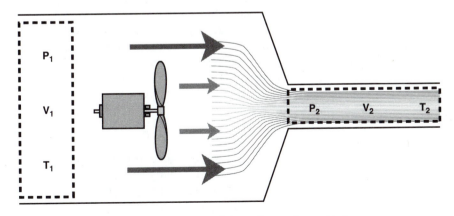

Figure 13–1. Simple equipment can perform a thermodynamic change.

A compression of a gas in a cylinder by a piston is a *discrete or batch process*. The gas to be compressed is first captured, then worked on during compression, and later released. To perform the process again, a new volume of gas must be captured and the process repeated.

In direct contrast to this type of operation, the thermodynamic process shown in Figure 13–1 does not occur on a specific quantity of gas. Rather, gas is continuously being fed into the duct inlet and gas is continuously being compressed as it moves through the duct. This kind of process is called a *continuous process*. The device shown in Figure 13–1 is open at the left and the right; therefore, the thermodynamic processes are called *open processes*. On the contrary, processes we have studied in Chapter 6 through 11 are called *closed processes* because they occur in closed systems.

How will open processes be analyzed? Figure 13–1 follows a certain volume of gas indicating that its pressure, specific volume, and temperature are those at state 1. As the volume goes through the fan compressor, it changes in form and arrives at state 2. Can the perfect gas law and the energy equation be used to calculate the properties of the gas?

Surely the perfect gas law describes the thermodynamic state at point 1 and 2. The perfect gas law for changing situations will be valid depending on whether the open process is constant volume, constant pressure, constant temperature, adiabatic, or polytropic.

The energy equation is an accounting equation that predicts that heat energy absorbed by a system can be turned into internal energy, WORK, latent heat or chemical energy, or a variety of other forms of energy. In *open processes*, energy in the form of heat will be converted into internal energy, WORK, or latent heat. In *open systems*, however, we may have to consider kinetic energy. Why did we not consider this important energy before, in closed systems? Kinetic energy implies that a system *as a whole* has motion. In systems we have studied in closed processes, the gas does not move from place to place, although the molecules of the gases move randomly within the system. This random motion energy we have accounted for as internal energy, not kinetic energy. Since there is no overall motion of the gas in closed systems, there is no kinetic energy. In an open process, however, gas flows through a compressor which results in an overall motion. The gas velocity can change through the process. This change in speed causes a change in kinetic energy, which is accounted for in the energy equation with the term:

$$KE = \frac{1}{2} m \ (avg. \ vel.)^2$$

As the gas goes from point 1 to point 2 in Figure 13–1, the velocity and kinetic energy changes, but for the types of systems that we are going to study, the kinetic energy changes are very insignificant compared to the other energy changes (heat addition or removal, internal energy, latent energy, and so on). For the time being, therefore, the energy equation derived in Chapter 5 will be

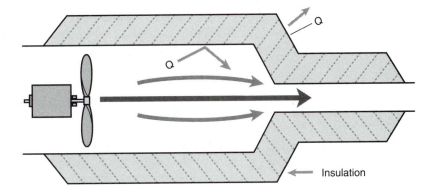

Figure 13–2. No heat gets in or out during an adiabatic compression.

appropriate for open processes; we will not include the change in kinetic energy. The analysis of open processes, therefore, will be identical to that of closed processes, except in a few small ways that we will see.

13.2 OPEN COMPRESSION PROCESSES

What kind of a thermodynamic process does the compressor in Figure 13–1 represent? Certainly it's not a constant pressure process, since the pressure is going to increase. Could it be an adiabatic process? It appears that as the gas goes through the compressor, there is no heat addition from explosions of a fuel, nor are there any hot plates around to add heat. Therefore, the heat process may be adiabatic. This would be absolutely true if the compressor was wrapped with insulation so no heat transfer could take place through the walls. (Figure 13–2) On the other hand, the gas could be maintained at a constant temperature by wrapping the compressor with water coils that

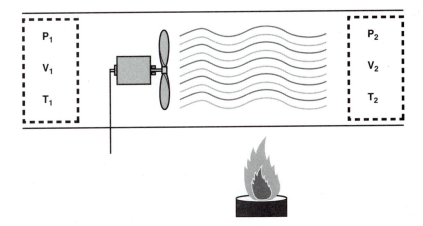

Figure 13–3. A constant pressure continuous process.

would cool or heat the gas to maintain it at the temperature of the water. This addition of water coils would change the compressor from an adiabatic one to a constant temperature compressor.

What type of process would maintain a constant pressure? Figure 13–3 shows such a device. In a continuous process in which there is no change in duct area, there is no change in the pressure of the gas in that duct. Therefore, heat can be added with a hot plate or candle or an explosion in a constant cross-section-area duct and the heat addition will take place at constant pressure.

The only process that is difficult to imagine in this type of situation is a constant volume process. Constant volume processes are not important in continuous applications, and we will not study them.

13.3 CALCULATING WORK IN AN OPEN PROCESS

As we well know, thermodynamic processes often do WORK on a gas, or the gas does WORK itself. In the piston-and-cylinder situation, the WORK done on the gas was done by the piston. What device does the WORK on the gas in the open process? In order to compress the gas, there must be some force pushing the gas through. This is done by a fan, as shown in Figures 13–1 through 13–3; therefore, the fan is the WORK-generating device in an open process.

Open processes are useful in many everyday thermodynamic devices, but the most recognizable example might be the jet engine of Figure 13–4. In this figure, the fans that are familiar to us now for open cycles are mounted directly in the compressor section itself. It's no wonder this engine might be called a fan-jet.

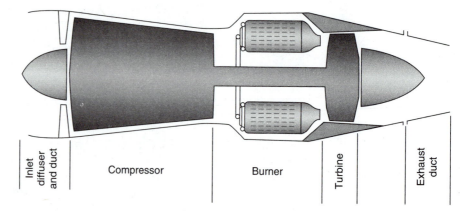

| Inlet diffuser and duct | Compressor | Burner | Turbine | Exhaust duct |

Figure 13–4. A fan-jet engine.

SAMPLE PROBLEM 13–1

Figure 13–5 shows a constant temperature compressor. The compression ratio for this compressor is 5:1 ($v_1/v_2 = 5$). The constant temperature is maintained by blowing outside air over the compressor to keep it cool. The compressor is fed with atmospheric room air, at a temperature of 70°F and a pressure of 14.7 lb per in^2. Calculate the temperature and the specific volume of the air after the compressor. Determine also the heat rejected and the WORK done during the process.

SOLUTION:

$$\frac{V_1}{V_2} = 5 = \frac{v_1}{v_2}$$

$$T_1 = T_2 = T_{air} = 70°F = 530°R$$

$$P_1 = 14.7 \text{ psia} = 2116 \text{ psfa}$$

$$v_1 = \frac{RT_1}{P_1} = \frac{53.3 \times 530}{2116} = 13.4 \text{ ft}^3/\text{lb}$$

$$v_2 = \frac{13.4}{5} = \frac{2.68 \text{ ft}^3}{\text{lb}}$$

$$P_2 = P_1\left(\frac{v_1}{v_2}\right) = 10{,}580 \text{ psfa}$$

$$WORK_{1-2} \text{ (per lb)} = \frac{P_1 V_1}{J\,m}\ln\left(\frac{v_2}{v_1}\right) = \frac{P_1 v_1}{J}\ln\frac{v_2}{v_1}$$

$$= \frac{2116 \times 13.4}{778}\ln(.2) = -58.4\,\frac{\text{Btu}}{\text{lb}_{air}}$$

$$Q \text{ (per lb)} = -58.4\,\frac{\text{Btu}}{\text{lb}_{air}}$$

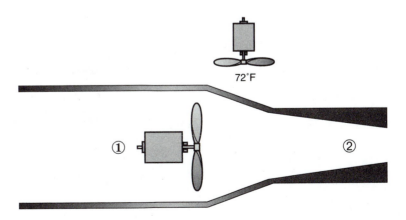

Figure 13–5. Sample Problem 13–1.

This sample problem illustrates one subtle but significant change in calculations in open processes from those of the piston-and-cylinder type. Notice that the volume occupied by the gas was never calculated; we never found V_1 or V_2. In fact, the gas never actually occupies a certain volume. Instead it continuously flows in and out of the device. Certainly the gas occupies a volume, but without hypothetical partitions around a specific amount of it, such as those in Figure 13–1, it doesn't make sense to determine the volume occupied by the gas.

This is particularly embarrassing when we recall that the formulas for WORK in Table 9–1 often require the volume occupied by the gas before and after the process. Fortunately the *specific volume* of a gas does make perfect sense in open processes. As shown in Sample Problem 13–1, both v_1 and v_2 are easily calculated. To use this fact, the formula for WORK has been rewritten in Sample Problem 13–1 in terms of the specific weight of the gas. This implies that the WORK will be calculated *per pound of gas* flowing. Therefore, when we present formulas for WORK on a per pound basis, we will use lowercase letters: *work* = WORK/m. Similarly, when we present formulas for HEAT on a per pound basis (Sample Problem 13–1), we will use q = Q/m.

SAMPLE PROBLEM 13–2

Figure 13–6 shows a compressor that allows for only thermodynamic changes that occur adiabatically. Suppose that the compression was to begin with air at the same state as in the outcome of Sample Problem 13–1 and finish at a temperature of 200°F. Find the pressure on the vapor after the compression and determine the WORK done on or by the gas (per lb$_m$).

SOLUTION:

$$P_3 = P_2\left(\frac{T_3}{T_2}\right)^{\frac{k}{k-1}} = 10{,}580\left(\frac{660}{532}\right)^{3.5}$$

$$= 22{,}799 \text{ psfa}$$

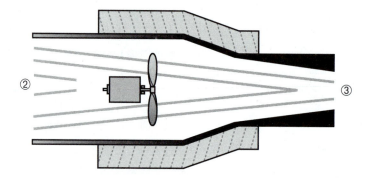

Figure 13–6. Sample Problem 13–2.

$$v_3 = v_2\left(\frac{P_2}{P_3}\right)^{\frac{1}{k}} = 2.68\left(\frac{10{,}800}{22{,}968}\right)^{\frac{1}{1.41}}$$

$$= \frac{1.54\ \text{ft}^3}{\text{lb}}$$

$$work_{3-2} = \frac{P_3v_3 - P_2v_2}{J(1-k)} = \frac{22{,}968 \times 1.54 - 10{,}580 \times 2.68}{778(1-1.41)}$$

$$= -\frac{22.2\ \text{Btu}}{\text{lb}_m}$$

WORK is done on the gas and, therefore, fans must be placed in the system to supply the WORK.

13.4 OPEN PROCESSES THAT EXPAND

Looking at the compressor from another angle turns it into an expander. If the flow is in the opposite direction through the device of Figure 13–1, then the gas realizes an expansion rather than a compression. The thermodynamic process is directly related to the piston moving down in the cylinder, increasing the captured volume.

SAMPLE PROBLEM 13–3

Figure 13–7 shows a constant temperature expander. This device maintains a constant temperature by hot water coils that are wrapped around the device itself, transferring heat to the gas inside as it is expanded. If the temperature of the inlet air is 660°F and the hot water bath is at the same temperature, calculate the pressure of the gas (air) after the diffuser (expander). Consider that the inlet gas properties are the same as the gas coming off the adiabatic compressor in Sample Problem 13–2 and that the water transfers 73 Btu/lb$_\text{air}$. Also calculate the WORK done by the gas during the expansion.

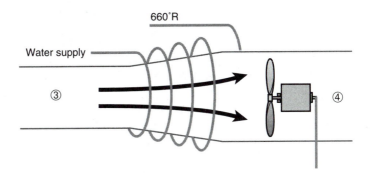

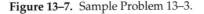

Figure 13–7. Sample Problem 13–3.

SOLUTION:

Checking the energy equation for a constant temperature process:

$$q = \frac{P_3 v_3}{J} \ln\left(\frac{v_4}{v_3}\right)$$

indicates that every quantity is known except for v_4, or:

$$73 = 22{,}968 \times 1.56 \ln\frac{\left(\dfrac{v_4}{1.56}\right)}{778}$$

$1.58 = \ln\left(\dfrac{v_4}{1.56}\right)$. Solving for v_4 requires taking each side to the exponential 'e'.

$$e^{1.58} = \frac{v_4}{1.56}$$

$$4.94 = \frac{v_4}{1.56}$$

$$7.71 \ \text{ft}^3/\text{lb} = v_4$$

Then

$$P_4 = P_3 \times \frac{v_3}{v_4} = 22{,}968\left(\frac{1.56}{7.79}\right) = 4559 \ \text{psfa}$$

and

$$work_{3-4} = 73 \ \text{Btu/lb}$$

A great amount of WORK is done by the gas during this process. Where is the WORK used? In the cylinder-and-piston arrangement of Chapter 12, the WORK done by the gas is used to force the piston up and turn a shaft that allows the WORK to be taken off as mechnical WORK. In this device, there is no such mechanism to take the WORK away; therefore, a fan must be placed inside the expansion device that will be rotated by the gas flowing through it. This is the mechanism for taking the WORK away. Fans that are fitted in such an expansion device are called turbines. Figure 13–8 shows a turbine at the outlet of the expander that allows the gas to do WORK. Figure 13–9 shows a more realistic set of turbines placed in the expansion device. Now that the turbines have been placed in the expander, the device again resembles a portion of a jet engine—this time, the outlet portion, or thrust section.

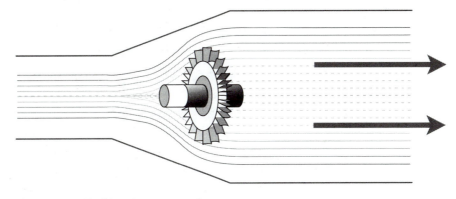

Figure 13–8. Turbines in an expander.

PROBLEM 13–4

Figure 13–10 shows an adiabatic expansion device (an adiabatic expander or adiabatic diffuser). This device takes a gas and expands the gas, with no heat addition, to a lower pressure and a lower temperature. Consider that the intake gas (air) to this device will be at 660°R and that it is desired to discharge the air out of this device at standard temperature and pressure (STP). What must the expansion ratio v_4/v_5 be? What must the initial pressure (P_4) and initial specific volume (v_4) be? During the process, calculate the WORK/lb done and specify whether it is done *on* or *by* the turbine.

SOLUTION:

$$\frac{v_4}{v_5} = \left(\frac{T_5}{T_4}\right)^{\left(\frac{1}{k-1}\right)} = \left(\frac{530}{660}\right)^{2.5} = .58$$

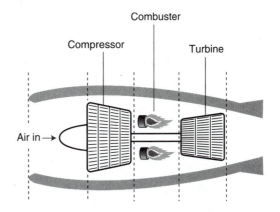

Figure 13–9. A turbojet engine.

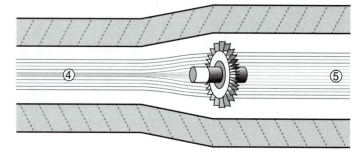

Figure 13–10. Sample Problem 13–4.

$$v_4 = .58, \quad v_5 = .58 \times 13.4 = \frac{7.7\,\text{ft}^3}{\text{lb}} \quad (v_5 \text{ is for air at STP})$$

$$P_4 = P_5\left(\frac{v_5}{v_4}\right)^k = 2116\left(\frac{13.4}{7.7}\right)^{1.41} = 4559\,\text{psfa}$$

$$work_{4-5} = \frac{P_5 v_5 - P_4 v_4}{J(1-k)} = \frac{28{,}354 - 35{,}420}{778(-.41)} = 21.7\,\text{Btu/lb}$$

WORK is done against the turbine, by the gas.

13.5
THE OPEN CYCLE

The devices that have been investigated in the previous sections of this chapter are in use in many ways in thermodynamics today. Their biggest application is when they are put together in sequence to form a cycle. As in any cycle, WORK is done in some of the processes, and WORK is taken off in other processes.

Figure 13–11 shows all the devices put in sequence: first the constant temperature compressor, then an adiabatic compressor, then a constant temperature expansion diffuser, then an adiabatic diffuser. The figure shows that the air traveling through the cycle begins at certain conditions entering at position 1 and regains those same conditions after process 4. This describes a thermodynamic cycle. Figure 13–12 shows the pressure versus specific volume diagram for this cycle. Notice that this cycle does WORK. The net WORK is the area on the inside of the cycle line, and the WORK is accomplished by the addition of heat at the hot water coils. The cycle shown is not the same as that of the automobile engine (Otto cycle) in that this cycle consists of an adiabatic compression, a constant temperature compression, an adiabatic expansion, a constant temperature expansion, and an adiabatic expansion. Although it's a different process than that of the automobile engine, it certainly qualifies as a legitimate cycle.

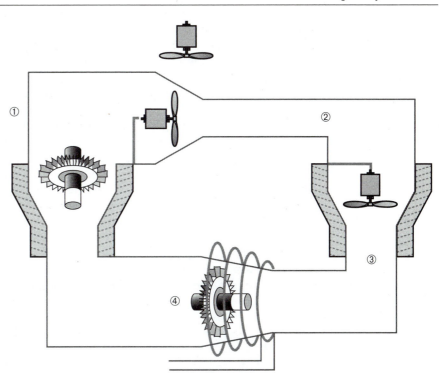

Figure 13–11. Putting continuous processes together to make an open cycle.

SAMPLE PROBLEM 13–5

Find the net WORK per pound of air and the efficiency of the open cycle shown in Figure 13–11.

SOLUTION:

The properties of each cycle have been previously calculated in Sample Problems 13–1 through 13–4. Collecting information from those problems yields:

$Net\ work$ (per lb) $= work_{12} + work_{23} + work_{34} + work_{45}$

$= -58.4 - 22.2 + 73 + 22.2 = 14.3\ \text{Btu/lb}$

$$Efficiency = \frac{Net\ work}{Heat\ Input} = \frac{14.3\ \dfrac{\text{Btu}}{\text{lb}}}{73\ \dfrac{\text{Btu}}{\text{lb}}} = .19 = 19\%$$

13.6 APPLICATION: SOLAR POWER

Such a device as Figure 13–11 illustrates might be a jet engine or a steam turbine used by a power plant to generate electricity, but this cycle may, in fact, best describe an ingenious device to convert solar energy into mechanical WORK: a solar engine. The 200°F water that provides the heat is typical of the water temperatures that might be generated by a water-tube solar collector. This hot water fed to the constant temperature expander means that a great deal of WORK can be done in this device. So much WORK is done that some of the WORK is diverted to power the compression fans in the other portions of the cycle to keep gas (air) circulating and pressurized to its proper point.

Figure 13–11 shows the configuration of the parts of the cycle, but it is Figure 13–12 that truly illustrates how the solar engine works. The solar hot water adds heat at an appropriate time (3–4) to keep the pressures high while the gas expands through a turbine to a much greater specific volume. The adiabatic expander (4–1) drains the last bit of useful WORK from the system while dropping the pressures. This allows the first compressor (1–2) to expend a minimum amount of WORK to make the gas much more compact and dense by reducing the specific volume. Taking away heat at the same time keeps the pressure down during this compacting of the gas medium. To get the pressures needed to begin the power portion, a short adiabatic compressor section (2–3) performs this job at little WORK because very little compacting needs be done. Poised now before the power section is a high-pressure, dense gas, one that can do some WORK on its own but will realize far more WORK by being heated as it is let go through the expander.

A little more reality can be added to this solar engine by changing the configuration of some of the components. For example, Figure 13–13 demonstrates a much more miniaturized adiabatic compressor/expander than the

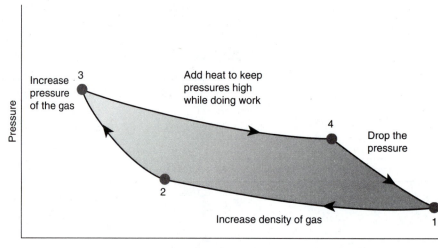

Figure 13–12. Operation of a solar engine.

converging or diverging duct sections of a jet engine compressor/turbine. This unit is called a *rotary compressor* and consists of two metal cylinders, one hollow and stationary and the other solid and rotating by a shaft at its center. As a compressor, the center cylinder rotates counterclockwise by the action of an external power source (WORK). Two large grooves are cut into the rotating cylinder fitted with metal blocks called *vanes.* These vanes can slide in and out of the grooves in a manner that allows them to make contact with the outer barrel at all times. As the vane rotates down past the intake port of the compressor, it sweeps out a continuously larger volume, and gas is swept into the chamber. The second vane sweeps past the inlet and captures the gas volume. Continued turning allows the first vane to pass the discharge valve. The gas rushes out, being pushed by the trailing vane, which pressurizes while it pushes (Figure 13–14).

This device is an expander when worked in the opposite direction. High-pressure gas comes in from the left and pushes on the vane, rotating it and supplying shaft WORK on the rotor. Soon the trailing vane shuts off the port and captures the pressurized gas. Further rotation moves the gas into a continuously larger volume where its pressure is dropping. Finally, the first vane sweeps past the right-hand port, and the gas is discharged.

Figure 13–15 shows the solar engine using rotary vane compressors and expanders for the adiabatic portions of the cycle and turbine-type compressors and expanders for the constant temperature portions. Notice that the rotary vane compressors are directly coupled to each other so that the output power of one is the input power of another. Such solar engines have been built and

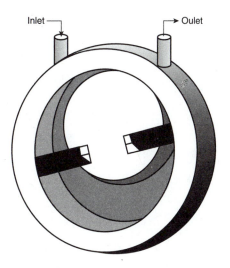

Figure 13–13. A rotary-vane compressor.

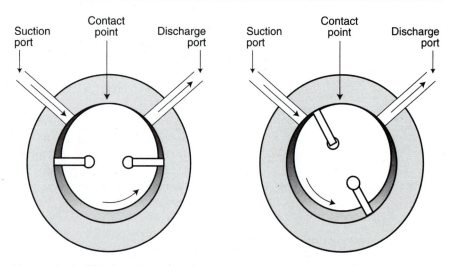

Figure 13–14. The operation of a rotary-vane compressor or expander.

perform as indicated in Figure 13–16. Their output is limited by the amount of solar heat that can be collected and the relatively low efficiencies of the cycle due to low water temperatures from the collector. We will learn more about that in Chapter 14.

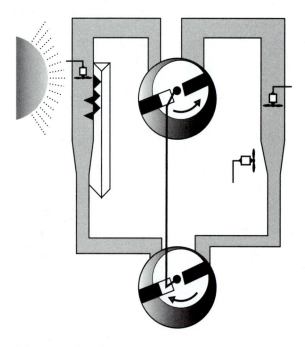

Figure 13–15. Schematic of a solar power generator.

SAMPLE PROBLEM 13–6

If the solar generator (Figure 13–15) circulates 15 lb_m/min of air inside and the other conditions are identical to Sample Problem 13–5, how much power is generated?

SOLUTION:

The formula for horsepower from Chapter 9 is only for pistons and cylinders:

$$Hp = \frac{Net\ work \times RPM}{2 \times 33{,}000}$$

of the four-stroke variety. The general calculation is:

$$Hp = \frac{\dfrac{Net\ work}{min}}{33{,}000}$$

To calculate net work/min, multiply:

$$\frac{Net\ work}{min} = \frac{Net\ work}{lb} \times \frac{lb}{min}$$

Figure 13–16. Mirrors of a solar electrical generating plant focus sunlight on a black boiler. (*Courtesy of the Department of Energy.*)

$$= 14.3 \frac{\text{Btu}}{\text{lb}} \times 778 \frac{\text{ft}-\text{lb}}{\text{Btu}} \times 15 \frac{\text{lb}}{\text{min}} = 167,231 \frac{\text{ft}-\text{lb}}{\text{min}}$$

$$Hp = \frac{167,231}{33,000} = 3.9 \text{ hp}$$

13.7 WORK DELIVERED

In the last two sections, many problems were solved, including open constant temperature and pressure situations, and adiabatic compressions and expansions. The WORK done by or on the gas was calculated in all of the solutions. Can it be assumed in these cases that the WORK done by or on the gas is *delivered* to the turbine or fans?

Remember that in the piston-and-cylinder arrangements, the delivered WORK could be affected by the pressure surrounding the piston. Is there an effect of the surroundings on the continuous process? Consider one pound of air as it goes through a continuous process. Imagine that the gas is enclosed in a piston-and-cylinder arrangement so that when the compression by the fan is accomplished, we see the piston move in to demonstrate the compression. With this model, it is clear that the pressure surrounding the piston before the process is P_1 and after the process is P_2. The WORK done on the gas does not have to come from the compression alone but is helped by the existing pressure in the process, both P_1 and P_2. This amount of WORK supplied by the surrounding pressure is calculated to be $P_1 v_1 - P_2 v_2$. Therefore, the WORK done by the piston itself (actually the fan) is:

$$work = Wd - \frac{(P_1 v_1 - P_2 v_2)}{J} = Wd + \frac{(P_2 v_2 - P_1 v_1)}{J}$$

Equation 13–1

or

$$Wd = work + \frac{(P_1 v_1 - P_2 v_2)}{J}$$

SAMPLE PROBLEM 13-7

In Sample Problems 13–1 through 13–4, determine the amount of WORK actually delivered to the fan or turbine during each process.

SOLUTION:

Sample Problem 13–1

$$Wd = work + \frac{(P_1v_1 - P_2v_2)}{J} = -58.6 + 0 = -58.6 \frac{Btu}{lb}$$

Sample Problem 13–2

$$Wd = work + \frac{(P_2v_2 - P_3v_3)}{J} = -21.7 - 8.9 = -31.8 \frac{Btu}{lb}$$

Sample Problem 13–3

$$Wd = work + \frac{(P_3v_3 - P_4v_4)}{J} = 73 + 0 = 73 \frac{Btu}{lb}$$

Sample Problem 13–4

$$Wd = work + (P_4v_4 - P_1v_1) = 21.7 + 8.9 = 31.8 \frac{Btu}{lb}$$

In all these cases, the pressure of the gas itself either helps the turbine or fans do their WORK or must be overcome by the fan or turbine. The difference between the WORK *done by the gas* and the WORK *delivered* can be seen by simplifying the equation for *Wd* (Equation 13–1) for the special cases of constant temperature, constant pressure, adiabatic, and polytropic processes. (Table 13–1) In the case of a constant temperature process, the perfect gas law states that $P_1v_1 = P_2v_2$, and therefore, the second term in the *Wd* equation is always 0 for this case (see sample problem for processes 1–2 and 3–4) and therefore *Wd* = *work* for constant temperature processes. For the constant pressure case, $P_1 = P_2$, the second term in the *Wd* equation, therefore, becomes $P(v_1 - v_2)/J$, which is equal but of opposite sign as *work* and, therefore, *Wd* = 0 for this case. For the adiabatic case:

$$Wd = \frac{(P_2v_2 - P_1v_1)}{J(1-k)} + \frac{(P_1v_1 - P_2v_2)}{J}$$

or by writing with common denominators:

$$Wd = \frac{(P_2v_2 - P_1v_1)}{J(1-k)} - (1-k)\frac{(P_2v_2 - P_1v_1)}{J(1-k)}$$

and collecting like terms:

$$Wd = k\frac{(P_2v_2 - P_1v_1)}{J(1-k)}$$

Table 13–1 collects these *Wd* formulas.

Table 13–1

	SPECIAL CASES	
	WORK/lb	**Wd**
Constant T	$\dfrac{\left[P_1 v_1 \ln\left(\dfrac{v_2}{v_1}\right) \right]}{J}$	$\dfrac{\left[P_1 v_1 \ln \dfrac{v_2}{v_1} \right]}{J}$
Constant P	$\dfrac{P(v_2 - v_1)}{J}$	0
Adiabatic	$\dfrac{(P_2 v_2 - P_1 v_1)}{J(1-k)}$	$\dfrac{k(P_2 v_2 - P_1 v_1)}{J(1-k)}$
Polytropic	$\dfrac{(P_2 v_2 - P_1 v_1)}{J(1-n)}$	$\dfrac{n(P_2 v_2 - P_1 v_1)}{J(1-n)}$

Notice the result for the constant pressure *Wd*. It states that in a constant pressure heating process, although the gas expands and does WORK, *all of the WORK is absorbed by the flow itself* and is not available to the turbine. *Flow WORK,* or *Flow Energy,* is a term often used to describe this effect. A constant pressure open heating process increases the flow WORK but will not push a turbine. A constant pressure open cooling process reduces flow energy, but no fan is needed to keep the process running. Therefore Figure 13–3 is incorrect in showing a turbine in the section because there will be no WORK available for a fan or turbine.

**13.8
THE JET ENGINE
CYCLE**

An interesting modification of continuous cycles may be made if the working medium gas is air. In such a case, air from the atmosphere might be continuously fed into a thermodynamic heat engine and exhausted directly back into the atmosphere after it has made one pass-through. When this is done, the working medium gas is not circulated over and over through the device but, instead, makes only one pass through the device, which can be an open cycle.

Figure 13–17 illustrates such a continuous heat engine. In this instance, air is taken into the front of the machine, compressed adiabatically to a high pressure, and fed directly to a constant pressure section where a fuel is exploded, causing an appreciable addition of heat. Then the hot air passes through an adiabatic turbine where WORK is done. Finally, the gas is exhausted back into the atmosphere.

Figure 13–18 is the P-v diagram for this device, and much can be learned from it. First notice that the cycle consists of only three processes, not four. The last process is missing, since hot gas at atmospheric pressure is exhausted and no process is needed to prepare it to be injected into the inlet of the cycle. Net WORK for this cycle, then, is:

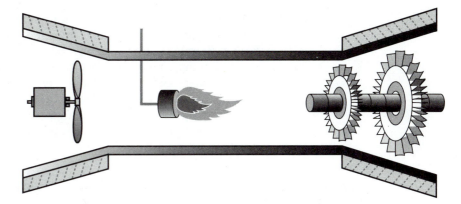

Figure 13–17 The jet engine does not recirculate the working medium but, instead, exhausts it.

$$Net\ work = work_{12} + work_{23} + work_{34}$$

One important machine that utilizes a three-process open cycle is the jet engine. Gas-turbine jet engines are widely used to power aircraft because they are light and compact and have a high power-to-weight ratio. Aircraft gas turbines operate on an open cycle called a jet-propulsion cycle. The ideal jet-propulsion cycle differs from the simple ideal Brayton cycle in which the gases are expanded to ambient pressure in the turbine and then exhausted hot to the atmosphere. Instead, the gases are expanded to a pressure such that the power produced by the turbine is just sufficient to drive the compressor and the auxiliary equipment, such as a small generator and hydraulic pumps. That is, the net WORK output of a jet-propulsion cycle is zero. The gases that enter

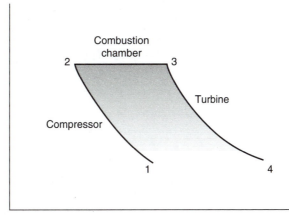

Figure 13–18. The P-v diagram for the jet-propulsion cycle.

the turbine at a relatively high pressure are subsequently accelerated in the turbine to provide the thrust to propel the aircraft (Figure 13–17). Aircraft gas turbines operate at high pressure ratios (typically between 10 and 25).

The P-v diagram of the ideal turbojet cycle is shown in Figure 13–18. Air is compressed in the compressor. It is mixed with fuel in the combustion chamber, where the mixture is burned at constant pressure. The high-pressure and high-temperature combustion gases expand in the turbine, producing enough power to drive the compressor and providing enough net WORK to propel the aircraft.

SAMPLE PROBLEM 13-8

A turbojet aircraft flies with a velocity of 260 m/sec at an altitude where the air is at 33 kPa (absolute) and –40°C. The compressor has a pressure ratio of 10, and the temperature of the gases at the turbine inlet is 1100°C. Air enters the compressor at a rate of 45 kg/s. Determine the following:
a. the temperature and pressure of the gases at the turbine exit
b. the heat input/lb$_m$
c. the net work/lb$_m$
d. the efficiency of the cycle
e. the power output of the engine (k/sec)

SOLUTION:

a. $V_1 = \dfrac{RT_1}{P_1} = \dfrac{.287 \times (-40+273)}{33} = 2.03 \dfrac{m^3}{kg}$

$V_2 = \dfrac{V_1}{10} = .203 \dfrac{m^3}{kg}$

$P_2 = P_1\left(\dfrac{V_1}{V_2}\right)^k = 33\left(10^{1.41}\right) = 848 \text{ kPa}$

$Wd_{12} = \dfrac{k(P_2V_2 - P_1V_1)}{(1-k)} = -361 \dfrac{kJ}{kg}$

b. $q_{23} = c_p(T_3 - T_2) = 1 \times (1373 - 233) = 1140 \dfrac{kJ}{kg}$

$v_3 = .287 \times \dfrac{1378}{33} = .46$

$T_4 = T_3\left(\dfrac{P_4}{P_3}\right)^{\frac{k-1}{k}} = 534 \text{ K}$

$v_4 = .287 \times \dfrac{534}{33} = 4.6 \dfrac{m^3}{kg}$

$Wd_{34} = \dfrac{k(P_4v_4 - P_3v_3)}{(1-k)} = 826 \dfrac{kJ}{kg}$

c.   *Net WORK* = −361 + 826 = 465 $\dfrac{kJ}{kg}$

d.   *EFF* = $\dfrac{465}{1140}$ = .41

e.   *Power = Net WORK* × $\dfrac{kg}{sec}$ = 20,925

TECHNICIANS IN THE FIELD

VENTILATION SYSTEM TECHNICIAN

On the shores of Lake Erie, behind the beaches where sunbathers rest, behind the marinas that hold the fleet of recreational fishing boats poised for an assault on walleye and perch, behind the condos with water access, a leviathan cement cylinder rises from the ground, the telltale clue that a nuclear power plant is nearby. To the local population, the Davis-Besse Power Plant is a mysterious but inconspicuous neighbor. To the technicians who work inside the plant, nuclear power is not mysterious, but very conventional. They know it as a simple power cycle, just like that in Figure 13–16 but with a different heat source—nuclear reaction rather than solar radiation.

Most of the technicians at this plant, in fact, do not deal with the nuclear fuel section itself but with all the thermodynamic requirements. For instance, three technicians are responsible for the ventilation of the nuclear containment building. This building houses the reactor core and the steam lines that run in and out of it. It is a thick cement building designed to confine a nuclear incident, and it is in the shape of a sphere to minimize the amount of space in the building that is not reactor core. This space must be conditioned for human occupancy so that nuclear technicians will have access to the reactor core. If the air in the building were not ventilated and conditioned, the heat given off by the core, even though confined by several feet of insulation, would slowly drive the temperature in the containment building to uninhabitable levels.

The ventilation system for the containment building is a below-floor recirculating type, meaning that the air removed from the building is conditioned and recirculated back into the building. This type of system is required to confine any small amount of radioactivity to the building itself and avoid discharging it to the atmosphere. Notice that it includes a fan, a heat exchanger with chilled water, and a set of filters. The technicians are responsible for maintaining the air coming off the heat exchanger coil at 60°F. (Figure 13–19) It is expected that this air temperature will keep the confinement building at 80°F, an acceptable level for the maintenance crews.

SAMPLE PROBLEM 13–9

Some nuclear scientists from the Nuclear Regulatory Commission are coming to inspect the core at Davis-Besse. Management has sent a memo to the ventilation technicians to have the containment building at 74°F on this day, for the comfort of the visitors. The technicians run a computer program that tells them that at 74°F, the containment building will pick up 1,260,000 Btu/hr through the core wall and through the outside walls. If the air flows across the coil at 50,000 ft³/min (cfm), what must the off-the-coil air temperature be to maintain the building at 74°F?

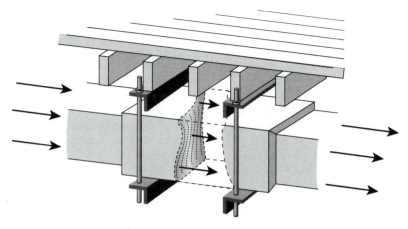

Figure 13–19. Air going through a cooling coil is a constant pressure process.

SOLUTION:

As air goes through the cooling coil, its specific volume changes, but since there is no piston confining the air, the pressure cannot change. Therefore, the cooling process in the ventilation system is a constant pressure process. (Figure 13–19) The technicians use an energy equation that has been simplified for their purpose. Since the process is constant pressure:

$$Q \ (/\text{hr}) = c_p \dot{m} \ (T_2 - T_1)$$

The mass flow rate can be determined from volume flow rate (cfm) by using $\dot{m} = \rho \dot{v}$. For $c_p = .24$ and $\rho = .075 \ \text{lb/ft}^3$, the energy equation becomes:

$$\dot{Q} = 1.05 \times cfm \times (T_2 - T_1)$$

This is a fundamental working equation of ventilation (see chapter 22). Therefore:

$$\dot{Q} = -1,260,000 = 1.05 \times 50,000 \times (T_2 - 74)$$

or

$$T_2 = 59°\text{F}$$

CHAPTER **13**

PROBLEMS

1. Air expands in an air turbine from a pressure of 50 psia and a temperature of 600°F to an exhaust pressure of 20 psia. Assume the process is reversible and adiabatic with negligible changes in kinetic and potential energy. Calculate the WORK per pound of air flowing through the turbine, both *work* and W_D.

2. Air enters a constant temperature compressor at 400°F and 75 psia and is compressed to 135 psia. What is the density of the air as it leaves?

3. Suppose that the continuous cycle of Figure 13–11 were filled with sulphur dioxide rather than air but otherwise operated with the same specifications of:

$$P_1 = 2116 \text{ psfa}, \quad T_1 = 530°R, \quad \frac{v_1}{v_2} = 5, \quad T_3 = 200°F$$

   Calculate the net WORK per pound of sulphur dioxide flowing and the efficiency of the system. Compare these results with those in Sample Problem 13–5, which uses air as the working medium. Is one more efficient than the other?

4. Figure 13–20 is a typical centralized comfort heating system showing air being pulled from a room, mixed with fresh outside air, and then heated by a set of hot-water heat-exchanger coils. The outside air and the return air are at two different temperatures; therefore, heat is exchanged between them as they mix. This constitutes a thermodynamic process.

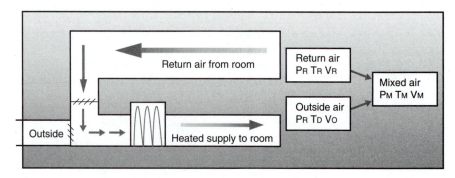

Figure 13–20. Problem 4.

a. Consider the relationship between the temperatures of the premixed air samples (o = outside air, r = return air) with the final mixed sample (m = mixed air). Which of the following is true?
   (1) $T_r = T_m$
   (2) The specific volume (or the density) of the premixed samples is different from the final specific volume
   (3) $v_r = v_m$ or $v_o = v_m$, $P_r = P_m$
   Look closely at the mixing device in the top picture of Figure 13–20. When you figure out which statement is true, you will know the type of fundamental thermodynamic process under which this mixing would be classified.
b. Write the energy equation for each of the two air samples.
c. Use the fact that when these two air samples are mixed, the heat gained by one is lost by the other (just like the horseshoe and water), and write the equation to calculate the temperature of the mixture. What is the resultant air temperature of 300 cfm return air at 72°F mixing with 100 cfm of outside air at 30°F?

5. Automobile air conditioners are made up of compressors, condensors, hoses, belts, clutches, and a lot of other things that can malfunction. Another design that appears to be far superior has no moving parts. The design is a set of continuous thermodynamic processes in a converging and diverging section of pipe mounted between the bumper and the firewall (dashboard) of the car.

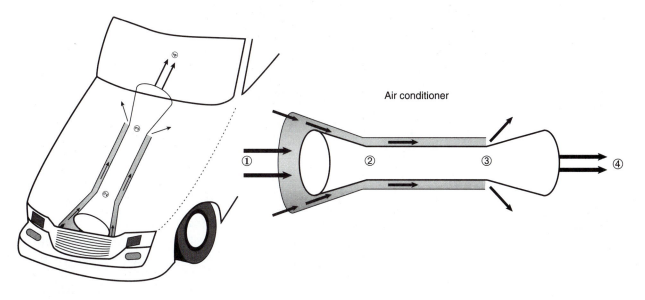

Figure 13–21. Problem 5.

  a. To see how the device works, consider the blown-up diagram in Figure 13–21. The device is based on adiabatic compression, constant pressure cooling, and adiabatic expansion. Find the temperature of the air-conditioned air as it enters the automobile cabin if the initial compression ratio is $v_1/v_2 = 1.5$, the outside entering air is at 100°F, and the adiabatically compressed air cools to 120°F (T_3) in the constant pressure section. How much net WORK is required to drive one lb_m of air through the device? How much heat is taken out of each pound of air? What is the efficiency of the system?

  b. The above problem illustrates how the device would work, but the output temperature might not be the desired temperature. Rework the problem, this time finding the initial compression ratio that will output air at a temperature of 55°F into the auto cabin if the constant pressure cooling process again reduces the air temperature to 120°F before the final expansion. Compute net WORK, heat lost, and efficiency.

6. Figure 13–22 shows a P-v diagram for a cycle operating with continuous process equipment.

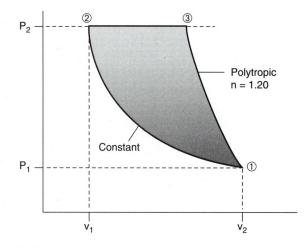

Figure 13–22. Problem 6.

  a. Find the state properties at each of the points.
  b. Determine q for each process q_{12}, q_{23}, and q_{31}.
  c. Calculate *work* for each process.
  d. Calculate Wd for each process.
  e. Show that net WORK is the same calculated by Σ *work* or Σ Wd.
  f. Determine the Carnot efficiency of the cycle that compares to this.

7. Show from Sample Problems 13–1, 13–2, 13–3, 13–4, and 13–6 that for a cycle, $\Sigma work = \Sigma W_D$.

CHAPTER
14

Carnot's Magnificent Machine: The Second Law of Thermodynamics

PREVIEW: Is there a cycle that is better than all other cycles—a set of processes that converts heat into WORK more efficiently than any other cycle? If so, wouldn't this be the cycle of choice for any power-generating device? Shouldn't the automobile manufacturers be told about it, or do they already know? Here is the story of that cycle and how it developed the second law of thermodynamics.

OBJECTIVES:
- ❑ Identify the processes of the Carnot cycle.
- ❑ Do some fast cycle analysis using Carnot cycle mathematics.
- ❑ Discover the second law of thermodynamics.

14.1 ENTER SADI CARNOT

The Otto cycle is surely one of the most common means of converting heat into WORK, but there are many others. Sometimes these machines look much like the typical automobile engine, but the series of thermodynamic processes going on inside them are much different. We often think of a diesel truck as being powered by just a large automobile engine. A diesel mechanic will tell you, however, that there is a substantial difference. A jet airplane is powered by an engine that differs from both the Otto engine and the Diesel engine. A steam engine is different from all three.

Of all the engines, which one is best? What do we mean by *best*? To some, this might mean the most compact, or the lightest. To others, it might mean the most fuel efficient. Still others might call the most powerful engine the best. To a thermodynamicist, *best* means specifically the engine that has the highest thermodynamic efficiency.

The question "Which is best?" was answered around 1824 by a British engineer, Sadi Carnot. (Figure 14–1) At that time the first heat engine had already been developed and was literally pulling England into the Industrial Revolution. It was the steam engine—the device that triggered the mechanized age. Carnot's studies resulted in an amazingly readable, thermodynamically oriented study of the steam engine quaintly titled "Reflections on the Motive Power of Fire and on Machines Fitted to Develop That Power."

Carnot introduces his topic by pointing out the importance of the steam engine to the economy of England:

> Already the steam engine works our mines, impels our
> ships, excavates our ports and our rivers, forges iron, fashions
> wood, grinds grains, spins and weaves our cloths, transports

Figure 14–1. Sadi Carnot explained the steam engine and more.

the heaviest burdens, etc. It appears that it must someday serve as a universal motor, and be substituted for animal power, waterfalls, and air currents.

To better put the steam engine in perspective, he claimed, "To take away from England her steam engines . . . would be to dry up all of her sources of wealth, to ruin all on which her prosperity depends, in short, to annihilate that colossal power."

Carnot was surely correct in predicting that the steam engine and other heat engines would take over from the horse and ox in transportation and take over from the grist mill and waterwheel in supplying industrial power. He declared, "If someday, the steam engine shall be so perfected that it can be set up and supplied with fuel at a small cost, it will combine all of the desireable qualities, and will afford the industrial arts a range the extent of which can scarcely be predicted."

How prophetic! Carnot not only predicted that the heat engine would dominate the industrial scene, as it has for almost a century and a half, but he also suggested that the criteria by which it would flourish would be "fuel at a small cost." Could he have foreseen that someday fuel would be so expensive that animal power (bicycles and so on), waterfalls, and air currents all would again have their day?

The steam engine was invented long before Carnot's time. The first such engine was patented in 1704. James Watt, for his developmental work on the device, was titled the "Father of the Steam Engine" before Carnot was born. A locomotive was powered by steam first in 1785, and the first steam boat operated twenty years later. Carnot took no credit in developing the steam engine. But he suggested that "notwithstanding the satisfactory condition to which (steam engines) have been brought today, their theory is very little understood, and the attempts to improve them are still directed almost by chance."

It was left to Carnot to describe in an orderly, mathematical manner the thermodynamic operation of the steam engine. The primary questions that he was striving to answer were twofold:

1.   Is the motive power of heat unbounded, that is, could the steam engine design be continually improved to develop more and more power from one unit of heat?
2.   Is steam the best medium to use to develop power, or is there another substance, perhaps atmospheric air, that would be an even better medium?

In answering these questions, Carnot accomplished much, as we are about to see. It can be said that Carnot put the "dynamics" into the science of thermodynamics.

14.2 CARNOT DREAMS UP A MACHINE

Carnot began his scientific investigation by scrapping any notions of analyzing the steam engine component-by-component. In fact, he simply lumped all steam engines into the class of "heat engines" and went about trying to discover the perfect engine. He asked, "What series of thermodynamic processes best converts heat into WORK?"

He began the analysis with a very fundamental observation about work-generating cycles. A thermodynamic cycle basically has two phases to it. The most obvious phase is the power stroke, that portion of the cycle in which a cylinder is moved by high pressures to generate WORK. The remainder of the cycle, the second phase, is that which resets the piston in a position to again do WORK and regenerates the high pressures needed behind the piston to provide for a great amount of WORK.

To determine which thermodynamic process should be used for the power stroke, we must ask which process is the most efficient in converting heat into WORK. Surely the constant temperature process fits this description because it converts all the heat entirely into WORK. It is one hundred percent efficient. Carnot's perfect "paper" engine uses a constant temperature power stroke (Figures 14–2 and Figures 14–4, process 3–4).

Next, what is the most efficient way to reset the piston and raise the pressure again to reactivate the power stroke? Surely the best way to move the piston back to its original position without requiring a great deal of WORK to be done is to first lower the resisting pressure behind the piston. This can be done without the cost of additional heat and without losing any of the precious heat that has been put into the machine by incorporating an adiabatic (no heat loss) expansion. So the next step in the perfect cycle is to reduce the pressure in the system by allowing the piston to expand adiabatically to a lower pressure (Figures 14–3 and Figures 14–4, process 4–1).

The next step is to move the piston back toward its original position, preferably without the use of WORK and without the loss of precious heat. The diagram in Figure 14–4 makes it clear that there is no way to get from point

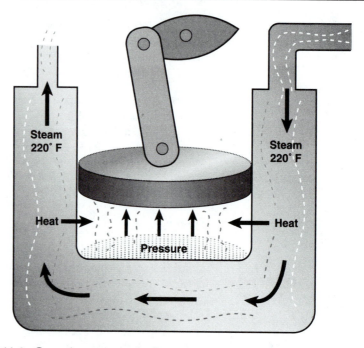

Figure 14–2. Carnot's power stroke is constant temperature: all heat is converted to WORK.

1 to point 3 without providing some WORK. There is no way to move between these two points without generating some area underneath the process curve. Of the many paths possible, some require far less WORK than others.

For instance, an adiabatic process will take the cycle from 1 back to 4, but a great deal of WORK will be expended doing it. We must use a process that will keep the restricting pressure low. This means either a constant temperature or constant pressure process, both of which require cooling to keep the pressures low. Therefore, we cannot avoid some loss of precious heat—heat that will have to be replaced during the next power stroke. A constant temperature process requires less cooling than a constant pressure one; therefore, it is the best compromise if we must lose heat and do WORK in driving the piston back to the original location.

If we follow the piston back to its original location with a constant temperature process, however, the pressure behind the piston will not be high enough to drive the power stroke the next time around. Carnot's philosophy, therefore, was to stop short of completing the reset phase with this process (Figure 14–4, process 1–2) and change to another thermodynamic process that will allow the piston to arrive at its original location with a sufficiently high pressure. Selecting the fourth and final process is easy. We need the one that increases pressure with no additional heat input: adiabatic compression (Figure 14–4, process 2–3).

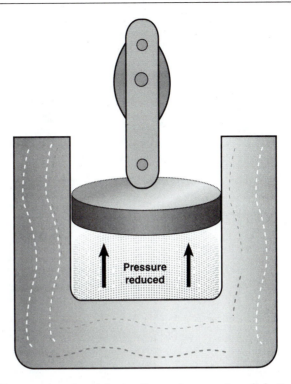

Figure 14–3. Pressure is reduced during the reset process by adiabatic expansion.

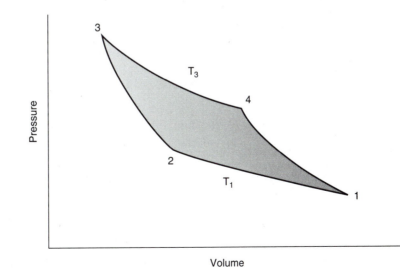

Figure 14–4. The P-V diagram of a Carnot cycle.

Now the cycle touted to be the best is on paper, although it has been neither built nor tested. The Carnot cycle can be summarized in four steps:

1. Constant temperature compression with heat loss
2. Adiabatic compression
3. Constant temperature expansion with heat gain
4. Adiabatic expansion

Figure 14–4 is a graphic illustration of the pressure-volume diagram that results from a Carnot cycle.

14.3 PITTING CARNOT'S "AUTOMOBILE" AGAINST OTTO'S ENGINE

Is the Carnot cycle better than the other types of heat engines? We begin this analysis by supposing that we have constructed a Carnot automobile and we are going to compare it with the Otto engine in Sample Problem 12–1, the Ford Taurus.

As a guideline to keep both engines competitive, we require that they be equally large: $V_1 = 258$ in^3. Furthermore, both have intake of atmospheric air: $T_1 = 70°F$ and $P_1 = 14.7$ psia. Both burn the same amount of gasoline except, because of the differences in design, the Otto cycle burns 5.4 Btu between points 2 and 3 and the Carnot cycle burns 5.4 Btu between points 3 and 4. Both engines are required to reach the same hottest temperature: $T_3 = 4133°R$. For now we can imagine that this requirement allows both engines to be constructed of the same material. The goal of our comparative analysis is to determine which engine is more efficient.

SAMPLE PROBLEM 14–1

Suppose an automobile engine has been built using the Carnot cycle with the same characteristics of the Ford Taurus in Sample Problem 12–1. Find the state properties at points 1, 2, 3, and 4; then compare the thermodynamic efficiency of the Carnot cycle with the Otto cycle.

SOLUTION:

This problem is best solved by working backward through the cycle, starting with process 4–1 instead of 1–2. (See Figure 14–4.)

Specifications:

$V_1 = 258$ in^3 = .149 ft^3

$P_1 = 14.7$ psia = 2116 psfa

$T_1 = 70°F = 530°R$

$m_{air} = .011$ lb$_m$

$T_4 = 4133°R$

$Q_{3-4} = 5.4$ Btu

Process 4–1 Adiabatic Expansion

$$V_4 = V_1\left(\frac{T_1}{T_4}\right)^{\frac{1}{k-1}} = .149\left(\frac{530}{4133}\right)^{2.5} = .000876 \text{ ft}^3$$

$$P_4 = P_1\left(\frac{V_1}{V_4}\right)^k = 2116\left(\frac{.149}{.000876}\right)^{1.41} = 2,808,475 \text{ psfa}$$

$$Q_{4-1} = 0$$

$$WORK_{4-1} = \frac{(P_1 V_1 - P_4 V_4)}{J(1-k)} = \frac{(315 - 2460)}{778(1-1.41)} = +6.89 \text{ Btu}$$

Process 3–4 Constant Temperature

$$Q_{3-4} = P_3 V_3 \ln\frac{(V_4/V_3)}{J} = P_4 V_4 \ln\frac{(V_4/V_3)}{J}$$

or

$$5.4 = 2,808,475 \times .000876 \ln\frac{(.000876/V_3)}{J}$$

Solving for V_3:

$$V_3 = .000158 \text{ ft}^3$$

$$P_3 = P_4\left(\frac{V_4}{V_3}\right) = 2,808,475\left(\frac{.000876}{.000156}\right) = 15,770,667 \text{ psfa}$$

$$WORK_{3-4} = Q_{3-4} = 5.4 \text{ Btu}$$

Process 2–3 Adiabatic Compression

$$V_2 = V_3\left(\frac{T_3}{T_2}\right)^{\frac{1}{k-1}} = .000158\left(\frac{4133}{530}\right)^{2.5} = .02683 \text{ ft}^3$$

$$P_2 = P_3\left(\frac{V_3}{V_2}\right)^{1.41} = 15,571,038\left(\frac{.000158}{.02683}\right)^{1.41} = 11,578 \text{ psfa}$$

$$Q_{2-3} = 0$$

$$WORK_{2-3} = \frac{(P_3 V_3 - P_2 V_2)}{J(k-1)} = \frac{(2460 - 315)}{778(-.4)} = 6.89 \text{ Btu}$$

Process 1–2 Constant Temperature

$$Q_{12} = P_1 V_1 \ln \frac{(V_2/V_1)}{J} = 2116 \times .149 \ln \frac{\left(\frac{.02683}{.149}\right)}{778} = -.69 \text{ Btu}$$

$$WORK_{1-2} = Q_{1-2} = -.69 \text{ Btu}$$

Summarizing:

$$Net\ WORK = -.69 - 6.89 + 5.4 + 6.89 = 4.71 \text{ Btu}$$

$$\eta = \frac{4.71}{5.4} = .87$$

Comparing this calculation for a Carnot engine with that for the Otto engine in Sample Problem 12–1 indicates that the Carnot automobile is much more efficient, 87% compared to 57%.

14.4 BUILDING A CARNOT ENGINE

It is often repeated at thermodynamic parties that "a working model of the Carnot cycle has never and can never be built." In a strict sense, this is true. For one thing, there is no such thing as a totally adiabatic process. No matter how much the process is insulated, some heat will escape. It may be limited to one percent, but in the real world some heat is going to escape. The same is true of a constant temperature process. Since the Carnot cycle consists of these processes, no "perfect" Carnot cycle can be built.

These processes, however, can be approximated very closely. A Carnot engine can be built that is very close to the theoretical device. After all, this is all that we expect from the Otto cycle. An internal combustion engine is by no means a "perfect" Otto cycle. It is not really fair to say that an Otto cycle can be built, but a Carnot cycle cannot.

One difficulty with the Carnot cycle is that it is bulky and cumbersome when designed as a piston-and-cylinder engine. It is almost inconceivable to think of building a working model of the processes in Figure 14–4 with ice packs applied to the cylinder at certain moments and insulation applied a fraction of a second later. But this doesn't eliminate it as a possible WORK-producing device if we employ something different than pistons and cylinders. If the processes involved were the open processes of Chapter 13, a Carnot cycle device such as that illustrated in Figure 14–5 would not be difficult to build.

Now we are faced with two clear facts: (1) the Carnot cycle is the most efficient way of converting heat into WORK, and (2) a prototype version of the device can be built. Then why don't we have Carnot cycles propelling our automobiles, or at least our airplanes, at a higher efficiency than that of the present devices?

Consider the maximum pressure in the Carnot cycle of Sample Problem 14–1. The pressure at P_3 is 15,770,667 psfa, a fantastically high pressure. What

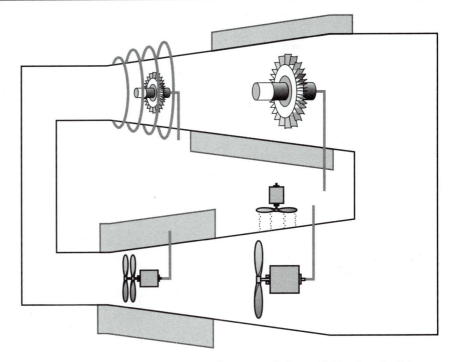

Figure 14–5. An open-process Carnot cycle is not difficult to build.

material would hold such a pressure? Such practical engineering considerations have doomed the Carnot engine as a possible prime mover.

14.5 CARNOT CYCLE AS AN ANALYSIS TOOL

Carnot's idea of creating this "paper engine" was a tool of analysis, so let's return to the two questions he proposed to answer: "Is there a best medium (gas) to use to drive such a device?" and "Can a heat engine be improved to generate more and more power?" Both questions can be answered by investigating the detailed calculations of the efficiency of the best-ever cycle of Carnot's imagination.

Before continuing to the next important point, some preliminary properties of the Carnot cycle can be established that will have future use. First, consider the adiabatic relationship between points 2 and 3 (Figure 14–4):

$$\frac{T_2}{T_3} = \left(\frac{V_3}{V_2}\right)^{k-1}$$

Similarly, between points 4 and 1:

$$\frac{T_1}{T_4} = \left(\frac{V_4}{V_1}\right)^{k-1}$$

We have seen several times that in the Carnot cycle:

$$T_2 = T_1$$

$$T_3 = T_4$$

so

$$\frac{V_2}{T_3} = \frac{T_1}{T_4} = \left(\frac{V_3}{V_2}\right)^{k-1} = \left(\frac{V_4}{V_1}\right)^{k-1}$$

or

$$\frac{V_4}{V_3} = \frac{V_1}{V_2}$$

This states that the constant temperature compression ratio is the same as the constant temperature expansion ratio.

SAMPLE PROBLEM 14–2

Using the results of Sample Problem 14–1, compute the adiabatic compression ratio V_3/V_2 and the adiabatic expansion ratio V_4/V_1.

SOLUTION:

$$\frac{V_3}{V_2} = \frac{.000158 \text{ ft}^3}{.02683 \text{ ft}^3} = .00589$$

$$\frac{V_4}{V_1} = \frac{.000876 \text{ ft}^3}{.149 \text{ ft}^3} = .00589$$

Both ratios are the same.

SAMPLE PROBLEM 14–3

Using the results of Sample Problem 14–1, calculate the constant temperature process volumetric ratio V_1/V_2. Repeat for the constant temperature expansion ratio.

SOLUTION:

$$\frac{V_1}{V_2} = \frac{.149 \text{ ft}^3}{.0268 \text{ ft}^3} = 5.54$$

$$\frac{V_4}{V_3} = \frac{.000876 \text{ ft}^3}{.000158 \text{ ft}^3} = 5.54$$

These ratios are the same for a Carnot cycle.

Now we are prepared to discover a very useful concept from Carnot's Cycle. Thermodynamic efficiency is defined as:

$$\eta = \frac{Net\ WORK}{heat\ input} = \frac{(WORK_{12} + WORK_{23} + WORK_{34} + WORK_{41})}{Q_{34}}$$

But a simplification of this formula can be made, since for the Carnot cycle, $WORK_{23} = -WORK_{41}$, as can be seen by reviewing Sample Problem 14–1. Efficiency, therefore, becomes:

$$\eta = \frac{(WORK_{12} + WORK_{34})}{Q_{34}}$$

But the equations for $WORK_{12}$ and $WORK_{34}$ can be written (constant temperature processes):

$$WORK_{12} = P_1 V_1 \ln \frac{\left(\dfrac{V_2}{V_1}\right)}{J} = mRT_1 \ln \frac{\left(\dfrac{V_2}{V_1}\right)}{J}$$

$$WORK_{34} = P_3 V_3 \ln \frac{\left(\dfrac{V_4}{V_3}\right)}{J} = mRT_3 \ln \frac{\left(\dfrac{V_4}{V_3}\right)}{J}$$

$$Q_{34} = P_3 V_3 \ln \frac{\left(\dfrac{V_4}{V_3}\right)}{J} = mRT_3 \ln \frac{\left(\dfrac{V_4}{V_3}\right)}{J}$$

where use of the perfect gas law has been made.

$$P_1 V_1 = mRT_1$$

$$P_3 V_3 = mRT_3$$

Now the formula of efficiency becomes:

$$\eta = \frac{(WORK_{12} + WORK_{34})}{Q_{34}}$$

$$= \frac{\dfrac{mRT_1}{J}\ln\left(\dfrac{V_2}{V_1}\right) = \dfrac{mRT_1}{J}\ln\left(\dfrac{V_4}{V_3}\right)}{\dfrac{mRT_3}{J}\ln\left(\dfrac{V_4}{V_3}\right)} = \frac{\left[T_1\ln\left(\dfrac{V_2}{V_1}\right) + T_3\ln\left(\dfrac{V_4}{V_1}\right)\right]}{T_3\ln\left(\dfrac{V_4}{V_3}\right)}$$

For further simplicity, remember that

$$\frac{V_4}{V_3} = \frac{V_1}{V_2}$$

or

$$\ln\left(\frac{V_4}{V_3}\right) = \ln\left(\frac{V_1}{V_2}\right) = -\ln\left(\frac{V_2}{V_1}\right)$$

Then the equation for efficiency takes the most simplified form:

$$\eta_{carnot} = \frac{(-T_1 + T_3)}{T_3} = \frac{(T_3 - T_1)}{T_3}$$

With this simple formula, we can calculate the efficiency of any Carnot cycle without going through the complete cycle calculations. Note that in this formula, temperature *must* be in °R (K).

From Sample Problem 14–1, calculate:

$$\eta = \frac{(T_3 - T_1)}{T_3}$$

and compare it with the efficiency you computed for the Carnot cycle in that sample problem.

SOLUTION:

$$\eta = \frac{(4133 - 530)}{4133} = .87$$

This single formula answers Carnot's first question. It clearly states that the substance used to drive the cycle does not change the performance of the engine; the formula for efficiency contains no factor that is related to the substance used. The only quantity that determines the efficiency of the Carnot cycle is the temperatures of the medium during the power stroke (T_3) and during the reset stroke (T_1).

SAMPLE PROBLEM 14–5

How would the formula for efficiency have to be changed if it were to show that indeed the efficiency is dependent on the type of medium used?

SOLUTION:

It would have to contain a term that is special to different types of gases, such as R, c_v, c_p, or k. For example:

$$\eta = \frac{k(T_3 - T_1)}{T_3}$$

This formula would show that the gas with the highest value of k is the best for use in the Carnot cycle (and possibly all heat engines). If this were so, with two Carnot cycles operating between the same source and sink temperatures, the one using a gas with the highest k would be the best. This would imply that some gases are better than others. Carnot proved this is not the case.

Figure 14–4 clearly illustrates that the Carnot cycle operates from only two different temperatures. T_3 is the temperature at which heat is put into the cycle. It is the high temperature in the cycle, and the heat provided to the engine at this point must be generated by some heat source. The low temperature in the cycle, T_1, is the temperature at which heat must be rejected from the cycle in order to make the reset process easier. Some cold substance, such as a large body of water, must be available to absorb this heat. This substance is referred to as a *heat sink*.

The Carnot cycle can be described very simply as a device that absorbs heat from a high-temperature source and rejects some of it into a lower-temperature sink, generating WORK continuously with the remainder. Anytime there are two areas of different temperatures, a Carnot cycle can be connected between them to generate useful WORK. In Carnot's words, "Wherever there exists a difference of temperature, motive power can be produced." The formula for Carnot efficiency, however, states that if the temperatures are not substantially different from each other, the efficiency of the operation will be extremely low. For example, if you tried to operate a Carnot engine from a source of warm water at 100°F and an atmospheric sink at 70°F, the efficiency of conversion into WORK would be:

$$\eta = \frac{\left[(100 + 460) - (70 + 460)\right]}{(100 + 460)} = \frac{30}{560} = 5.5\%$$

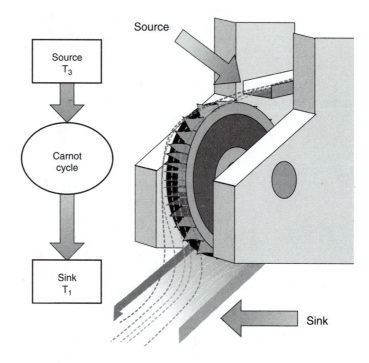

Figure 14–6. Heat goes through a carnot cycle like water over a grist mill.

And since Carnot's engine is the most efficient type of engine, you can be guaranteed that any other engine hooked up to this heat source and sink would be even less efficient in operation.

Figure 14–6 shows Carnot's concept of the manner in which a heat engine converts heat into WORK. He likened it to a waterwheel that takes water in at a high elevation and drops it to a low elevation, doing WORK along the way. The Carnot engine takes heat in at a high temperature, converts it to heat at a low temperature, and does WORK along the way.

SAMPLE PROBLEM 14–6

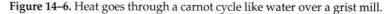

An inventor claims to have a device that connects to a chimney of a home furnace to convert the waste heat of the fire into electricity. He claims that it operates at 80% efficiency. Can you substantiate his claim?

SOLUTION:

Heat is converted to electricity by a generator, which is a motor that is turned by a thermodynamic engine and converts rotary motion (WORK) into electrical energy. To create the motion, the invention would be most efficient with a Carnot engine. The heat source for the engine is the chimney, which can achieve a maximum temperature of 700°F. The sink is the atmospheric air at a temperature of 0°F on a cold day. The Carnot efficiency is:

$$\eta = \frac{(700+460)-(0+460)}{(700+460)} = \frac{700}{1160} = 63\%$$

At best the device can work at 63% efficiency, so the inventor's calculations or claims are in error.

Carnot's second question is answered now as easily as the first. Can a steam engine be modified and improved to put out more and more WORK? The answer is *no*. The best the engine can do is the Carnot efficiency, which is limited by the source temperature (steam) and the sink temperature (atmosphere). The actual steam engine can be improved only to the point at which it achieves its Carnot efficiency.

The equation for Carnot efficiency demonstrates one more interesting fact. The only way to convert heat into WORK at 100% efficiency is either to have a very hot source (T_3 = infinite) or a very cold sink (T_1 = 0°R). Practically speaking, neither of these limits can be achieved, so it is safe to say that all the energy received as heat in a heat-engine cycle cannot be converted to WORK.

SAMPLE PROBLEM 14–7

Just like Carnot's engine, all power cycles must have a source of heat and a sink for the heat to be rejected. The high-temperature heat source in the Otto cycle is the ignition of the fuel. This is why it is called an internal-combustion engine; the high temperature is generated internally rather than coming from an external heat source. The cold-temperature sink is the atmosphere, where heat is rejected. Is dumping the heat into the sink an important requirement for an automobile engine? Would the conventional automobile work properly on the planet Mercury, where outside temperatures may be 2000°F or greater?

SOLUTION:

No, it would not work. After the explosion of gas in the cylinder, the temperatures and pressures inside would be no greater than those on the outside, so the piston would not be blown down the cylinder. Without the cold heat sink, the cycle would not operate.

The Carnot formula can be used in many ways for cycle analysis. For example, it can be shown that for any cycle, the difference between the heat input and the heat rejected is equal to the net WORK produced by the cycle. It sounds logical that if 10 Btu of heat are put into a cycle and 4 Btu are rejected, then the other 6 will be converted to net WORK. With this in mind, efficiency can be written:

$$\eta = \frac{Q_{in} - Q_{out}}{Q_{in}}$$

The Carnot efficiency can help predict the heat requirements of a cycle.

SAMPLE PROBLEM 14-8

A Carnot engine produces 13 hp at 2000 cycles per minute and exhausts heat to a 60°F water bath at a rate of 500 Btu/min. Find the temperature of the heat that drives the cycle, the rate at which heat is added, and the efficiency of the cycle.

SOLUTION:

$$Hp = \frac{Net\ WORK \times RPM}{33,000}$$

$$Net\ WORK = \frac{Hp \times 33,000}{RPM} = \frac{13 \times 33,000}{2000} = 858\ ft-lb_f = 1.1\ Btu$$

$$Q_{out} = \frac{500\dfrac{Btu}{min}}{2000\dfrac{cycles}{min}} = .25\ Btu$$

$$Q_{in} = Net\ WORK + Q_{out} = 1.1 + .25 = 1.35\ Btu$$

$$\eta = \frac{Net\ WORK}{Q_{in}} = \frac{1.1}{1.35} = .81$$

$$\eta = \frac{(T_3 - T_1)}{T_3}$$

or

$$T_3 = \frac{T_1}{(1-\eta)} = \frac{(460+60)}{(1-.81)} = 2736.8°R$$

14.6 THE SECOND LAW OF THERMODYNAMICS

If we summarize the concepts developed in this chapter, we would have a statement of what is known as the second law of thermodynamics. Whereas the first law is stated as a clear-cut equation, the second law is a concept that can be stated in many ways:

1. A thermodynamic cycle cannot convert all of the heat it absorbs into WORK.
2. A thermodynamic cycle operates from heat transferred from a region

of high temperature to a region of lower temperature, with that heat passing through the cycle.

3.   The maximum efficiency of a WORK-generating cycle is calculated from $\eta = T_3 - T_1/T_3$ (°R) where T_3 is the highest temperature of the cycle and T_1 is the lowest.

4.   When it comes to generating WORK, some Btu's are better than others (see Technicians in the Field).

TECHNICAL WRITER

The office seems bigger now. Ten years ago, there were thirty people in a room; now there are six. Everyone has more space. Soon, however, the office will be closed. There will be no one working here. The job is done.

The Solar Energy Utility Consortium, Inc. (SEUC) in Boulder, Colorado, was founded in 1972 as a central point of research into the commercial uses of solar energy. Funded by governmental agencies and public power utilities, its goal was to search out and develop the many ways that solar energy could replace or augment fossil fuels and nuclear power to meet the energy needs of an energy-hungry country. By the mid-1980s, it employed thirty engineers, urban development technicians, and business planners—but mostly engineering technicians. The technicians were responsible for a variety of projects throughout the country on the many facets of solar energy. Some were involved in solar hot water heating for cooking and cleaning. Others were researching the making of DC electricity by shining solar rays on ceramic crystals called photovoltaic cells. Most, however, were helping to develop a solar-powered, open-cycle thermodynamic power plant in the Mohave Desert east of Los Angeles. This plant operates by concentrating solar rays onto a tube that carries a working medium gas, almost identical to that in Figure 13–17 and in the discussion of Sections 13–5 and 13–6.

In the early days, when projects were getting started, technicians spent most of their time at the project site. Once the project was operating, however, their emphasis switched to writing the detailed operating manuals that would be used by operations personnel to get the most performance out of the plant. Technical writing is a very exacting skill. Plant operators who would be using the manuals had only high school educations; therefore, the reading level had to be no greater than a ninth-grade level. Yet the manuals included sections on technical concepts, operating procedures, control schematics, and troubleshooting. Words became the technician's tools, and it took a while to learn how to use them.

Once the sites were operational, all personnel trained, and the systems running smoothly, the technicians found themselves writing again, this time to present the interim results of the projects. Since these reports were targeted for influential individuals such as sponsors, politicians, businesspersons, and so on, the technicans improved their report presentation through the use of sophisticated computer software. Their CAD packages were linked with desktop publishing software so they could integrate words and diagrams. The technicians learned to use digitizing scanners for processing photographs and to modify the photos once they were in the computer. They worked with advanced spreadsheets, mostly to generate clear and pleasing tables and graphs. In 1984, the first interim report on the Mohave project was distributed, and each subsequent annual report has been much anticipated.

In their reports on the projects, the technicians sought to set aside predjudices and provide all the information and analysis necessary for others to make decisions on the future of solar energy. The performance criterion they used was *feasibility*. This is an elusive criterion, not nearly as precise as efficiency or capacity. For a project to be feasible, it must be competitive in terms of cost, both cost per kilowatt-hour and cost of construction. There is also feasibility of scale; what might work for small-power applications may not work for large power plants. There are ecological considerations as well—the impact of the project on nature.

Technicans have long realized that one of the major drawbacks to solar engines is the "curse of Carnot," or as one technician wrote: "Some Btu's are better than others." A Btu heats one lb_m of water one degree regardless of whether that Btu comes from a source that is solar, nuclear, natural gas, wood, or any other fuel, even though each of these sources presents that Btu to the water at a different temperature. Solar heat, as in making sun tea on a porch, presents Btu's at 90–100°F, unless they are concentrated in a solar collector or by using lens-focusing, in which case temperatures of 200–500°F are possible. Natural gas gives off its heat at ignition temperatures at 1800°F, and nuclear energy reacts to 10,000°F. Still, no matter what the temperature of the Btu, every Btu raises water the same amount in accordance with the first law of thermodynamics.

When the Btu is used to drive a power-generating device, Carnot's theory says that the efficiency of that Btu in creating WORK depends on the temperature of the Btu:

$$\eta = \frac{(T_3 - T_1)}{T_3}$$

This means that 1 Btu of solar energy at 200°F will generate .197 Btu of WORK (see Sample Problem 13–5), whereas 1 Btu from natural gas will generate much more WORK.

SAMPLE PROBLEM 14-9   Compare the number of Btu's of heat it takes to make 1 Btu of WORK, using a Carnot engine with a sink temperature of 70°F, for six popular heat sources.

SOLUTION:

Source	Max Temp (°F)	Carnot Efficiency	Btu's of Heat / Btu's of WORK
Solar (water panel)	200	.197	5
Solar (concentrating lens)	500	.45	2.1
Natural gas	1800	.76	1.33
Oil	2500	.82	1.24
Coal	3000	.84	1.22
Nuclear	10,000	.99	1.01

The technicians tried to work around the Carnot curse through the years by employing concentrating collectors and by installing solar tracking systems that maximized the solar radiation collected. But these advances created their own problems. They drove up the cost of the installation, and it was found that lenses had to be cleaned frequently or their reflectivity dropped appreciably.

This year the SEUC project is complete and the last interim report will be written. It will point out that there are some products that have achieved commercial success, such as the hot water heater project. Today, homes in high-solar areas like Florida, Texas, and Arizona typically are built with small solar hot water heaters installed in their roofs. The photovoltaic cell project proved not to be feasible on a large scale, but there have been many small solar cells installed that are powering traffic lights and security lights in situations in which the high cost of the cells is offset by the elimination of power lines.

The solar thermal power plant proved to have little commercial value, and even some environmental disadvantages. Large tracts of land are required to collect the solar energy, which retires the land from plant and vegetable growth. And, as the table above shows, solar energy, with its low T_3, is very inefficient in producing power.

The SEUC will close soon. The last of the technicians will leave. They have done an excellent job in representing solar energy. They will move on to new jobs; there are always jobs for technicians. Solar energy now will take its proper place in the energy future of the world.

CHAPTER **14**

PROBLEMS

1. A steam power plant consists of the following essential equipment:
   (1) The furnance provides the source of heat
   (2) The boiler actually contains the vapor (steam)
   (3) The prime mover is the engine or turbine that generates the WORK
   (4) The condenser cools the steam before returning it to the boiler
   (5) The circulator brings the spent steam (liquid and condensate) back to the boiler
       a. Carnot found that all cycles consist of the following necessary components: a high-temperature heat source, a low-temperature heat sink, and a WORK generating device. Identify each of these from the diagram in Figure 14–7.
       b. Assume that the boiler and condenser are constant pressure devices and that the prime mover and circulator pump are adiabatic processes. Draw the schematic of the continuous cycle that is described.
       c. Draw the P-v diagram for the cycle.
       d. Using the P-v diagram, describe the function of the condenser.
       e. What is the function of the circulator?

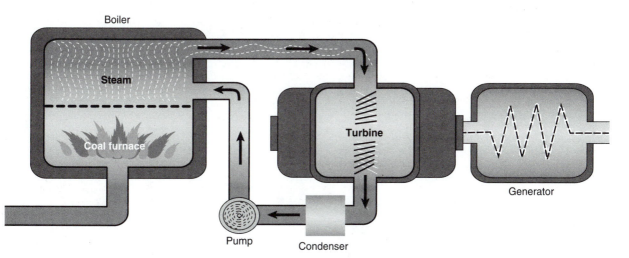

Figure 14–7. Problem 1.

2. An inventor claims to have discovered a device that will produce 293.25 watts of output (1 watt = 3.4 Btu/hr) from 1 cu ft of natural gas fuel. Is this possible?

3. What is the maximum output (ft-lb)/min of any air-cooled heat engine that uses natural gas fuel at a rate of .1 cu ft/min (1 cu ft of natural gas will release 1000 Btu of heat).

4. A Carnot cycle operates between temperatures of 50°F and 1200°F. If the engine absorbs 860 Btu/min, how much heat does it reject?

5. Calculate the efficiency of the Carnot solar engine in Sample Problem 13–5 using the Carnot formula and compare it to the efficiency calculated in that sample problem.

Problems in SI Units

6. Calculate the Carnot efficiency for an engine working between $T_3 =$ 3500°F and $T_1 = 70$°F using both °R and K and show that both calculations give identical results.

7. Find the maximum output (kJ/min) of an air-cooled (21°C) engine that uses coal as a heat source at a rate of 100 kg/hr.

CHAPTER 15

Application: The Grand Prix Race Car

PREVIEW: There are many thermodynamic power-generating machines that have thrilling developmental histories, but none more popular or exciting than the automobile engine. Its increased performance over the last one hundred years has been accomplished in the public eye because of the sport of automobile racing, where most of the new refinements have been first tested, indeed even conceived. This chapter reviews the thermodynamic history of the automobile engine through the exciting history of the premier racing circuit in the world, the Grand Prix Formula 1 circuit.

OBJECTIVES:

❑ Discover how calculations previously derived can describe the increased performance of the automobile engine from its beginnings in 1895 to the current fleet.

❑ Learn the vocabulary of high-performance automotive engines and become familiar with specific components of the engine that accomplish the thermodynamic processes studied.

❑ Introduce other components besides the engine that changed racing and road vehicles through the years.

Figure 15–1. Early Grand Prix racing.

15.1
THE EARLY YEARS

Years before Henry Ford imagined the manufacture of assembly line automobiles, the motor car was not only a proven invention but was so far advanced in design that cars were actually being raced. In 1894, a race was completed over seventy-nine miles of French countryside where the winning car, a steam engine, averaged just short of twelve miles per hour.

Auto racing is well entrenched today as an exciting and worthwhile sport. The story of the Grand Prix race car is the story of thermodynamics in action, for within one century, such racing machines have increased their speed twentyfold to over two hundred miles per hour. The changes in the engines that have generated these fantastic speeds can be analyzed only through the processes and calculations of thermodynamics.

In 1895, the brief era of steam engine automobile racing was ended. (Figure 15–1) The winning car was an internal combustion (Otto cycle) vehicle with an internal cylinder displacement of 1.2 liters, or about 73 cubic inches (1 liter = 61 cubic inches). The complete specifications for this engine are given below.

SPECIFICATIONS

Displacement = 73 in^3
Compression ratio = 4:1
Weight of captured air per cycle = .003 lb
$Q_{combustion}$ = 1.66 Btu/cycle
Speed = 1000 RPM

With this information, a complete cycle analysis can be made (see Sample Problem 12–1) with the following results:

CYCLE COMPUTATION SUMMARY

State Pressures psfa	State Volumes ft^3	State Temperatures °R	Process WORK
P_1 = 2116	V_1 = .042	T_1 = 530	$WORK_{1-2}$ = −.21
P_2 = 14, 947	V_2 = .0107	T_2 = 935	$WORK_{2-3}$ = 0
P_3 = 63,368	V_3 = .0107	T_3 = 4012	$WORK_{3-4}$ = .925
P_4 = 9115	V_4 = .042	T_4 = 2023	$WORK_{4-1}$ = 0

Net WORK = .715 BTU/cycle
Efficiency = .426
Mean effective pressure = 13,152 psfa
Horsepower = 8.37 hp

Even in the early days, winning was the name of the game. The first attempts at increasing engine horsepower were simple: enlarge the size of the cylinders. This was so much the trend that in 1905 a famous name in racing, Fiat, made its debut at the French Grand Prix with a gigantic leviathan, a 16.2-liter engine, and swept the race.

**SAMPLE
PROBLEM 15–1**

Assuming that the 1905 Fiat had the same engine specification as the 1895 engine except for the increase in displacement and a corresponding increase in heat input per cycle to make the same BTU/lb$_{air}$, find the efficiency, horsepower, and mean effective pressure of the Fiat.

SOLUTION:

The procedure for solving this problem might be to compute the pressure, temperatures, and volumes at each point of the Otto cycle. This may not be necessary, however, since everything is identical to the specifications of the original 1895 engine except for one thing, V_1. Notice, by inspecting Sample Problem 12–1, that V_1 enters the solution only in the calculation for WORK. It is true that V_2/V_1 is used, but this is a compression *ratio* that has not changed from the 1895 version. So everything is identical except WORK (and volumes). Therefore, we only need to recalculate these quantities.

Calculate:

$$V_1 = 16.21 \text{ liter} = 988.2 \text{ in}^3 = .572 \text{ ft}^3$$

$$V_2 = .572 \frac{\text{ft}^3}{4} = .143 \text{ ft}^3$$

$$V_3 = V_2 = .143 \text{ ft}^3$$

$$V_4 = V_1 = .572 \text{ ft}^3$$

$$WORK_{1-2} = \frac{P_2V_2 - P_1V_1}{J(1-k)} = \frac{(14{,}947 \times .143) - (2116 \times .572)}{778(1-1.4)} = -2.90 \text{ Btu}$$

$$WORK_{2-3} = 0$$

$$WORK_{3-4} = \frac{P_4V_4 - P_3P_3}{J(1-k)} = \frac{(9115 \times .572) - (64{,}368 \times .143)}{778(1-1.4)} = 12.59 \text{ Btu}$$

$$WORK_{4-1} = 0$$

$$Net \ WORK = WORK_{1-2} + WORK_{2-3} + WORK_{3-4} + WORK_{4-1}$$

$$= -2.90 - 0 + 12.59 + 0 = 9.68 \text{ Btu}$$

$$Q_{combustion} = \frac{.572}{.042} \times 1.66 = 22.7 \ \frac{\text{Btu}}{\text{cycle}}$$

$$MEP = \frac{Net \ WORK}{displacement} = 9.68 \times \frac{778}{.572} = 13{,}152 \text{ psfa}$$

$$Hp = \frac{Net \ WORK \times RPM \times 778}{2 \times 33{,}000} = 114$$

$$Efficiency = \frac{9.68}{22.7} = .426$$

Figure 15–2. Race cars grew to astronomical proportions from 1885 to 1910.

Notice that making the engine bigger (Figure 15–2) did not improve the efficiency, but it did increase the horsepower in the same ratio as the increase in displacement. The fact the mean effective pressure remains the same indicates that this engine is no more clever than the 1895 one, although it is more powerful (bigger).

Imagine this tank of a race car with twelve or sixteen cylinders packed into an enormous engine. Surely it would not have been very good in turns, but turns were rare in the cross-country races. Engines continued to grow in size after the turn of the century, well past the 17-liter level.

In 1912, a mere 7.6-liter Peugeot engine beat the giants. How could such a small engine (comparatively) beat the larger ones? This engine ran almost twice the speed of previous engines, making up for the lower volume with a higher RPM. The engine had the advantage of high speed because it was lighter.

SAMPLE PROBLEM 15–2

Assuming that the 1912 Peugeot had the same specifications as the 1895 engine except that the displacement was 7.6 liters and the RPM was 2000, find the Peugeot efficiency, horsepower, and mean effective pressure.

SOLUTION:

As seen in Sample Problem 15–1, changing the displacement does not change efficiency but does change the net WORK in the same ratio as the displacement changes. Therefore,

Figure 15–3. The 1915 Mercedes.

$$Net\ WORK_{1912} = \frac{7.6\ \text{liters}}{1.2\ \text{liters}} \times .71 = 4.5\ \text{Btu}$$

$$MEP = \frac{Net\ WORK}{displacement} = 13{,}152\ \text{psfa}$$

$$Hp = \frac{Net\ WORK \times RPM}{2 \times 33{,}000} = \frac{778 \times 4.5 \times 2000}{66{,}000} = 106.1$$

Even though the behemoth race car was on its way out by 1914, an engine capacity limit of 4.5 liters was belatedly placed on racing engines to maintain a spirit of competition. With this restriction began the regulation of racing by a sanctioned governing body, a practice that is continued today. Regulation according to size is as necessary in racing as it is in wrestling or boxing, since competition according to weight makes for better sport.

With the initiation of the 4.5-liter limit, Mercedes outstripped the Peugeot high-RPM technology (Figure 15–3) and became the vehicle to beat. It dominated for the next three years with a 4.5-liter, 2800-RPM engine that boosted the speed record to 110 mph. The advancement to high RPM required many technical breakthroughs in spark timing and spark plugs that would allow the number of spark ignitions per minute to be doubled. It also required that engine exhaust and intake occur over a much shorter time, meaning that intake and exhaust valves had to be increased in size. Larger valves were unwieldy for opening and closing at high speed, so Mercedes increased the number of valve sets to two, that is, four valves per cylinder instead of two. This meant that the engine could take in its complete engine displacement of atmospheric air and better achieve its theoretical horsepower.

SAMPLE PROBLEM 15-3

The engine of 1895 design suffered from obstruction by small intake and exhaust valves. This meant that the cylinder did not fill with its full quota of intake air. Assume that the intake stroke of the 1895 engine pulled in only eighty percent of the calculated weight of air but maintained all other conditions the same, including the same Btu/lb$_{air}$ heat input. Find the efficiency, horsepower, and mean effective pressure of this more realistic engine.

SOLUTION:

Since the compression ratio is the same, conditions at state 2 will be unchanged. Since the Btu/cycle is decreased to match the decrease in air ingested, conditions at state 3 will be the same. All state conditions are the same except the volume is decreased to 80% and, therefore, net WORK will be reduced to 80%.

$$Net\ WORK = .8 \times .71 = .57\ \frac{Btu}{cycle}$$

$$Efficiency = .426\ (stays\ the\ same)$$

$$Hp = \frac{net\ WORK \times RPM}{2 \times 33,000} = \frac{.57 \times 778 \times 1000}{2 \times 33,000} = 6.7\ (decreased)$$

$$MEP = \frac{1.70 \times 778}{.042} = 10,558\ psfa$$

15.2 SUPERCHARGING

In the years between 1909 and 1911, a closed circuit automobile track was developed in Indianapolis, Indiana, for testing vehicles made in the local area. At first the track was made of crushed stone and asphalt, but this rapidly broke up. Eventually it was covered with three million ten-pound bricks, the best paving material available. While it never was used as a test track because the seat of American automobile manufacturing moved to Detroit, the track was used in 1911 to run one race at a distance of 500 miles. This proved so successful that the race was repeated the following year and has been repeated again every year to the present. It is now known worldwide as the Indy 500. The race was added to the international Grand Prix circuit in 1921. In the early years, Indy spawned classic American racing teams: Stutz from Milwaukee, the Duesenberg that made its appearance in 1922 with the first use of hydraulic brakes, and Gaston and Louis, the French Chevrolet Brothers, driving their American car.

Meanwhile, Grand Prix racing continued to reduce the engine limit capacity, down to 1.5 liters by 1922. But the reduction in engine size did not reduce speeds because of a development that might be called the most significant technological advancement in racing. In 1922, Mercedes beat the racing world to the punch with a powered air injection system that utilized a small air pump mounted next to the exhaust manifold to force air into the cylinders. (Figure 15–4) It was called a supercharger and could easily and immediately add thirty to forty percent more horsepower to an engine by packing more air in the cylinder, and therefore more gasoline could be burned on each cycle. The 1922 Mercedes was the first supercharged racing car.

**SAMPLE
PROBLEM 15–4**

If a supercharger is applied to the 1895 winner so that the initial pressure is twice the normal pressure, and the heat added is adjusted to maintain the Btu/lb_m ratio the same, find the efficiency and mean effective pressure.

SOLUTION:

Since the efficiency of the Otto cycle depends only on compression ratio, and since the supercharger does not affect the engine size, bore, or stroke, the efficiency for the supercharged engine is the same as the original engine, 43%, even though the initial pressure is twice the atmospheric pressure and other state pressures are elevated. The amount of air ingested

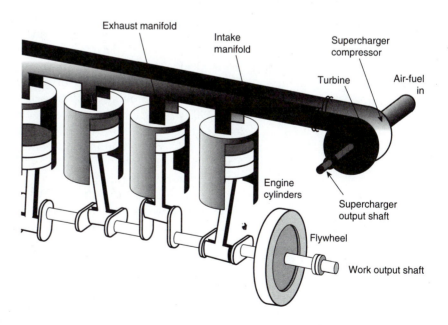

Figure 15–4. The supercharger forces compressed air into the cylinders.

in the supercharged engine is twice that of the original ($m = PV/RT = .006$ lb$_m$), and the amount of heat added during combustion is also doubled ($Q_{23} = 3.33$ Btu). Since

$$Efficiency = \frac{Net\ WORK}{Q_{23}} = \frac{Net\ WORK}{10} = .426$$

Then

$$Net\ WORK = 1.42\ Btu$$

and

$$MEP = 1.42 \times \frac{778}{.042} = 26{,}303\ psf$$

and

$$Hp = 16.7$$

Increasing the mean effective pressure indicates that this supercharged engine uses the piston cylinder volume effectively, but for the drivers, the horsepower is the significant number, and this almost immediately more than doubles.

The sample problem points out that a supercharger does not improve efficiency, but it does increase horsepower by making more effective use of the engine volume (look at the improvement of MEP). It does this not by increasing the initial pressure (this effect is minor), but by packing more air and oxygen in the cylinder so that more fuel can be burned per cycle with a corresponding increase in net WORK.

Supercharged engines became an immediate sensation. In 1924, Indy was won by a supercharged car, an American Duesenberg, and the following year there wasn't a car in the race that wasn't supercharged. The supercharger drove some great names out of racing—those teams that did not quickly respond to the technology. Bugatti had been in the winner's circle consistently since 1905 with creative and novel machines such as the "barrel body," but his team scorned the use of superchargers and tried to make up the difference with superior road handling and braking to offset their power disadvantage. Their hopes were not fulfilled, and by 1930, they were no longer a serious threat in Grand Prix racing.

Top speeds increased drastically in the 1920s. The sanctioning committee tried to limit them first by lowering the engine displacement to 1.5 liters, then by limiting the weight of the total vehicle, then the overall dimensions, and then a combination of weight and displacement. There were so many requirements that engine designers claimed it was a "delicate formula" the sanctioning body was brewing each year, and Grand Prix races soon became known as the Formula races, or Formula One. In order to foster creativity in regularly aspired engines, the formula took on a dual nature: one formula of displacement and size for nonsupercharged and, therefore, underpowered engines, and another formula for supercharged engines. The sanctioning body played

Figure 15–5. The driver and navigator often acted as mechanic.

with the formula yearly in order to bring a balance between the two concepts, and by the early thirties, regularly aspired engines again began making trips to the winner's circle.

In some years, the formula was inspiring to designers. It gave them freedom to develop novel concepts, such as the inherently balanced V-engine and new aerodynamic shapes for the body. Other formulas devastated the sport until they were changed. In 1926, the formula was so strict that the French Grand Prix was run with only three entries, all Bugattis.

Overall, the years 1925 to 1935 saw many changes. One was extremely evident. Race cars had always carried two occupants—a driver and a navigator, who doubled as repairman. (Figure 15–5) As more races were run on confined tracks, the repairman became a pit crew. Furthermore, from the Indy track came the rearview mirror, and a navigator was no longer needed. Even though two drivers in the cockpit was no longer necessary, the formula required it. In the early 1930s, Alfa Romeo introduced its 3P model as a single-seater, or "montoposto," configuration. It was fully sanctioned in 1935, ending the era of two-seaters.

In this era, too, the search for more power turned to exotic fuels, primarily nitromethane with 250,000 Btu/gal. Immediately, the horsepower developed from an engine took a quantum leap.

SAMPLE PROBLEM 15– 5

To show the effect of the increased power of exotic fuels, compare the efficiency, horsepower, and mean effective pressure of the 1895 winner against this same engine using nitromethane.

SOLUTION:

Increasing the heat value of the fuel by sixty-seven percent increases Q_{23} to 2.77 Btu/cycle, but since it does not affect the compression ratio of the engine, it does not affect the efficiency. Therefore

$$Efficiency = \frac{Net\ WORK}{Q_{23}} = .426$$

$$Net\ WORK = .26 \times 2.77 = 1.18$$

$$MEP = 1.18 \times \frac{778}{.042} = 21,858\ psf$$

$$Hp = 13.9$$

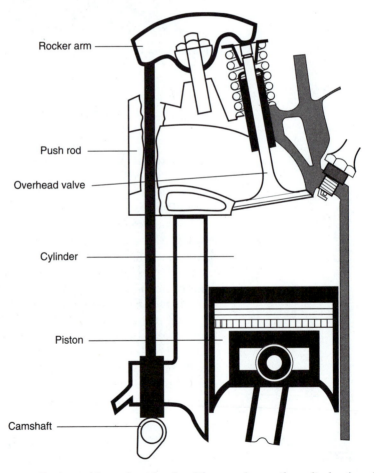

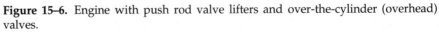

Figure 15–6. Engine with push rod valve lifters and over-the-cylinder (overhead) valves.

The search for power also turned to less obvious, but more technical, changes in engine design. The intake and exhaust valves of the cylinder must be timed perfectly, even for low-RPM engines. To accomplish this, a rotating shaft driven by the driveshaft, which is at the base of the engine, is acommodated with a series of cam lobes that, as they rotate, push up on a straight rod that is mechanically tied to the valves, which are located at the top of the engine. This complete mechanism of camshaft, push rod, and valve is called the valve train. In the 1895 engine, the intake and exhaust valves were located at opposite sides of the cylinder, necessitating two complete valve trains. To eliminate one of the drivetrains, the intake and exhaust valves were moved to the same side of the cylinder (Figure 15–6), but a great deal of resistance to the gas flow was created to reach the side pocket where the valves were located. Valves located directly in the cylinder head were a big advantage, with a valve pivot connecting the push rod to the valve that allowed both valves to be reached from the same side.

This configuration is used in most automobile engines today. The valve size is limited, however, by the size of the cylinder bore. Therefore, an angled head was developed to allow the valves to be larger, which has the advantage of giving a more favorable shape for the combustion chamber.

In a later era, Ford Motor Company pioneered the V-engine with its famous V-8. This was an advancment over the straight, or I-block, engines because it was more mechanically balanced (opposing forces instead of additive forces) and because it moved the camshaft above the driveshaft, making the push rods shorter and, therefore, more rigid.

All these modifications were attempts to make the exhaust and intake phases of the Otto cycle more efficient—to make the engine a "better breather."

SAMPLE PROBLEM 15–6

The 1895 engine was so inefficient on its intake phase (volumetric inefficiency) that it did not completely fill up with air on the intake stroke. Because of this, its compression ratio was effectively reduced by twenty-five percent. Find the efficiency, mean effective pressure, and horsepower, of this engine assuming a volumetric efficiency of seventy-five percent.

SOLUTION:

The compression ratio is actually 3, rather than the compression ratio of 4 computed from the bore and stroke, therefore

$$Efficiency = \eta = 1 - \left(\frac{V_1}{V_2}\right)^{k-1} = 1 - \left(\frac{1}{3}\right)^{.41} = .35$$

$$Net\ WORK = \eta \times Q_{23} = .35 \times 1.66 = .58$$

$$MEP = .58 \times \frac{778}{.042} = 10{,}743 \text{ psfa}$$

$$Hp = 6.83$$

These performance indicators imply that the volumetric efficiency severely robbed power from the early engines. (Compare to Sample Problem 15–1.)

15.3 ADVANCES DURING THE WAR YEARS

By the early 1940s, it appeared that every advantage and modification to the Otto cycle had been taken, that engine technology was almost completely developed, and that maximum power had been achieved from this cycle. But one factor—RPM—had yet to be fully developed. The RPM is a direct factor in the calculation of horsepower. Increasing the RPM of an engine definitely creates more horsepower. The speed at which a cycle takes place is determined by two factors. The maximum pressure in the cylinder has an effect (P_3). In particular, the switch to the alcohol fuels increases P_3 and correspondingly increases RPM. But a bigger factor determining the speed of the power stroke adiabatic expansion is the load on the driveshaft that the piston must over-

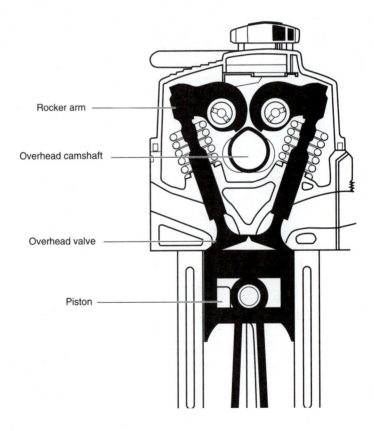

Rocker arm

Overhead camshaft

Overhead valve

Piston

Figure 15–7. An overhead cam engine.

come. Reducing the load that the piston pushes against makes this expansion take place faster. This is accomplished in the gearbox to the tires. Increasing the rearend gear ratio automatically increases the RPM of the engine, meaning that more cycles can take place per second.

Speeding up the engine, however, took its toll. Volumetric efficiency went down, calling for even more attention to intake and exhaust valve configuration. But even more important, the higher RPM stressed the valve train and particularly the push rods beyond their capability. The push rods represented far too clumsy an arrangement for engines destined to rev at over 10,000 RPM. In the 1940s, the overhead cam was introduced, which moved the camshaft directly over the cylinder to drive the valve rocker arms directly without push rods. (Figure 15–7) Overhead cams had immediate response and had such short linkages that they stood up at any RPM. The problem now was that the mass of the valves could not move fast enough, so lighter valve materials were sought. The eventual solution was to reduce the size of the valves and increase their numbers. Therefore, multivalve engines, mostly four-valve (two intake, two exhaust), were produced.

World War II suspended many sporting activities, but not automobile racing because it was a war of motorized technology, and racing cars were in the forefront of this technology. One name, more than any other, sparked creativity during this period: Ferdinand Porsche. Under the graces of Hitler (and with a Jewish partner), yet unimpressed with the communist mystique, this man effected many changes that are still evident today. Working under the company name of Auto Union, Porsche and his engineers designed a vehicle that could win under a formula that severely restricted overall weight. To eliminate the heavy driveshaft that delivered power from the engine in

Figure 15–8. Auto Union put the motor in the rear.

front to the wheels in the rear, Porsche moved the engine to the rear, directly over the wheels. (Figure 15–8) At the same time, he eliminated the heavy leaf-spring suspension and replaced it with a torsion-bar design. Auto Union ruled the racetrack in the late 1940s and even spawned a car for the average driver called Volkswagen.

15.4 AERODYNAMICS TAKES OVER

Auto Union created its own demise, after almost a decade-long grip on the winner's circle, because the rear-engine design opened the door to new concepts in aerodynamic design. Auto Union's P-Wagen and later models looked exactly like their front-engine brothers except the driver was positioned closer to the front of the vehicle. Taking advantage of the relative uselessness of the front-end body, Alfa Romeo made streamlined changes that reduced drag and gave the race car body a radically modern profile. (Figure 15–9) Other auto manufacturers followed the trend, and wind tunnels became a critical component of the design process.

Figure 15–9. An Alfa (rear) duels a Ferrari in 1951.

It was Mercedes in the 1950s that brought racing back to the engine designers with the introduction of fuel injection. Fuel injection eliminates the carburetor, the primary function of which is to mix air and fuel on the way to the cylinders. The air flow is restricted using a tapered orifice called a venturi, which in turn creates a vacuum that pulls fuel into the airstream. This means the pistons have to pull harder against the venturi to fill the cylinder with air, reducing the volumetric efficiency. Fuel injection allows small pumps to inject precise amounts of fuel directly into the cylinders after they have been filled with air through an unrestricted manifold.

Most of the advancements of the sixties and seventies belonged to the body designers. Speeds increased to the point that the cars began to look more like airplanes than land-bound machines. Radiators, which had always been such a prominent part of the front end, began to shrink. The high speeds of racing meant there was plenty of air available for cooling, and Alfa Romeo made it clear that bulky front ends were a thing of the past. By 1972, the radiator was totally hidden (Figure 15–10) in the side of the vehicle on the Lotus-Ford Model 72. The wedge-shaped nose did more than reduce drag; it also put aerodynamic down forces on the vehicle to help keep it on the road, for the velocities of the vehicles were closing in on airplane speeds and the vehicles had a tendency to try to take off and fly. Inverted wings became the craze: wings in the front, in the rear, and on the sides. Aerodynamics were being used to keep the vehicle manageable.

Figure 15–10. By 1972, the front-end radiator was gone, moved behind the tire in the body.

Aerodynamics were used by the engine designers, too. In the early 1970s, small protrusions thrust out the top of Formula One cars, much like aerodynamic telescopes. These air scoops provided sufficient quantities of air for the ever-more-hungry engine and used the motion of the vehicle to pressurize and ram the air into the engine in a modified supercharging effect.

SAMPLE PROBLEM 15-7

The laws of aerodynamics indicate that an airstream traveling at a speed S, when rammed into a scoop and brought to rest in the engine, will deliver a pressure equal to $\rho\, S^2/2g$. If the winning car of 1895 could have achieved 200 mph (292 ft/sec) as modern cars do, how much supercharge boost would the ram air (Figure 15–11) give it (in psig, psia, and inches of mercury)? What improvement in efficiency, mean effective pressure, and horsepower would result?

SOLUTION:

Since the density of air is .072 lb/ft³, the boost is calculated from the aerodynamic formula

$$P = 0.72 \times \frac{(292)^2}{(2 \times 32.2)} = 95.32 \text{ psfg} = .66 \text{ psig} = 15.36 \text{ psia}$$

Figure 15–11. The aerodynamic scoop changed the look of the Formula One race car.

Designers prefer to discuss boost in terms of the deflection of a mercury manometer: 1 psi = 2 in of boost. Therefore, the boost is 30.72 inches. The boost in a nonsupercharged engine is 14.7 psia, or 29.4 inches. The aerodynamics ram increased the boost by 1.33 inches or 4.5%. The scoop supercharger has no effect on efficiency; therefore, it increases MEP and hp by 4.5%.

The benefit of the ram air was not to replace the far more effective supercharger but rather to give 4.5% more power to an engine that was running under the formula as a regularly aspired engine. The formula by now was published in a book, as it was so lengthy. Because auto racing's original goal involved promoting the development of all components of the automobile, the sanctioning committee attempted to keep all engine and vehicle concepts competitive. Therefore, in the same race, regularly aspired engines would be given an advantage in engine displacement over supercharged engines (typically sixty percent greater displacement). Push rod engines were allowed thirty-three percent more cubic-inch displacement over multivalve, overhead cam engines. Somewhere in the formula, even the 1895 winner might still be competitive.

15.5 THE TURBINE CAR

The attention of Grand Prix racing was temporarily diverted from beautiful body designs in 1967 by a sensational new car that represented the greatest thermodynamic revolution since the supercharger. Andy Granatelli's STP Corporation backed a car powered by a gas turbine engine rather than a reciprocating power plant. In fact, the engine came out of a Pratt & Whitney helicopter. The engine worked on the principle of a Brayton cycle (Section 13–7). The motor had one rotating shaft with an air compressor on one end and a turbine on the other. (Figure 15–12) In operation, the shaft spun, turning the compressor that pulled outside air into the engine and forced it into a burner in which fuel was mixed with the air and the mixture was ignited. After initial ignition, the mixture burned with a continuous flame as long as fuel and air were supplied. The heat developed in the burner gave increased energy to the gases. As the gases left the burner, they were ducted to the compressor turbine. The compressor turbine, its shaft, and the air compressor were driven by these high-energy gases. The remainder of these gases were directed to a second turbine called the power turbine. It was geared to a transmission and a drive line to power the vehicle.

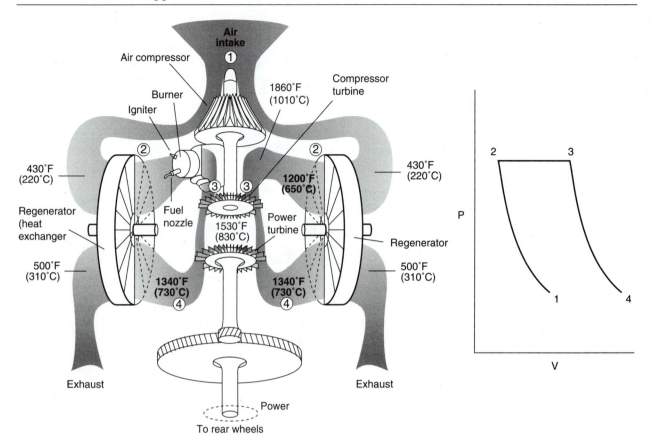

Figure 15–12. The Granatelli turbine engine.

SAMPLE PROBLEM 15–8

The specifications for the STP Indy turbine engine are below. Calculate the state points and its efficiency.

$$P_1 = 14.7 \text{ psia}, \ T_1 = 70°\text{F}, \ v_1 = 13.36 \ \frac{\text{ft}^3}{\text{lb}_m}$$

T_3 (combustion temperature) = 1345°F

Compression ratio $\dfrac{v_1}{v_2} = 4$

SOLUTION:

The cycle diagram is shown in Figure 15–12, with the state points marked.

a.   Adiabatic Compression
b.   Constant Pressure Heat Addition

c. Adiabatic Expansion
d. Constant Pressure Heat Rejection

Process 1–2

$$v_2 = \frac{v_1}{4} = 3.34 \text{ ft}^3/\text{lb}_m$$

$$P_2 = P_1 \left(\frac{v_1}{v_2}\right)^{\frac{1}{k}} = 14{,}958 \text{ psfa}$$

$$T_2 = T_1 \left(\frac{v_2}{v_1}\right)^{k-1} = 989° \text{R}$$

$$Wd_{12} = \frac{k(P_2 v_2 - P_1 v_1)}{(778 \times (1-k))}$$

$$= -95.76 \frac{\text{Btu}}{\text{lb}_{air}}$$

Process 2–3

$$v_3 = v_2 \left(\frac{T_3}{T_2}\right) = 6.09 \text{ ft}^3/\text{lb}_m$$

$$P_3 = P_2 = 14{,}958 \text{ psfa}$$

$$q_{23} = c_p (T_3 - T_2) = 195.8 \text{ Btu}/\text{lb}_m$$

$$Wd_{23} = 0$$

Process 3–4

$$P_4 = P_1$$

$$T_4 = T_3 \left(\frac{P_4}{P_3}\right)^{k-1} = 1023° \text{R}$$

$$v_4 = v_1 \left(\frac{P_1}{P_4}\right)^{\frac{1}{k}} = 30.22 \frac{\text{ft}^3}{\text{lb}_m}$$

$$Wd = \frac{k(P_4 v_4 - P_3 v_3)}{(778(1-k))} = 120 \frac{\text{Btu}}{\text{lb}_m}$$

$$Efficiency = \frac{net \ work}{q_{23}} = \frac{24.24}{170} = .142$$

SAMPLE PROBLEM 15–9

With an inlet area of 30 in^2, how many lb$_m$/min of air is ingested by the turbine moving at 150 mph? What is the horsepower?

SOLUTION:

To find the air flow:

$$\frac{lb_m}{min} = \frac{ft}{min} \times area \times \frac{lb_m}{ft^3} = velocity \times area \times \rho_{air}$$

$$= 150 \text{ mph} \times 5280 \frac{ft}{mi} \times \frac{1 \text{ hr}}{60 \text{ min}} \times 30 \text{ in}^2 \times \frac{1}{144} \frac{ft^2}{in^2} \times .072 \frac{lb_m}{ft^3}$$

$$= 198$$

$$Hp = \frac{net\ work}{lb_{air}} \times \frac{\frac{lb_{air}}{min}}{33,000} = 24.24 \times 778 \times \frac{198}{33,000} = 326.8$$

Parnelli Jones led at Indy in the turbine right from the start and extended its lead throughout the race. It appeared to have the clear victor, but with three laps to go a bearing failed in the transmission and it was out of the race, finishing sixth.

The turbine had met the rules of the formula and showed it was a more compact power plant than a reciprocating engine. The ruling body of racing quickly recognized that turbines would take over racing. They therefore restricted the inlet area of the compression section. Instead of simply taking away the advantage of turbines, this restriction dealt so harshly with them that they were no longer competive and almost within the year this great thermodynamic revolution became obsolute.

15.6 THE CASE OF COLIN CHAPMAN

There were a great many successes in Grand Prix racing in the 1970s and 1980s, but none overshadowed the achievements of designer Colin Chapman and his Lotus Car Company. Chapman started this London firm in 1952 to supply racing enthusiasts and sports car hobbyists with a sleek, simple, and lightweight racing vehicle dubbed the "Mark 3 Lotus." For ten years his company expanded as an aggressive manufacturer of racing sport cars and eventually even built production road cars, all with Chapman's creative mark of design: light weight. His expertise was primarily in the mechanical aspects of race car design rather than the thermodynamic components. Specifically, he made advances in the chassis, which is the backbone of the vehicle on which all the other components are hung; in the suspension systems that provide the excellent road-handling characteristics of quality race cars; in the braking systems, and in the use of fiberglass as a material for the body skin. In fact, he

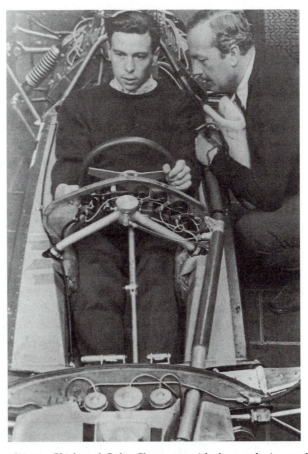

Figure 15–13. Jimmy Clark and Colin Chapman with the revolutionary Monocoque.

popularized the monocoque concept which uses the vehicle skin and engine as the structure of the vehicle, eliminating the frame. (Figure 15–13) "It's illegal!" the competition shouted, "There's no chassis! The driver is sitting right on the gas tank!"

In 1962, Dan Gurney introduced the smooth line of monocoque Lotus 25 at the Dutch Grand Prix. This car took full advantage of the rear-mounted engine that allowed the front end to be formed into a knife-edge to reduce aerodynamic resistance. Gurney and Chapman approached Ford and asked them to build an engine to fit the Lotus vehicles so that they could be competitive at the Indianapolis 500. Ford desperately wanted to be part of a winning Indy vehicle and agreed. Their engine delivered 350 hp on normal gasoline, while the popular Offenhauser engine, which developed 400 hp, used methanol and nitromethane.

In 1963, Lotus entered its first Indianapolis race, and in 1965, Jim Clark piloted his Lotus 25 to victory, due in part to the amazing light weight of the vehicle which required only one change of tires throughout the race instead of the customary three. But the other credit for his victory goes to the high fuel efficiency of the engine, 7 mpg rather than the normal 4 mpg, which meant two gas stops rather than the traditional three or four. Lotus finished not only first but also second, fifth, and eighth!

Colin Chapman's main imprint on Grand Prix racing was yet to come. In 1966, the formula changed, and the engines with which Lotus had won now had to be reengineered. Ford came through again by sponsoring two ex-Lotus employees, Keith Duckworth and Mike Costin, to design a 400-hp engine for a Ford V-8 engine block. The engine, called the Cosworth DFV (double-four valve), was a superior engine for many reasons. First, it was designed and the engine block cast with racing in mind. It had all of the auxiliary apparatus designed into the block rather than hung on afterward. It was lightweight yet designed to carry stress and torsion, which made it perfect as a "load-bearing member" in a monocoque chassis design. The four-valve concept was essentially a doubling of the number of intake and exhaust valves on a cylinder, from one each to two sets. This made the engineer a better breather, improved its volumetric efficiency, and increased overall power by twelve percent.

The Cosworth engine achieved such great success that Chapman realized his fortune could be made by selling this engine to other racing teams. Indeed, many tried the engine and achieved success to the point that twenty years later, the Cosworth engine, modified many times from the initial design, was still a consistent winner on the Grand Prix circuit. Today, many races are competed with the same engine, a striking contrast to the completely unique entries of the past.

On a sad note, the superiority of the Cosworth engine took away much of the thermodynamics of Grand Prix racing. Most of the power plants of the vehicles were standardized on the Cosworth engines, and races were won in the 1980s and 1990s more by the aerodynamic configuration and driver expertise than the creative ability of the engine designer. The careers of many popular race drivers such as Roger Penski, Johnny Rutherford, Al Unser, Danny Sullivan, and others were made behind, or rather in front of, a Cosworth engine.

One innovation during the Cosworth years that did not come from the Cosworth team (although they quickly incorporated it) was the development of the turbocharger, a new type of supercharger. Conventional superchargers used the power of the driveshaft to perform the compression of inlet air gases. The power requirement decreased the available net WORK to the wheels. Therefore, although the supercharger could almost double the horsepower of the engine, some of this was used to drive the supercharger itself.

SAMPLE PROBLEM 15–10

Sample Problem 15–4 illustrates a supercharger that doubles the net WORK of the engine to 1.42 Btu/cycle and hp to 16.7, with a boost of 29.4 psia. What is the WORK required to perform this boost (per cycle) if the efficiency of the supercharger is 50%? What is the actual available hp of the engine when this WORK is subtracted from the net WORK?

SOLUTION:

Assuming the supercharger compression is adiabatic, then the volume of atmospheric air that the supercharger compresses per cycle is

$$V_{\text{free air}} = V_1 \left(\frac{P_1}{P_{\text{atmospheric}}} \right)^{\frac{1}{k}} = .042 \left(\frac{29.4}{14.7} \right)^{\frac{1}{1.41}} = .0686$$

The WORK done during the adiabatic compression is

$$WORK = \frac{(29.4 \times 144 \times .042 - 14.7 \times 144 \times .0686)}{(778 \times (-.41))} = .102 \text{ Btu}$$

Since the supercharger is only 50% efficient, it actually uses twice this net WORK (.2 Btu/cycle). Reducing the theoretical net WORK by this amount gives 1.42 − .2 = 1.22 Btu, or hp = 14.38, which is a 13% reduction.

The new supercharger design that emerged, called the turbocharger, performed the compression with no cost to the net WORK by utilizing the energy in the waste heat of the exhaust. In a way, this is the guarantee of the Carnot cycle: wherever there is a temperature source and a temperature sink, WORK can be done between them. Race cars at the time typically exhausted at over 2000°F, and a Carnot engine could run with that exhaust temperature.

SAMPLE PROBLEM 15–11

What is the efficiency of a Carnot turbocharger running off of 2000°F waste heat?

SOLUTION:

$$\eta = \frac{(T_3 - T_1)}{T_3} = \frac{(2460 - 530)}{2460} = .78 = 78\%$$

Although actual turbochargers do not achieve the Carnot efficiency, they easily provide the supercharger boost and restore the 4.5% lost horsepower. Most Formula One supercharged vehicles today are actually turbocharged.

15.7 PRESENT-DAY ENGINES

In 1994, something very nostalgic happened at the Indy track. The push rod, two-valve, V-8 engine was still written into the formula to compete against the high-speed, multivalve, overhead cam engines that were so popular, but seldom was the push rod engine a factor in a race. The sanctioning committee quietly increased the edge of the push rod engine in the fall of 1993 by allowing a displacement of 3.43 liters compared to the overhead cam 2.65 liter, and an

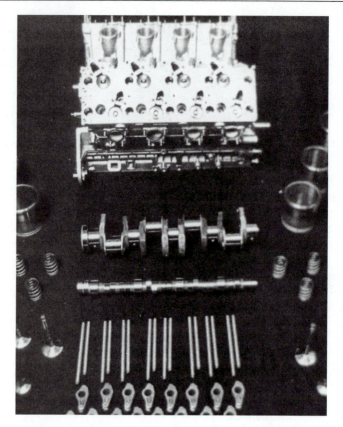

Figure 15–14. The Ilmor-Benz push rod engine.

increase in the amount of supercharger boost to 55 inches compared to 45 inches. The British team of Ilmor Engineering and Mercedes-Benz took the challenge to use the advantage to compete using this old-fashioned technology. They completely redesigned the V-8 engine using the latest in materials technology, employed a Garrett turbocharger, used computer modeling to reshape the cam lobes, and added fuel injection and a computerized energy-management system. (Figure 15–14) Once completed, the engine developed 1024 hp revving at 10,000 RPM and competed against engines that revved to 13,500 RPM but developed only 815 hp.

Mercedes entered three push rod, V-8 vehicles at Indy in 1994, driven by Paul Tracy, Emerson Fittipaldi, and Al Unser, Jr. During the race, Tracy's turbocharger failed and Fittipaldi crashed, but Al Unser, Jr., swept the field to win. Some said it was a clever manipulation of the rules, and in the off-season, the sanctioning body reacted with a reduction of the boost pressure to 48 inches. In 1995, the "all-American" push rod V-8, with a British twist, could not compete.

CHAPTER **15**

PROBLEMS

1. Using the specifications in Section 15–1 for the 1895 winning race car, calculate the thermodynamic state points of the cycle involved and compare them to those presented in that section.

2. Otto cycle calculations such as we have done assume that the combustion of gasoline occurs instantaneously. In real engines, combustion occupies a finite time interval. Ignition, therefore, begins slightly before top dead center, and combustion is not complete until the piston has moved beyond top dead center. Which of the P-V diagrams below best represents this real effect? (Figure 15–15)

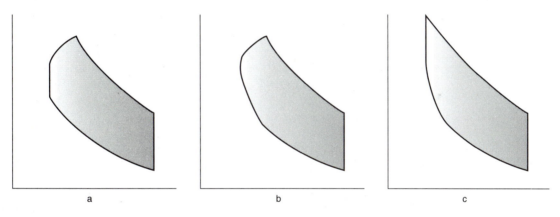

Figure 15–15. Problem 2.

3. According to the results from Problem 2, and by analyzing the P-V diagram only, tell whether the WORK put into the compression is greater or less than for the ideal case. Repeat for efficiency.

4. How much gasoline (gal) was burned in one hour in the 1895 Grand Prix winner if the heating value of the fuel was 100,000 Btu/gal?

5. It is interesting that the replacement of cast iron engine blocks and cylinder heads with comparable aluminum parts allowed engine designers to use higher compression ratios. This doesn't seem possible, since surely cast iron is a stronger metal than aluminum to withstand the

higher internal pressures. But when cast iron gets hot inside the cylinder head, it expands evenly and will crack. If the 1895 winner could have raised his compression ratio to 6.5, what would the vehicle efficiency have been?

6. Aluminum cylinder heads conduct heat out of the cylinder much faster and do not allow large temperatures to build. Therefore, they do not have the problem of temperature-related stress. But on the other hand, it is difficult to consider processes 1–2 and 3–4 as adiabatic. They can be described as polytropic processes. If the 1895 winner used an aluminum engine with a compression and expansion stroke polytropic index of 1.15, compute the state points of that cycle, Net WORK, and efficiency.

CHAPTER

16

Cycle Analysis by Gas Tables: The Gas Turbine

PREVIEW:
Calculations necessary to perform a complete analysis of a thermodynamic cycle are long and tedious. Most thermodynamic designers use a different method to drastically cut down on calculations, a method that is actually more accurate. This method is based on precalculated tables of thermodynamic data: the gas tables. This procedure is applied to cycles that introduce the gas turbine engine, a device used in power plants for the generation of electricity.

OBJECTIVES:
❑ Use carefully constructed tables of thermodynamic data rather than equations to make process and cycle analysis easier.
❑ Apply tabular calculation methods to the Brayton cycle and to sophisticated modifications to this cycle that increase net WORK.

16.1
THE POWER OF THERMODYNAMICS

If there is one fact that explains the importance of thermodynamics it is that nothing moves in today's modern world without the utilization of heat. Cars, boats, and airplanes are just some of the machines that employ the principles of thermodynamics to create their motive power. The "super fast" trains of France and Japan owe their speed to thermodynamics. Aircraft manufacturers seek out the best of the thermal engineers, and the same is true of automotive manufacturers, food processors, metal formers, and air-conditioning manufacturers. People are movers, too. We walk, crawl, and run with the aid of the Btu and the useful conversion of it to WORK.

Electric motors, the most prolific movers of all, also have their roots in thermodynamics, even though this doesn't appear to be so. There are no boilers, no burners, and no pistons moving in a fan motor or in a motor that pumps water out of a dishwasher. The source of this power is further downstream from the plug in the wall, all the way back at the generating plant. Electricity is a pure form of thermodynamic WORK, and electrical power plants operate by exciting and creative thermodynamic processes, all of which are described in the previous chapters of this text.

The basic elements of an electrical power plant and distribution system can be seen in the flow chart in Figure 16–1, which illustrates the process in reverse—from the user back to the source. The electrical generator is the mysterious element that creates the electrical current. This device is essentially an electric motor in reverse. A shaft that carries a heavy magnetic core is rotated, and with the rotation, the magnetic field cuts across bundles of heavy wire coils. This effect pushes electrons in either an alternating or direct current sense, depending on the configuration of the device. It takes a great deal of power to maintain the motion of the rotor. For example, one kilowatt of continuous electricity requires about 1.34 hp of power turning on the shaft. This power comes from an engine much like an automobile engine, or possibly a continuous device (see Chapter 13). In either case, the rotating WORK is created in a heat engine from a thermodynamic cycle. In a power plant, this engine is called a *prime mover*, whether it be coal-fired, oil-fired, or powered by natural gas or nuclear power.

It is no wonder that back at the power-generating station, the emphasis is on thermodynamics more than electricity. Prime movers that operate on the principles of pistons and cylinders would be ineffective in creating the magnitude of WORK needed in a power plant. They would be too large and inefficient for such a job. Rather, steam turbines or gas turbines, forms of the continuous cycle, are most often employed. Turbines and rotating compressors can rotate at one hundred times the speed of a reciprocating engine and, thereby, can produce tremendous power from an extremely small package or frame. These high-speed components are not without penalty, however, since there are limits to which they can be pushed—limits of stress and efficiency.

The gas turbine can be easily studied with the tools that have been developed in this text. It is a relative newcomer to the power-generation field. Gas turbines have been the subject of experimental development for decades, but their performance and cost have made them commercially acceptable only in recent years. The research has focused on one of the basic principles of the Carnot cycle: for best efficiency, source temperatures must be high. These high temperatures create high pressures in gases that are the working medium for the engine and push against the rotating machinery. Material for turbine

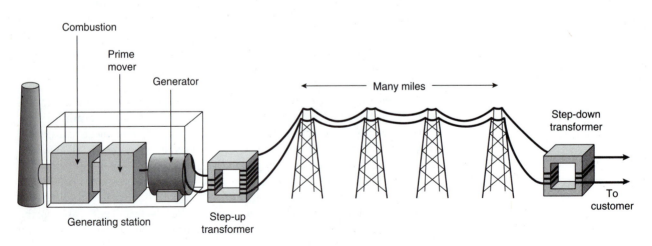

Figure 16–1. Elements of a power-generating and distribution system.

blades that can withstand these temperatures, in excess of 1800°F, has only recently been developed. The loss in power in converting high-velocity gases into rotating power also has meant a series of component redesigns in order to achieve high (80 to 90%) compressor and turbine efficiencies.

Leaving these design matters to the technicians who study "turbomachinery," we can fully understand the thermodynamics behind the unique cycle that generates power for gas turbines through a straightforward application of thermodynamic principles.

16.2 THE BRAYTON OPEN FLOW-THROUGH CYCLE

The gas turbine is unique from most cycles in this text, since it is an open, flow-through cycle. As Figure 16–2 indicates, it receives a continuous supply of fresh air that becomes the working medium. This air is exhausted directly back into the atmosphere, and not returned into the system. There are three components to this device, each representing its own thermodynamic process. Most of the cycles previously studied involve four processes, but due to the flow-through nature of this system, one of the processes is eliminated.

Air is first compressed at the inlet by a rotating compressor fan. The flow through the compressor is so fast that no heat escapes even though the air temperature and pressure rises. The compressor air then goes to a combustion chamber, where fuel is burned, raising the temperature and increasing the specific volume at constant pressure. The gas enters the power turbine at high pressure and expands through the turbine blades, rotating the turbine impeller to generate useful WORK. This expansion takes place so fast that again the process is adiabatic, with no heat loss.

The cycle can best be described in graphic form. Referring to Figure 16–3, air existing at state 1 in the atmosphere is compressed to state 2 where it is delivered from the compressor to the combustion chamber. The compression WORK is represented by the area filled with left-sloping lines. Between points 2 and 3 is the combustion process. The amount of fuel burned in the combustion chamber is such as to produce the maximum temperature of the cycle. The upper limits run between 1500°F and 2300°F in present turbine practice. The products of combustion, now at P_3 and T_3, are expanded in the turbine to

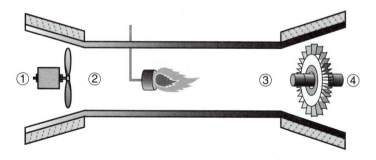

Figure 16–2. The Brayton open flow-through cycle.

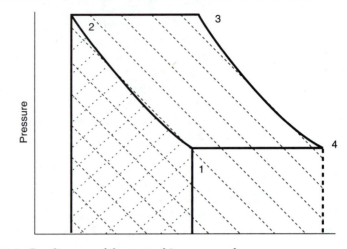

Figure 16–3. P-v diagram of the gas turbine open cycle.

state 4, which is the discharge state. From this diagram, it can be seen that P_3 = P_2 and P_4 = P_1. The processes 1–2 and 3–4 are adiabatic.

The WORK output from processes 2–3 and 3–4 is shown in right-sloping lines on Figure 16–3. Notice that the WORK out is much greater than the WORK in, thus qualifying this cycle as one that generates net WORK. Notice also that in order to close the cycle on the P-v diagram, another process would be needed, that from state 4 to 1. The gas turbine cycle is more accurately called the Brayton cycle. It exhausts hot air and combustion by-products directly into the atmosphere, and no final cooling process is necessary.

SAMPLE
PROBLEM 16–1

Find the state points, net work, and heat input for the gas turbine with the following specifications:

   Pressure ratio (P_2/P_1) = 5.00
   Conditions at state 1 (Figure 16–3) are STP
   Combustion chamber temperature (T_3) = 1400°F
   k = 1.4 assumed constant with temperature

SOLUTION:

1–2 Adiabatic Compression

   R = 53.3, T_1 = 530°R

   P_1 = 2116 psfa

   $v_1 = \dfrac{RT_1}{P_1} = 13.4 \text{ ft}^3/\text{lb}_m$

$$T_2 = T_1 \left(\frac{v_1}{v_2} \right)^{k-1} = 840°\,\text{R}$$

$$P_2 = 5P_1 = 5 \times 2116 = 10{,}580 \text{ psfa}$$

$$v_2 = \left(\frac{P_1}{P_2} \right)^{\frac{1}{k}} v_1 = \left(\frac{1}{5} \right)^{\frac{1}{1.41}} 13.4 = 4.25 \text{ ft}^3$$

$$Work = \frac{(P_2 v_2 - P_1 v_1)}{J(1-k)} = -53 \text{ Btu/lb}_m$$

$$Wd_{1-2} = k\frac{(P_2 v_2 - P_1 v_1)}{J(1-k)} = -74.2 \text{ Btu/lb}_m$$

2–3 Constant Pressure

$$v_3 = v_2 \left(\frac{T_3}{T_2} \right) = 9.4 \frac{\text{ft}^3}{\text{lb}_m}$$

$$P_3 = 10{,}580 \text{ psfa}, \qquad q_{2-3} = c_p (T_3 - T_2) = 245 \text{ Btu/lb}_m$$

$$T_3 = 1860°\text{R}, \qquad WORK = \frac{P(v_3 - v_2)}{J} = 69.3 \text{ Btu/lb}_m$$

$$q_{2-3} = c_p (T_3 - T_2) = 245 \text{ Btu/lb}_m$$

$$WORK = \frac{P(v_3 - v_2)}{J} = 69.3 \text{ Btu/lb}_m$$

$$WD_{2-3} = 0$$

3–4 Adiabatic Expansion

$$P_4 = 2116 \text{ psfa}, \qquad v_4 = v_3 \left(\frac{P_3}{P_4} \right)^{\frac{1}{k}} = 29.7 \text{ ft}^3/\text{lb}_m$$

$$T_4 = T_3 \left(\frac{P_4}{P_3} \right)^{\left(\frac{k-1}{k} \right)} = 1175°\,\text{R}$$

$$Work = \frac{(P_4 v_4 - P_3 v_3)}{J(1-k)} = 117 \text{ Btu/lb}_m$$

$$Wd_{3-4} = k\frac{(P_4 v_4 - P_3 v_3)}{J(1-k)} = 164 \text{ Btu/lb}_m$$

4–1 Constant Pressure

$$Work = \frac{P(v_1 - v_4)}{J} = -44 \text{ Btu/lb}_m$$

$$WD_{4-1} = 0$$

$$q = c_p(T_1 - T_4) = .24(530 - 1175) = -155 \text{ Btu/lb}_m$$

Cycle Summary:

$$Net \ work = -53 + 69 + 117 - 44 = 89 \text{ Btu/lb}_m$$

$$Efficiency = \frac{89}{245} = .36 \text{ or } 36\%$$

$$Net \ wd = -74 + 0 + 164 - 0 = 90 \text{ Btu/lb}_m$$

$$q_{1-2} = 0, \quad q_{3-4} = 0$$

$$q_{2-3} = 245 \frac{\text{Btu}}{\text{lb}_m}, \quad q_{4-1} = -155 \text{ Btu/lb}_m$$

16.3 ANALYSIS BY GAS TABLES

The analysis of the Brayton cycle in Sample Problem 16–1 can be made more easily with fewer calculations through the use of a table created in 1936 by Kennan and Keyes (see Appendix I). This table is a creative method of tabularizing the equations of Table 9–1 to eliminate the need to make such calculations. The table is particularly useful for continuous processes where the energy equation is readily written in terms of WORK delivered (wd) rather than WORK done by the gas.

In Section 13–6, the energy equation was rewritten in terms of WORK delivered:

$$q = c_v(T_2 - T_1) + W_D - \frac{(P_1v_1 - P_2v_2)}{J}$$

Equation 16–1

By rearranging terms:

$$q = c_vT_2 + \frac{P_2v_2}{J} - \left(c_vT_1 + \frac{P_1v_1}{J}\right) + W_D.$$

The term $c_vT + Pv/J$ is conspicuous in this equation and represents both the thermal or internal energy of the gas (c_vT) and the pressure energy or flow energy (Pv/J). This equation clearly demonstrates the two types of energy that are associated with the working medium itself. It is convenient to lump these two types of energy together and give them a name. That name is enthalpy, designated by the letter h.

❖ **KEY TERM:**   **Enthalpy (h):** The sum of the total internal energy and the total flow energy of a gas at a specific state; $h = c_v T + Pv/J$

Enthalpy has units of Btu/lb$_m$ and measures the energy inherent in the gas itself. In terms of enthalpy, the energy equation is written

$$q = h_2 - h_1 + W_D$$

Equation 16–2

Writing the energy equation in this way shows enthalpy as an important parameter, and since enthalpy is determined only by the state the gas is in, Kennan and Keyes included it as one of the important properties on the table in Appendix I. The table is easiest to use to find the enthalpy if the temperature of the air is known. The temperature is in the left-hand column of the table, and the row of data represents the state of the gas.

The table columns P_r and v_r are helpful indicators. These are called the "adiabatic pressure ratio" and the "adiabatic volume ratio." They are useful in determining the state of a gas *after an adiabatic process* has taken place. To find the state after an adiabatic process, multiply v_r by the specific volume compression ratio of the process (or P_r by the pressure ratio) and find the corresponding value of v_r (or P_r) on the table. Simply said, *for an adiabatic process:*

$$P_{r2} = P_{r1}\left(\frac{P_2}{P_1}\right), \qquad v_{r2} = v_{r1}\left(\frac{V_2}{V_1}\right)$$

Equations 16–3

In Sample Problem 16–1, the first process involved a pressurization of 5:1 adiabatically. The P_r of the original state was 1.29, therefore, the P_r of the final state is

$$P_{r2} = P_{r1}\left(\frac{P_2}{P_1}\right) = 1.29 \times 5 = 6.45$$

This corresponds on the table to a temperature of 839°R, which agrees with the sample problem.

By finding the enthalpy of the gas (air in Appendix I) before and after a thermodynamic process, much can be learned about the heat transferred (q)

and the work done. Specifically, the following relationships (Table 16–1) are a direct result of Table 13–1.

TABLE 16–1

RELATIONS INVOLVING ENTHALPY FOR OPEN PROCESSES

$q = h_2 - h_1$ if the process is constant pressure
$Wd = h_1 - h_2$ if the process is adiabatic
$Wd = work_{gas}$ if the process is constant temperature

SAMPLE PROBLEM 16–2

Reconstruct the solution to Sample Problem 16–1 using enthalpy and Appendix I.

SOLUTION:

An abbreviated form of Appendix I is presented below using only information that is important for this problem. To the right is the format for a table of state points that should be used for simplified cycle analysis—a place where the critical information from Appendix I may be tabulated.

Abbreviated Table from Appendix II

T	h	P_r	
°R	Btu/lb		
520	124.3	1.215	
(**530**	126.5	1.306	State 1)
540	129.1	1.386	
640	153.1	2.514	
740	177.2	4.193	
840	201.6	**6.573**	State 2
1220	296.4	**25.53**	State 4
1860	466.1	**129.954**	State 3
1220	296.4	25.53	
1860	466.12	129.95	

Table of State Points Created for Sample Problem 16–2

State	T	h	P_r
1	530	126.5	1.306
2	840	201.6	6.573
3	1860	466.1	129.95
4	1223	297.4	26.0

State 1 is known from its temperature, so we can pick it up directly from the table. Going to state 2 is an adiabatic process, so we can use the P_r column. The pressure ratio is 5:1, so the P_r of state 2 will be 5 times P_r:

$$P_{r2} = P_{r1}\left(\frac{P_2}{P_1}\right) = 5 \times 1.29 = 6.5$$

This P_r corresponds to the state at 840°F (see Appendix I). State 3 occurs at a temperature of $T_3 = 1860$, which defines that point on the table. State 4 is an adiabatic process from state 3 with a pressure ratio of 1:5 or

$$P_{r4} = P_{r3} \times \frac{P_4}{P_3} = 129.9 \times \frac{1}{5} = 26.00$$

To finish the analysis:

$$q_{add} = q_{2-3} = h_3 - h_2 = 264.5 \frac{\text{Btu}}{\text{lb}_m}$$

$$Wd_{12} = h_1 - h_2 = -75 \frac{\text{Btu}}{\text{lb}_m}$$

$$Wd_{23} = 0$$

$$Wd_{34} = h_3 - h_4 = 168.7 \frac{\text{Btu}}{\text{lb}_m}$$

$$Net\ work = Wd_{12} + Wd_{23} + Wd_{34} = -75 + 168.7 = 93.7 \frac{\text{Btu}}{\text{lb}_m}$$

$$Efficiency = \frac{Net\ work}{q_{2-3}} = \frac{93.7}{264.5} = .35$$

The results of Sample Problem 16–2 do not quite agree with those of Sample Problem 16–1. Which one is more correct? The table method is more accurate because it takes into consideration the variability of specific heat with temperature. We have utilized the value $c_v = .17$ Btu/lb$_m$ – °F throughout this book. However, c_v is actually not a constant and changes slightly at different temperatures. In fact, at 1400°F, the value of c_v is .21 Btu/lb – °F for air. Whereas the previous calculations do not take this into consideration, the tables do and, therefore, are more accurate.

**16.4
COMPRESSORS
AND TURBINES**

Thousands of engineers and scientists in schools of engineering and well-known manufacturing companies within the automobile, aerospace, and power utility industries have contributed to the development of rotating compressor fans and turbines used in the gas turbine cycle. This research area is the study of turbomachinery.

A compressor or fan in a gas turbine (see Figure 16–2) must perform two functions: first, it must pressurize, and second, it must move air through the system. These are counteracting goals. The higher the pressure ratio, the less the compressor can move the flow. In effect, the compressor must move gas against the pressure that it creates. Some fans, such as a household fan, can move great amounts of gas but cannot build up much pressure. Other fans can build up great pressures but do not move much air. It depends on the pitch, size, and number of the fan blades.

Aerodynamic and mechanical inefficiencies increase the power to drive the fan or compressor. For example, in Sample Problem 16–1, 53 Btu of WORK per lb$_m$ of air must be flowing to compress the gas. In actuality, however, the amount of energy needed to drive the fan might be twice that amount, since the compressor can be only fifty percent efficient.

Much of the inefficiency of a compressor is due to the friction and separation of the air as it passes over the blades. Turbulence and eddy currents in the airstream must be minimized to reduce inefficiency. Separation and recirculation can build to be so great at some speeds that a condition called surge may occur, and air may actually spill out the fan instead of going through. The first operation models of a gas turbine generator typically had compressor efficiencies below eighteen percent.

SAMPLE PROBLEM 16–3

Calculate the actual input WD of the gas turbine of Sample Problem 16–1 if the compressor efficiency is 70%.

SOLUTION:

The theoretical compression WORK delivered was calculated to be wd_{12} = –75 Btu/lb$_m$.

$$Compressor\ efficiency = \frac{wd_{12}}{WORK_{INPUT}} = .7$$

$$WORK_{INPUT} = \frac{wd_{12}}{.7} = \frac{-75}{.7} = -107 \frac{Btu}{lb_m}$$

Most compressors are axial flow compressors where the gas enters in the shaft direction (radially) but is thrown outward by the centrifugal force that performs the compression. Therefore, the gas does not travel through the compressor but instead exists around the outside disk. This change in direction is actually beneficial because it means that the next process, the combustion process, takes place on the periphery, far away from the central rotating shaft. This is ideal because the rotating shaft is not designed to take the high temperatures created in the combustion chamber.

Although the turbine process is almost the reverse of the compression process, the turbine is designed quite differently from the compressor. The turbine extracts energy from the expanding gases that flow from the combustion chamber, converting it into shaft horsepower to drive the compressor and provide some net WORK. On many gas turbines used for power generators, the turbine has a radial flow rather than an axial flow. Figure 16–4 shows this piece of turbo-machinery as having vanes mounted on a round disk. The figure shows that high-velocity gas exiting the combustor is projected against the turbine blades, imparting momentum to the rotating blade. Again, the conversion of pressurized flow to rotary motion can have inefficiencies.

SAMPLE PROBLEM 16–4

If the turbine of Sample Problem 16–1 were 60% efficient, how much of the 168.7 Btu/lb$_m$ of WORK output actually is available?

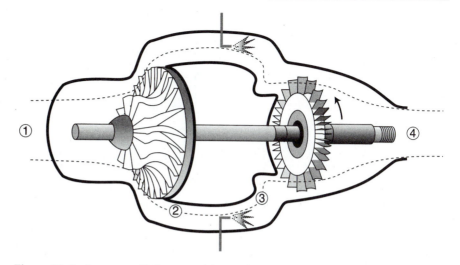

Figure 16–4. A more realistic gas turbine engine.

SOLUTION:

$$Efficiency = \frac{WORK\ available}{WD\ output}$$

$$.60 = \frac{WORK\ available}{168.7}$$

$$WORK\ available = 101.\ 2\ \frac{Btu}{lb_m}$$

16.5 COMBUSTORS

High-efficiency combustion chambers or combustors are crucial to the success of the gas turbine cycle. Problems arise from the nature of the combustion process itself. Combustion is a chemical reaction between oxygen and an organic substance called fuel. In order for this process to proceed in a proper manner, two factors must be considered. First, the proper amounts of fuel and air (which supplies the oxygen) must be brought together in the combustion chamber and mixed completely. In other words, the fuel-air ratio must be correct.

Mixing of fuel and air is important so that the chemical elements make good contact. The air is supplied as the working medium that flows through the cycle. Fuel is injected into the air in the combustion chamber by means of a burner head, either a single- or multiple-injector design. Burning takes place in a moving fuel-air stream, which means that the flame can be blown out if the velocity becomes too high. Maintaining the flame in the combustion chamber is called flame stabilization. (Figure 16–5)

To stabilize the flame, a baffle is often placed in front of the burner head. An even more sophisticated design is to completely encase the flame in a tube

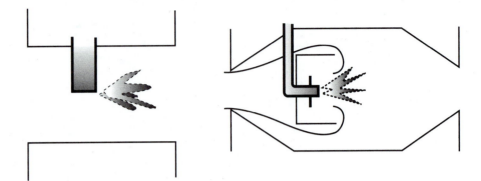

Figure 16–5. An unstable flame and a baffled injector.

called a flame tube. Through side wall ports, just enough air is fed to the flame to keep it properly fired. Temperatures inside get quite hot, so additional air is diluted into these burned gases downstream of the flame before the gas exits the combustion chamber. Typically twenty-five percent of the working medium feeds the flame while seventy-five percent of the air mixes with the gases after combustion.

SAMPLE PROBLEM 16–5

Consider that for the case of Sample Problem 16–1, a flame tube pulls one-fourth of the air input through the flame. What is the temperature of the gases in the flame tube after combustion?

SOLUTION:

In Sample Problem 16–1, it was calculated that 264.5 Btu of heat were added per pound of air flowing through the system. In the flame tube, this heat is released into one-fourth of this air; therefore, the flame tube gas absorbs 1058 Btu/lb$_{air}$. This constant pressure burning is governed by the equation:

$$q_{2-3} = c_p(T_3 - T_2)$$

$$T_3 - T_2 = \frac{1058}{.24} = 4408°R$$

or

$$T_3 = 4408 - 840 = 3572°R$$

With proper design of combustion chambers, combustion efficiency (defined as the amount of heat potential delivered divided into the amount of useful heat delivered to the working medium) is often as high as ninety-nine percent.

**SAMPLE
PROBLEM 16-6**

If the combustion efficiency of the gas turbine from Sample Problem 16–1 is 98%, how many Btu's of fuel are burned in the combustion?

SOLUTION:

We can define

$$Combustion\ efficiency = \frac{Heat\ output}{Heat\ input}$$

But the required heat output is 264.5 Btu/lb$_{air}$. So

$$.98 = 264.5\ \frac{Btu/lb_m}{Heat\ input}$$

or

$$Heat\ input = \frac{264.5}{.98} = 270\ Btu/lb_m$$

Now that the individual components of a gas turbine engine have been investigated, Figure 16–6 shows these components in an actual gas turbine engine. The engine shown is unique in that the gas does not go straight through the engine but goes through the compressor from left to right, then is transferred to the combustion chamber at the far right, where it changes direction to go through the turbine section from right to left and exhausts from the center of the engine.

**16.6
THERMAL
REFINEMENT OF
THE GAS TURBINE:
REGENERATION**

If practical gas turbine cycles have efficiencies as low as forty percent, they may waste fuel. Still, the cycle has advantages in weight, size, and vibration when compared to Otto and Diesel engines, and it has size and cost advantages over steam-operated engines. The efficiency handicap is surmountable at the expense of adding complexity to the power plant. These refinements are clever thermodynamic tricks. The principal ones are *regeneration, intercooling,* and *reheating.*

Regeneration is the transfer of heat energy from exhaust gases back into the system. Every engine designer at one time or another has looked longingly to the hot exhaust gases and wondered why they must be wasted. Why not take these gases and insert them directly into the intake and use this already hot working medium?

Figure 16–7 illustrates why this cannot be done. It is obviously impossible to directly recompress hot gases to the required combustion chamber pressures (state 4 to 3) because it costs so much in terms of input WORK that the turbine cannot keep up. Diverting the exhaust gases directly back to the intake serves no useful purpose.

Instead, consider trying to use the heat of the exhaust gas to provide some of the heat in the combustion chamber. Suppose the high-temperature, low-

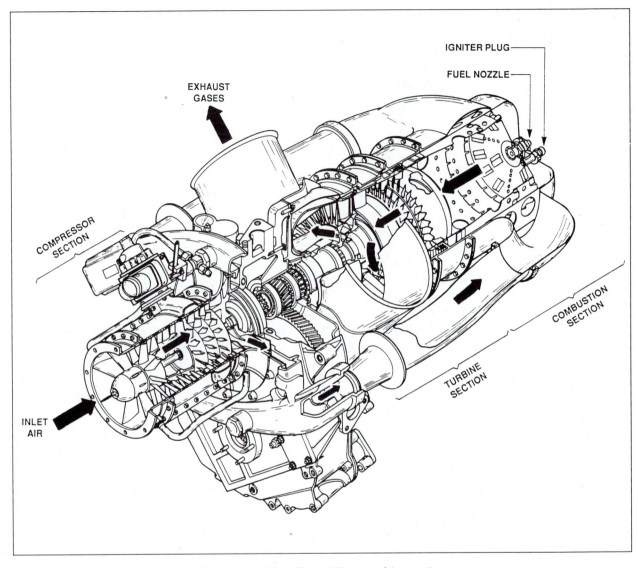

Figure 16–6. The Allison 250 gas turbine engine.

pressure exhaust gas (HTLPXG) is used to preheat the low-temperature, high-pressure gas entering the combustion chamber (LTHPCG). Is this possible? Would it be beneficial?

Beneficial it would be, since adding free heat at state 2 involves no additional WORK to the cycle and is, in fact, just what the original cycle calls for. Is it possible? The answer is maybe. Certainly the exhaust gas cannot be inserted directly into the combustion chamber because there is a mismatch of pressure. Instead the heat must be exchanged through the combustion cham-

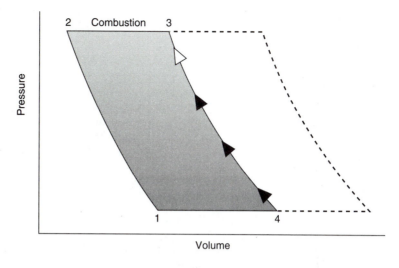

NOTE: Recompressing exhaust gases directly
does not start the cycle over again

Figure 16–7. Recompressing exhaust gases directly does not start the cycle over again.

ber walls (or rather the precombustion chamber walls). Figure 16–8 shows two
ways in which this can be done and illustrates when this heat exchange is
possible.

If the exhaust gas at state 4 is hotter than the combustion chamber entry
gas at state 2, then preheating of the combustion gas is possible. Sample
Problem 16–1 illustrates a case when such exhaust gas heating is possible. The
temperature of the exhaust gas is 1223°R, sufficient to warm the 840°R combustion chamber inlet air. The process of utilizing some or all of the exhaust
heat is called regeneration.

A most important question now can be answered. How much of the
combustor heat can be obtained from the exhaust gases? Figure 16–9 presents
a close-up of the regeneration heat exchanger. Each of the gases flow in the
same direction, and it can be expected that these flows will exchange heat until
their temperatures become equal. If both streams have the same amount of
gas, as they must, then both streams will reach the numerical average of their
initial temperatures. For the case of Sample Problem 16–1, this temperature is:

$$T_{\text{AVE}} = \frac{(1223 + 840)}{2} = 1031°\text{R}$$

The amount of heat absorbed by the precombustion air is

$$q = c_p (T_{AVE} - T_2) = 45.8 \frac{Btu}{lb_m}$$

Of the total heat required, $q_{2-3} = 264.5$, this represents 17.3%. Figure 16–10 shows a P-v diagram of the cycle, indicating the portion of the combustion heat that comes from regeneration.

Another method of heat transfer is even more effective. Figure 16–11 illustrates a flow pattern in which the exhaust gases and combustor gases travel in the opposite directions. In this configuration, the hot exhaust gas

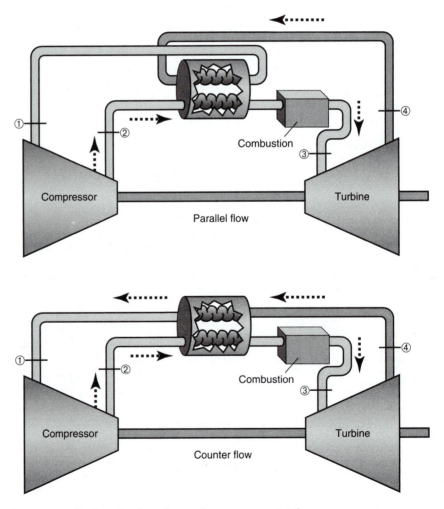

Figure 16–8. Exchanging heat from exhaust gases to intake gases.

continues to give up heat until it reaches the temperature of the *inlet* combustor gas. Similarly the precombustion chamber gases continue to warm until they reach the temperature T_4. For the case of Sample Problem 16–1, the combustor gas is preheated to a full 1223°R. This type of flow path in the heat exchanger is called a *counterflow* pattern and is more effective than the first heat exchange, which is called a *parallel flow* heat exchanger.

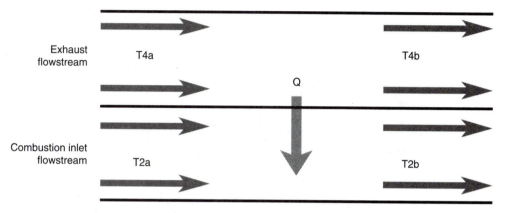

Figure 16–9. Parallel flow waste heat recuperator.

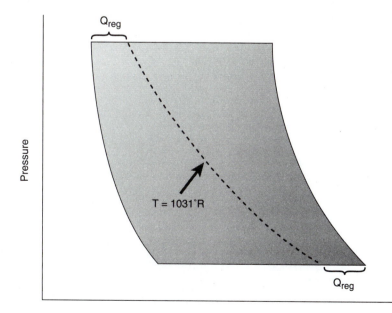

Figure 16–10. Regeneration on the P-v diagram.

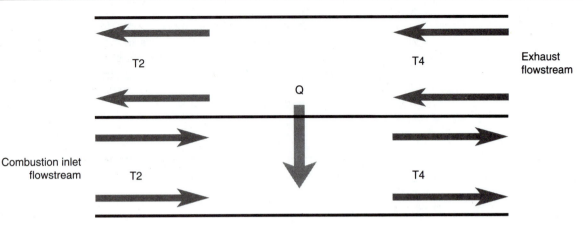

Figure 16–11. Counterflow waste heat recuperator.

SAMPLE PROBLEM 16-7

Apply the concept of a counterflow heat exchanger to Sample Problem 16–1 and determine how much heat can be transferred from exhaust gas to combustor gas in a counterflow heat exchanger. Calculate the overall thermodynamic efficiency of the cycle, considering that only a portion of Q_{2-3} is costly heat.

SOLUTION:

$$q_{\text{regenerated}} = c_p(T_4 - T_2) = 90.5 \frac{\text{Btu}}{\text{lb}_m}$$

$$Costly\ heat = q_{2-3} - q_{\text{regenerated}} = 174 \frac{\text{Btu}}{\text{lb}_m}$$

$$Efficiency = \frac{90.5}{174} = .54$$

16.7 INTERCOOLING

In some situations, major modifications to the cycle are needed to make additional improvements. In the Brayton cycle, the compression is an adiabatic one. However, a constant temperature compression will require less WORK. By changing the adiabatic compression to a constant temperature process, the cycle is changed and is called the Ericsson cycle. (Figure 16–12)

SAMPLE PROBLEM 16-8

Find the properties of an Ericsson cycle that operates on the same specifications of the cycle in Sample Problem 16–1.

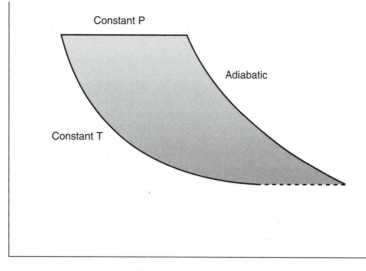

Figure 16–12. The Ericsson cycle.

SOLUTION:

This solution will be accomplished with the help of Appendix II.

We use the fact that

$$T_1 = T_2 = 530°R, \; T_3 = 1860°R, \text{ and } \frac{P_3}{P_4} = 5.000 \text{ (see Figure 16–16)}$$

Then the table of state points is:

State	T	h	P_r	v_r
1	530	126	1.30	150
2	530	126	1.30	150
3	1860	466	129.9	5.3
4	1223	297	26.0	17

The last row is found from

$$P_{r4} = P_{r3}\left(\frac{P_4}{P_3}\right) \text{ or } P_{r4} = 129.9 \times \frac{1}{5.0} = 26.0$$

Otherwise

$$P_2 = \frac{P_1 v_1}{v_2} = 2116 \times 5 = 10{,}580 \text{ psfa}$$

Then

$$q_{1-2} = P_1 v_1 \ln \frac{(P_1/P_2)}{J} = 2116 \times 13.4 \ln \frac{(1/5)}{J} = -58.4$$

$$Wd_{1-2} = q_{1-2} = -58.4$$

$$q_{2-3} = h_3 - h_2 = 340.4 \frac{Btu}{lb_{air}}$$

$$Wd_{2-3} = 0$$

$$Wd_{3-4} = h_3 - h_4 = 169 \frac{Btu}{lb_{air}}$$

$$Wd_{4-1} = 0$$

The efficiency of the Ericsson cycle is

$$\eta = \frac{111.6}{340.4} = .328$$

Notice that the Ericsson cycle did use less WORK of compression than the Brayton cycle (58.4 Btu to 75). The temperature of the air entering the combustion, however, is also lower and requires more costly heat to bring it up to 1860°F in the combustion chamber. The result is that the efficiency of the Ericsson cycle is actually lower than that of the Brayton cycle. The WORK delivered, however, is higher in the Ericsson cycle (111.6 to 93.7 Btu). Therefore, this engine has more horsepower than a Brayton cycle engine of the same size.

The Ericsson cycle does have a much greater efficiency if regeneration is included. Because of the lower combustion inlet temperatures, more exhaust heat can be reused. For example, the amount of reused heat in the Ericsson cycle of Sample Problem 16–8 with a counterflow heat exchanger is

$$q_{REJ} = .24 \times (1223 - 530) = 166 \frac{Btu}{lb_m}$$

Then

$$q_{Costly} = 340.4 - 166 = 174.4 \frac{Btu}{lb_m}$$

and

$$Efficiency = \frac{111.6}{174.4} = .64$$

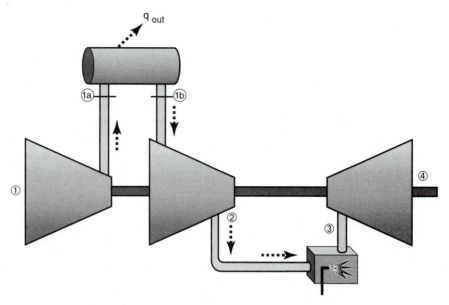

Figure 16–13. Gas turbine cycle with intercooling.

This is a nine percent improvement over the Brayton cycle of Sample Problem 16–8.

Compressors that raise the pressure of gases at constant temperature are hard to design. There must be significant cooling in the compressor to keep the temperature down. In the compressors of Section 16–4, there just isn't sufficient time for the gases to cool. One solution would be to make these compression units extremely long so the compression would take place slowly and there would be sufficient time to drive heat out of the gas. Unfortunately these units would be costly, extremely large, and probably have such a poor compressor efficiency that any benefit would be lost.

Another approach would be to perform the cooling separately from the compression process. This amounts to performing an initial compression through a conventional compressor, then ducting the gases to a constant pressure heat exchanger where enough heat is rejected to bring the gas temperature back down to its original value. From there, the gas is compressed in a second compressor (second stage) to its final value. This technique is called *intercooling* (Figure 16–13) and is an effective way of approximating a constant temperature process.

SAMPLE PROBLEM 16–9

Repeat Sample Problem 16–8 but use a two-stage compressor. (Figure 16–13) The first stage of compression has a pressure ratio of 3:1, and the second stage compression is 1.6:1. This gives an overall pressure of 5:1. Draw the P-v diagram first.

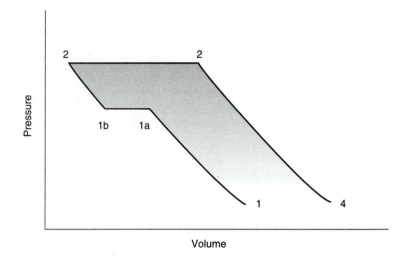

Figure 16–14. Brayton cycle with two-stage cooling.

SOLUTION:

State	Temperature	Enthalpy	P_r
1	530	126.5	1.30
1a	730	174.5	3.9
1b	530	126.5	1.30
2	603	144.5	2.08
3	1860	466.1	129.9
4	1223	297.2	25.8

Using

$$P_{r1a} = P_{r1}\left(\frac{P_1}{P_{1a}}\right) = 1.3 \times 3 = 3.9$$

Then

$$Wd_{1-1a} = h_1 - h_{1a} = -48.0 \ \frac{\text{Btu}}{\text{lb}_m}$$

$$Wd_{1a-1b} = 0$$

$$Wd_{1b-2} = h_{1b} - h_2 = -18.0 \ \frac{\text{Btu}}{\text{lb}_m}$$

$$Wd_{3-4} = h_3 - h_4 = 168.7 \ \frac{\text{Btu}}{\text{lb}_m}$$

$$Net \ work = -48 - 0 + 168.7 + 120.7 \ \frac{\text{Btu}}{\text{lb}_m}$$

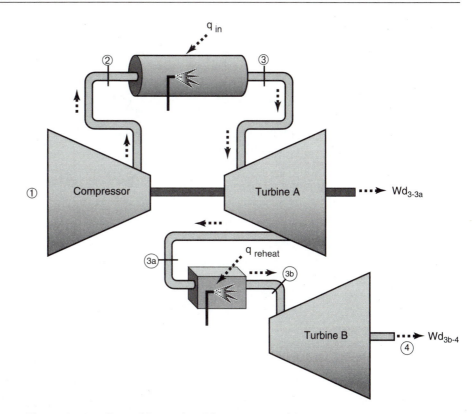

Figure 16–15. Gas turbine cycle with two-stage turbine.

16.8 REHEATING

As was true in the compression process, so it is true in the power-generating expansion process that performing it with constant temperature would create more horsepower. But maintaining a constant high temperature in the turbine compressor would require combustion within the turbine or multistaging. This could be hard on the turbine metal blades.

Therefore, just as for intercooling, an approximation is made. The gases are expanded partially, then ducted to a secondary combustion chamber where they are reheated, then ducted back to the turbine where the expansion is completed. (Figure 16–15)

SAMPLE PROBLEM 16–10

Reevaluate the engine of Sample Problem 16–1, but use reheat in the turbine (Figure 16–16) after an expansion of 3:1

$$\frac{P_3}{P_{3a}}; \ \frac{P_{3b}}{P_4} = \frac{5}{3}$$

to give overall

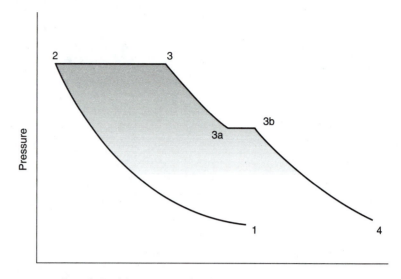

Figure 16–16. Sample Problem 16–1 with reheating.

$$\frac{P_3}{P_4} = 5$$

The reheating (3_a–3_b) continues until the temperature is increased back to T_3.

SOLUTION:

State	Temp	Enthalpy	P_r
1	530	126.5	1.30
2	840	201.6	6.57
3	1860	466.1	129.9
3a	1405	344.3	43.3
3b	1860	466.1	129.9
4	1657	409.1	81.18

$$Wd_{3-3a} = h_3 - h_{3a} = 121.82$$

$$q_{3a-3b} = h_{3b} - h_{3a} = 121.82$$

$$Wd_{3b-4} = h_{3b} - h_4 = 57$$

$$Net\ work = wd_{1-2} + wd_{2-3} + wd_{3-3a} + wd_{3a-3b} + wd_{3b-4}$$

$$= -75.0 + 121.82 + 57 = 104.8\ \frac{Btu}{lb_m}$$

$$Efficiency = \frac{104.8}{(264.5 + 121.82)} = .28\ or\ 28\%$$

Comparing Sample Problem 16–10 with Sample Problem 16–1 illustrates that the reheat engine has less overall efficiency but that the net work is increased for the same size engine, and therefore, there is more horsepower. Also, with $T_4 = 1657°R$, there is a great opportunity for regeneration.

POWER PLANT TECHNICIAN

There is a great deal of excitement at the Department of Facilities and Operations at the University of Florida in Tallahassee, Florida, these days. The university owns and operates a power plant that provides electricity for the whole UF campus. Years ago, the university bought electricity from Florida Power and Light (FPL), the electrical utility. Now they have joined a growing trend of industries and institutions that make their own power. This is due in large part to gas turbine technology, which has brought the price down for smaller power-generating stations. Whereas the university never could have considered building a nuclear plant or even a coal-fired plant on the campus, gas turbines pack a great amount of power in a small package—small enough to make it feasibile for the needs of the university. Natural gas has other advantages as a fuel. It is relatively plentiful, and it burns cleanly so that emission standards are easily met.

The University of Florida employs engineers and technicians to maintain and operate their 150-megawatt (MW) facility. Some of them operate it from

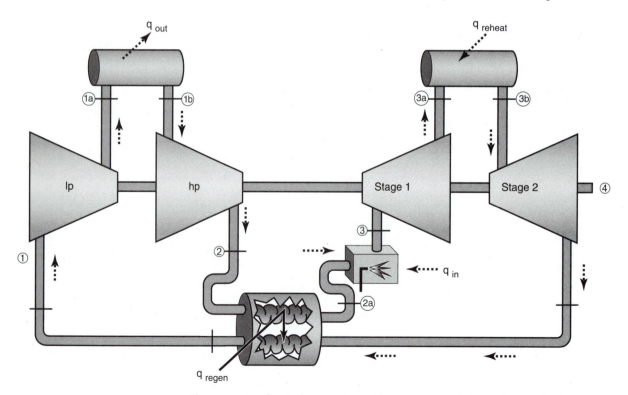

Figure 16–17. Gas turbine engine with regeneration, intercooling, and reheat.

the control room, analyzing data collected at different points of the gas turbine process to ensure maximum efficiency and sufficient power to meet electrical demands, to pinpoint potential component failures, and to monitor combustion gas emissions. Theirs is a complex job, since the three 50-MW gas turbines that they run have both compressor intercooling, which includes a low-pressure (l–p) compressor and a high-pressure (h–p) compressor, and two stages of turbine expansion with reheat in between. (Figure 16–17) One routine monitoring point of which they are continuously aware is the pressurization ratio of the l–p compressor. Fouling of the compressor blades by particulates in the entry airstream typically causes a reduction in pressurization ratio with a corresponding loss in efficiency and power. When this happens, the operating technicians alert the field service technicians to perform a compressor wash. A fine spray of water and detergent is injected into the inlet airstream to clean the compressor blades. Recently, technicians have generated specifications for an inlet air filter bank to be installed and are currently waiting for contractor bids to perform the work.

SAMPLE PROBLEM 16–11

To see the effect of reduced pressurization ratio on the performance of the Brayton gas turbine, repeat Sample Problem 16–2 using $P_2/P_1 = 4$ rather than = 5.

SOLUTION:

State	T	h	Pr
1	530	126.5	1.306
2	790	189.4	5.224
3	1860	466.1	129.95
4	1300	317	32.5

$$q_{2-3} = h_3 - h_2 = 276.7 \ \frac{Btu}{lb_m}$$

$$WD_{12} = -62.9 \ \frac{Btu}{lb_m}$$

$$WD_{23} = 0$$

$$WD_{34} = 149.1 \ \frac{Btu}{lb_m}$$

Net WORK = 87.8

$$\eta = \frac{87.8}{276.7} = .32$$

The field service technicians take the log books and service requests from the plant operators and service the many different types of equipment in the plant: pumps, controls, flow control valves, and so on. There are other technicians who also are critical to the UF power plant operation. They are specialists who work for Energy Management Consultants (EMC), a contracting firm in Orlando, Florida, that specializes in major overhauls of turbocompressors and turbines. When performance declines due to increased tolerances and changes to surface finishes and blade contours, the turbine is shut down to modify or replace compressor vanes and turbane blades. This precision work requires skilled technicians and often takes weeks to complete before the equipment is back on line. This is the reason that UF has three small gas turbines instead of one large one. Taking one-third of the power-producing capacity off-line during a nonpeak period (summer vacation) is much less disruptive than having to shut the complete system down.

A few technicians who work for the management of the facility currently are looking into the feasibility of using a different fuel than natural gas. This is a contingency measure in case the supply of natural gas is interrupted in the future. They could actually power their generators from the heat of nuclear reactors, but it would take a great deal of modification in their equipment. A better choice is synthetic coal gas, which has a lower Btu content than natural gas but can be burned with the same nozzles and burner heads that are already in place, although it would require the construction of a small coal gasification station at the plant.

Workdays are exciting at the University of Florida power plant, as technicians daily experience a technology that is literally changing as they drive to work in the morning.

CHAPTER **16**

PROBLEMS

1. If air is compressed adiabatically in a continuous process at 14.7 psia and 70°F to 120 psia, what is the final temperature and what is the WORK delivered to the gas (wd), using
    a. adiabatic equations
    b. gas tables for air
    Why might there be a slight difference between these results?

2. Consider that the Brayton cycle of Sample Problem 16–1 has been slightly modified so that the first process is constant temperature rather than adiabatic. Create the state table for this cycle as in Sample Problem 16–2 and compute the efficiency of the cycle (power generating). Require that $P_2/P_1 = 5.0$ as before and $T_3=1400°F$. Find the efficiency of the cycle. Is it better than the original cycle?

3. A gas turbine expansion section has an inlet temperature of 1200°F, and the air expands adiabatically to 400°F. Use the air tables to determine the net work produced by this turbine.

4. In an air-standard Brayton cycle, air enters the compressor at 1000 lb_m/hr, 14.7 psia, and 40°F, and it leaves the compressor at 70 psia. The inlet turbine temperature is 1440°F. Use Table 16–1 to determine the following:
    a. the temperature at the compressor outlet (°F)
    b. the work (ft-lb_f/lb_m) and power (hp) needed to run the compressor
    c. the rate of heat added (Btu/hr)
    d. the net work produced by the cycle (hp)
    e. the cycle thermal efficiency (%)

5. How much of the exhaust heat in Problem 4 is available for regeneration? What is the efficiency of this cycle with regeneration?

6. If the compression process of the cycle in Problem 4 is accomplished in two equal stages (each pressurization ratio is $\sqrt{70/14.7}$) with intercooling back to 70°F in between, determine the efficiency and net work of the cycle.

7. If the turbine expansion process of the cycle in Problem 4 is accomplished in two equal stages with reheat back to 1440°F in between, determine the efficiency.

8. One of the advancements being evaluated by University of Florida technicians is to replace their gas turbine burners with ones that can deliver higher temperatures (2350°F) from natural gas. Reevaluate Sample Problem 16–2 to see what would be the effect of this modification. In particular, determine the efficiency for the plant with this modification.

9. An aircraft is propelled by a reversible gas turbine, or turbojet engine. (Figure 16–18) The aircraft is flying at an altitude of 30,000 ft with a temperature of –20°F and a pressure of 1000 psfa. The compressor has a pressurization ratio of 16:1, and the kerosene fuel in the combustion chamber delivers 30 Btu/lb. *Using tables,*
   a. find the state points at 1, 2, 3, and 4
   b. find wd_{12}, wd_{23}, wd_{34}, wd_{41}, and q_{12}, q_{23}, q_{34}, q_{41}
   c. determine net work and efficiency

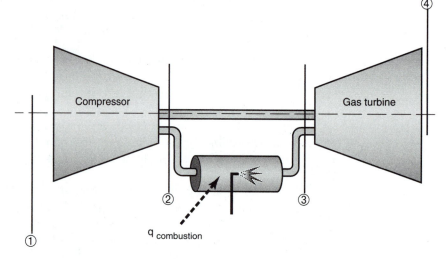

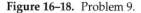

Figure 16–18. Problem 9.

10. Three lb_{mair}/sec are reversibly and polytropically ($n = 1.25$) expanded in a gas turbine from 2100°F. If the exhaust pressure is 15 psia, determine the WORK produced per lb_{mair}, the horsepower generated, and the rate of heat transfer. (Polytropic processes are not adiabatic and must be worked using the calculation method rather than the table method.)

11. Use air tables only to repeat the above problems, but assume the process to be adiabatic.

12. Problem 5 in Chapter 13 describes a novel auto air conditioner. In your homework you used conventional formulas to analyze the cycle. Now try it using Table 16–1.

    a. Using Table 16–1, prepare the state table for Problem 5 in Chapter 13.

State	Temperature	Enthalpy	vr	Pressure	Pr

    b. From the state table, find

    $q_{12} =$ $Wd_{12} =$

    $q_{23} =$ $Wd_{23} =$

    $q_{34} =$ $Wd_{34} =$

    $q_{41} =$ $Wd_{41} =$

    c. Find net WORK and efficiency

13. Sample Problem 15–8 analyzed the Granatelli Indy turbine engine. Rework this analysis using air tables (Appendix I) rather than calculations and compare the solutions.

14. The air-conditioning systems on aircraft are very similar to the system proposed in Problem 5 in Chapter 13. They cool air directly and inject

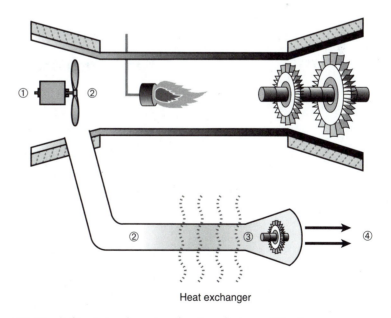

Heat exchanger

Figure 16–19. A direct simple system for aircraft air-conditioning.

the refrigerated air into the cabin. However, there is a modification from Figure 13–21 in that the first process of compression is done by the main jet engine in its normal compression process (Figure 13–17), and some air is bled off to be the compressed air for the air-conditioning system (Figure 16–19). This clean, hot, compressed air is then cooled through a heat exchanger, using outside air as the coolant, then run through a turbine expander to drop its temperature. If the bleed air enters the heat exchanger at 50 psia and 877°R, and the constant pressure heat exchanger cools the air to 615°R, what will be the resultant temperature of the air after the turbine expands out to atmospheric pressure?

15. The aircraft air-conditioning system in Problem 14 is called a simple system. A bootstrap system (Figure 16–20) has a slight modification that recompresses the cooled bleed air (state 3) to a higher temperature and pressure (state 3a) where it is fed to a secondary heat exchanger that cools the gas back to 615°R. If the compression process is required to use exactly the same amount of wd as the turbine process delivers (wd 3–3a $= -wd_{41}$), what is the state 3a and what is the conditioned air temperature T_4?

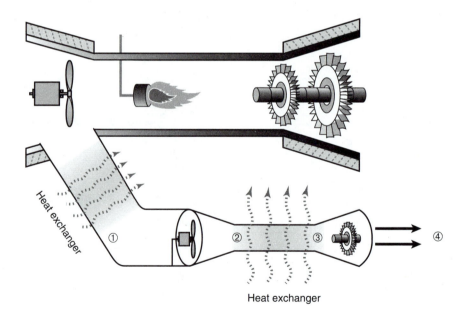

Figure 16–20. A direct bootstrap system for aircraft air-conditioning.

CHAPTER
17

Cycles in Reverse

PREVIEW: Not all thermodynamic cycles are used for power production. An air-conditioning system performs a much different function from a gas turbine, yet both are closely related. In fact, the equipment is (thermodynamically) identical, only the direction of flow of the working medium is reversed. This small difference makes for a whole new field of thermodynamics: reverse cycles.

OBJECTIVES:
- ❑ Analyze continuous cycles with the help of entropy.
- ❑ Distinguish whether a cycle proceeds in the forward (power) manner or reverse manner.
- ❑ Describe reverse cycles as devices that pump heat against its natural flow.
- ❑ Describe the primary components of a heat pump device.

17.1 INTRODUCING ENTROPY

Power generation is the *dynamics* in *thermodynamics,* and a great amount of the interest in thermodynamics deals with power cycles. The engineer's goal in power generation often is to use the lowest quality fuel in the most efficient cycle possible to generate power. Low-quality heat sources such as solar heat or dried grasses generate heat at low temperatures. High-quality fuels such as nuclear fuel, uranium, "burn" at high temperatures.

Low-quality fuels are more readily available than high-quality fuels, and for good reason. It takes nature a lot longer to create a high-temperature fuel like uranium than a lower quality fuel like coal, oil, or natural gas. Similarly, oil takes longer to create than wood, wood takes longer to grow than corn husks, and Buffalo chips and corn husks take longer to generate than collected sunlight. Table 17–1 shows the interesting relationship between the fuel source temperature (usually flame temperature) and the amount of time it takes to generate the fuel.

To generate power, both a source of heat and a sink of heat must be available. Usually the ambient conditions of the air are used as the sink. Sometimes in nature both a source and a sink are available. It is Carnot's principle that whenever a source and sink are provided, an engine can be coupled between them and WORK can be done. If the source and sink are provided by nature, a power cycle can be operated without any "cost" in terms of fuel.

An example of this, though a fleeting one, is a lightning bolt (a source) striking an iceberg (a sink). Somewhat more realistic would be a cold mountain

Table 17–1

Fuel	Maximum Temperature	Creation Time
Uranium	10,000°F	10,000 years
Coal	3000°F	1,000 years
Oil	2500°F	1,000 years
Natural gas	1800°F	500 years
Wood	1500°F	50 years
Corn husks	1100°F	1 year
Sunlight	200°F	Instantaneous

stream flowing through a Colorado valley in the middle of summer. (Figure 17–1) A black pan setting in the sun on the floor of the valley might reach 120°F while the stream next to it flows at 33°F. What a perfect opportunity to run a small machine! On what principle should this machine operate in order to get the best efficiency? Certainly it should be based on the Carnot cycle.

Figure 17–2 shows the inner workings of such a cycle, which was introduced in Chapter 13. Notice that the bottom portion of the cycle is the heat sink and must be bathed in cool water, implying that it must be in contact with the trout stream. Notice, too, that the top of the cycle must be surrounded by a high temperature bath of some sort—a source to heat the gas inside as it goes through this thermodynamic process. In the Colorado valley, this portion must be in contact with the hot air. What better application for such a device than an outboard boat motor? It would have access to both the stream and the heat of the air.

Figure 17–1. Hot and cold in a Colorado valley.

Before investigating the characteristics of such a cycle by calculating state properties and efficiencies, consider the computational techniques to be used. Since this is a continuous cycle, the calculations using gas tables shown in Chapter 16 will reduce greatly the number of calculations needed to complete the analysis. The gas tables in Appendix I, however, do have some minor drawbacks. First, they are only for air, and the medium inside the engine shown above is not air. Therefore, another table is necessary.

Another feature of the tables that can be improved upon is the use of the P_r and v_r columns to track adiabatic processes. Using this method requires only a few calculations (see Sample Problem 16–2). To eliminate even these calculations, the gas table in Appendix II uses a different approach. This table is for the gas called chlorodifluoromethane (R-22), a nontoxic, odorless, stable gas that is ideal for cycles like those we will analyze. This table eliminates the use of P_r and v_r for adiabatic processes and replaces them with a single parameter called *entropy*. This property is calculated for each state on the table, and it has the important characteristic that *all states that have the same entropy can be linked by a reversible, adiabatic processes.*

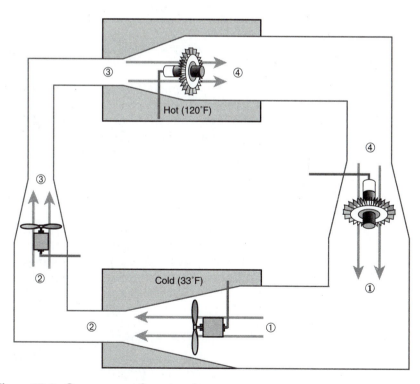

Figure 17–2. Open process Carnot cycle.

❖ **KEY TERM:**   **Entropy (s):** A property of a gas that does not change if the gas undergoes an adiabatic, reversible process.

Adiabatic processes, therefore, can be tracked on the table in Appendix II by matching states that have the same entropy. Notice that this table has a different format from that in Appendix I, since to find the proper state on this table, both the temperature and the pressure of the state must be known. The temperature determines the proper row, while the pressure determines the proper set of columns.

SAMPLE PROBLEM 17–1

The R-22 undergoes an adiabatic, reversible compression from a pressure of 40 psia and a temperature of 50°F to a final pressure of 120 psia. What is the resulting temperature? Use the adiabatic formula and compare the result to the result in Appendix II ($k = 1.16$).

SOLUTION:

$$T_2 = T_1 \left(\frac{P_2}{P_1} \right)^{\frac{(k-1)}{k}}$$

$$= (50° + 460°) \left(\frac{120}{40} \right)^{\frac{(1.16-1)}{1.16}} = 593.4°R \text{ or } 133.4°F$$

From Appendix II, the entropy for state 1 is $s = .2439$. Moving to the value of 120 psia for pressure and finding the state with the same entropy results in a temperature of 130°F ($s = .2437$). Both results are about the same, the table calculation being more accurate because the value of k for this gas over the range of temperatures in this process actually varies from 1.155 to 1.165, something that was not considered in the calculation technique.

Another inconvenience of the procedure using the Pr was with the evaluation of constant temperature processes. Sample Problem 16–8 demonstrated that there is no simple way to compute q and wd for a constant temperature process. Instead, we had to resort to the calculation method:

$$q = wd = P_1 v_1 \ln (P_1/P_2)$$

The problem of the lengthy computation for constant temperature WORK is also simplified with the concept of entropy. A property of this parameter is that *for a constant temperature process:*

$$q = wd = T\,(s_2 - s_1)$$

Equation 17–1

where T must be in °R. From this equation, it is seen that the units of entropy are Btu/lb$_m$ – °R. Entropy is quite a useful concept in eliminating lengthy calculations from cycle analysis.

17.2 APPLICATION: A CARNOT BOAT MOTOR

Colorado fishermen find that a special fishing boat motor works well in the trout streams during the summer months. It is called a Carnot troller and is purported to be an operating Carnot cycle with the sun as the heat source (Figure 17–3). One amazing thing about this motor is that it needs no gasoline or any other fuel. Rumor has it that it works using an "exotic" gas called chlorodifluoromethane.

SAMPLE PROBLEM 17–3

Evelute the Carnot trolling motor by determining the pressures and temperatures in the cycle if $P_1 = 40$ psia, air temperature is 120°F, and stream temperature is 33°F. Also find the net WORK/lb$_m$, the heat added/lb$_m$, the heat rejected/lb$_m$, and the cycle efficiency. (The overall compression ratio is $v_1/v_3 = 4$.)

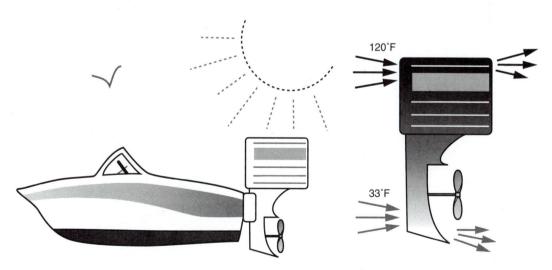

Figure 17–3. The Carnot troller.

SOLUTION:

Before beginning the calculations, we may want to investigate the mysterious chlorodifluoromethane (R-22) rumor. What is the mysterious nature of this gas? Carnot proved that there is none. Recall that he showed that the efficiency of a Carnot cycle does not depend on the nature of the working medium. The device using this working fluid, chlorodifluoromethane, is no more efficient than one filled with air.

How does the engine operate without fuel? Engines usually run on heat created by fuel, but in this case heat comes from another source. The trolling motor gets its heat from the sun and the 120°F air in the bottom of the canyon. It puts this heat to good use by having an extremely good cold temperature heat rejection sink: the 33°F stream.

The funnel of 120°F air at the top of the engine and the channel of 33°F water near the propeller are not part of the Carnot cycle itself. They represent the heat source and the heat sink that transfer heat into and out of the cycle. Still, we can identify the following properties of the system:

$P_1 = 4760$ psfa (40 psia)

$T_1 = T_2 = 493°R$

$T_3 = T_4 = 580°R$

$v_3 = \dfrac{v_1}{4}$

The state points have been defined in the familiar diagram of Figure 17–4. State 1 of the gas is readily found in Appendix II, and the properties are written in the cycle table on page 399. Very little information is known about state 2; therefore, state 4 may be easier to find, since it is linked to state 1 by a reversible adiabatic process with a final temperature of 120°F.

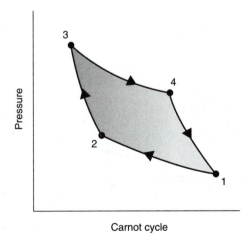

Figure 17–4. Labeling the state points of the Carnot troller.

Searching the $T = 120°F$ row on the table for the same entropy as state 1 ($.2390$ Btu/lb$_m$ – °R) identifies state 4, the properties of which are also listed in the table below. State 3 has a specific volume of $.36$ ft^3/lb$_m$ ($v_1/4$) and a temperature of 120°F, and this pinpoints its position on the table. Finally, states 3 and 2 are related by a reversible adiabatic process; therefore, their entropies will be same: $.2258$ Btu/lb$_m$ – °R.

State	Temperature °F	Pressure psia	v	Enthalpy Btu/lb$_m$	Entropy Btu/lb$_m$ – °R
1	33	40	1.44	109.90	.2390
2	33	60	.93	109.10	.2258
3	120	170	.36	118.6	.2258
4	120	110	.610	120.80	.2390

$$Wd_{1-2} = q_{1-2} = T(s_2 - s_1) = (460 + 33)(.2258 - .2390) = -6.5 \frac{Btu}{lb_m} \text{ (the}$$

entropy formula is used because this is a constant temperature process)

$$Wd_{2-3} = h_2 - h_3 = -9.9 \frac{Btu}{lb_m}$$

$$Wd_{3-4} = q_{3-4} = T(s_4-s_3) = (460 + 120)(.2390 - .2258) = 7.7$$

$$Wd_{4-1} = h_4 - h_1 = 10.9 \frac{Btu}{lb_m}$$

$$Net\ work = 2.6 \frac{Btu}{lb_m}$$

$$Efficiency = \frac{Net\ work}{q_{3-4}} = \frac{2.6}{7.7} = .33 = 33\%$$

Sample Problem 17–2 demonstrates how quickly and easily Carnot and other cycles can be analyzed using the entropy and enthalpy values.

17.3 THE REVERSE CYCLE

On one trip to the Colorado trout stream, a particular group with the amazing Carnot troller brought along a new man. The job of a rookie fisherman is always to provide the muscle power to start the boat. As great as the Carnot engine is, the device is not self-starting. Something has to push the air through the first compression so that WORK can be generated to keep it going. In fact, this engine is outfitted with a rope pull starter much like that on a lawn mower. It so happened that on this man's first day, he wound the rope the wrong way around the starter and pulled. The engine started up but ran terribly.

What happens to an engine in which the working medium is forced to circulate in the wrong direction? In Figure 17–2, the gas circulates clockwise, and everything works according to plan. What would happen if we stopped this engine, forced the R-22 gas to circulate counterclockwise, and made the operation work in reverse? (Figure 17–5) Would the boat go backward?

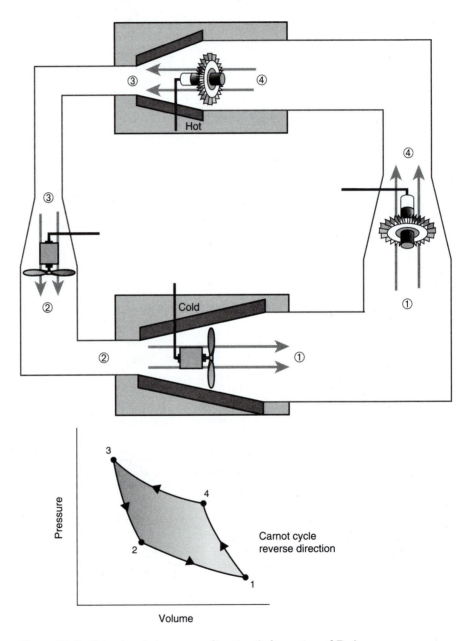

Figure 17–5. Carnot cycle in reverse direction (schematic and P-v).

SAMPLE PROBLEM 17–3

Suppose that every fan in the Carnot cycle is rotated in the opposite direction, causing the working medium inside to work in the opposite direction. In effect, every fan (WORK in) becomes a turbine, and every turbine becomes a fan compressor. Calculate the pressures, temperatures, and specific volumes throughout the cycle, assuming still that $T_3 = 120°F$, $T_1 = 33°F$, and $P_1 = 14.7$ psia. Notice that since the device does not change its configuration, all expansion and compression ratios stay the same.

SOLUTION:

The first thing to note for this solution is that the cycle equipment does not change, and even the numbering system needed on the P-V diagram need not change. The only change is that instead of going around the cycle in a 1-2-3-4 fashion, we go around in a 1-4-3-2 manner. This means that the 4–1 device that used to be an adiabatic expansion is now a 1–4 adiabatic compression.

Since the 4–1 device does not change, the specific volume ratio (v_4/v_1) does not change either, except now we look at it as a compression from .61 ft/lb$_m$ to .36 ft/lb$_m$ instead of an expansion from .36 ft/lb$_m$ to .61 ft/lb$_m$. As you look at the procedure of Sample Problem 17–2, you will realize that there is no change in the temperatures, pressures, and specific volumes for the reversed cycle from what they were in the forward cycle.

SAMPLE PROBLEM 17–4

Using Figure 17–5 as a guide and noting direction of flow and whether WORK is positive or negative, calculate the WORK of each process, paying special attention to the sign. Find the net WORK/lb$_m$, and identify where and how much heat is added and rejected.

SOLUTION:

At first glance, you might expect the work and net work to be the same as in Sample Problem 17–3. But notice that according to the figure, the work from 1–4 is negative, whereas for the original problem it was calculated to be positive. While before we were interested in wd_{1-2}, wd_{2-3}, wd_{3-4}, and wd_{4-1}, we are now interested in wd_{1-4}, wd_{4-3}, wd_{3-2}, and wd_{2-1}. Each has been calculated before, but the sign is now different. For example, $wd_{1-4} = h_1 - h_4 = -10.9$.

Summary

$wd_{4-3} = -7.7$

$wd_{3-2} = 9.9$

$wd_{2-1} = 6.5$

Net work $= -2.5$

Also from the formulas, you will find

$$q_{4-3} = -7.7 \ \frac{\text{Btu}}{\text{lb}_m}$$

$$q_{2-1} = 6.5 \ \frac{\text{Btu}}{\text{lb}_m}$$

There are many implications of the solution to Sample Problem 17–4. The fact that the net work is negative is disturbing. This means that more WORK must be put into the cycle than is gotten out. It is not much of a power-generating device if it takes more WORK to keep it running than the WORK it generates. The rookie fisherman soon discovered this when he realized that the engine would run only as long as he pulled the cord. When he stopped pulling, the engine ground to a hasty halt.

Another interesting result is that the cycle rejects heat during the 1–4 process into the 120°F air. We imagine the hot air and sunshine to be a source of heat, not a sink of heat. But as the Carnot cycle working medium goes through the adiabatic compression from 1–4, its temperature rises dramatically. As it is further compressed from 4–3, the only way to keep the temperature from getting higher is to reject heat into the surrounding air, even though that air is already extremely hot.

What is the effect of this heat on the 120°F surroundings? Although we have assumed that the air goes through the funnel on the top of the engine fast enough that its temperature does not measurably rise, we might now relax this assumption and admit that the air may exit the funnel at 121°F or 125°F or 130°F. This should not destroy our faith in the analysis of the cycle because the assumption that the process 4–3 is a constant temperature one is still very close to being true. The important point here is that the *hot surrounding air gets hotter*.

The fact that the hot surrounding air gets hotter is interesting only in regard to what happens at the opposite end of the cycle. As the working fluid of the cycle goes through the expansion from 3–2, its temperature is reduced to 33°F. Then as it expands further from 2–1, the only way to keep the temperature from falling in this constant temperature process is to absorb heat from the surroundings. Heat absorbed by the cycle means that heat is removed from the surroundings—from the stream water. Although we have assumed that the water temperature remains constant as it comes in contact with the Carnot engine, realistically it cannot. For this case, the water temperature drops slightly as it goes through, maybe down to 32°F or even below. The rookie fisherman found this to be true in a dramatic sense. He noticed that when he pulled the starter cord over and over in the wrong direction, a small chunk of ice would float up from the troller water outlet. The engine had actually reduced the temperature of the stream water below freezing. The other fishermen thought this was great and grabbed the ice to put into their orange juice container and make a cold, refreshing drink. One of the fishermen made a telling comment: "Say, fella, you just refrigerated the trout stream."

17.4 THE NATURAL FLOW OF HEAT: FOR AND AGAINST

The concept of natural flow of heat is a simple one: heat flows from high temperature to low. In the case of the Colorado trout stream, consider what would happen if the cold stream stopped flowing and the sun didn't shine so brightly. The trout stream water temperature would rise because the hot air would heat it. The air would cool because the trout stream would absorb heat from it. Sometime later, the whole valley would be at the same temperature. This is the natural flow of heat.

The natural flow of heat is also demonstrated in the operation of the Carnot troller. While it is powering around the lake, it is hastening the natural flow of heat. It absorbs heat from the air and rejects it into the stream. Heat flows from high temperature to low. This time, however, some WORK is accomplished as the heat moves through the cycle. Nevertheless, the end result is that the stream warms and the air cools.

Notice the difference in the situation if we position the rookie fisherman on his boat and allow him to continuously pull the rope in the wrong direction. Now the hot atmospheric air gets hotter, and the cold stream gets colder. If we wait long enough, the stream may become solid ice and the air flame hot. We have reversed the natural flow of heat. We have made heat flow from a low temperature to a high temperature.

The reason for this heat flow reversal is the man standing alone in the middle of the boat pulling the rope that operates the machine. He is somehow pulling Btu's out of the cold stream and forcing them into the hot air. He and his simple machine are absorbing Btu's at a temperature of 33°F or less, increasing the temperature of these Btu's, and forcing them out of the device into a hot airstream of 120°F to make the air hotter.

In a similar manner, the man could be pumping water, pulling water in from a low altitude and depositing it at a high altitude. But this time it is heat that is being elevated—elevated in temperature, not height. With this analogy, we can consider that the man is operating a *heat pump*.

Consider the function of a household freezer. If the ice cream in the freezer is not cold enough, the unit is turned on in the hope that it will draw some heat out of an already cold box. What will it do with the heat it collects? Reject it to the atmosphere. But the outside temperature is high. Somehow the low-temperature Btu's that were collected in the freezer must be "pumped up" so they may be rejected into the atmosphere. The device that will do this is a heat pump, a reverse thermodynamic cycle.

17.5 COMPONENTS OF A REFRIGERATION CYCLE

Heat pumps similar to the one discovered by the rookie fisherman are used in this country almost as much as internal combustion engines. Yet, they are far more misunderstood. Air conditioners, refrigerators, freezers, and refrigerated industrial processes utilize a cycle similar to the reverse Carnot. Chapter

19 will introduce significant differences that are incorporated in actual refrigerating equipment. But for now the reverse Carnot answers will give great insight into how a refrigeration system operates.

Figure 17–5 illustrates that there are four distinct processes in the cooling cycle, and this implies there must be four distinct pieces of equipment that make up the heat pump cycle. Table 17–2 lists the processes, names the component that performs the thermodynamic processes, and gives the equation that determines WORK delivered and heat transferred in that component.

Table 17–2

Process Number	Thermodynamic Process	Component Type	Equation
3–2	adiabatic expansion	valve	$q_{3-2} = 0$; $wd_{3-2} = h_3 - h_2$
2–1	heat rejection	evaporator	$q_{2-1} = T_2(s_1 - s_2)$; $wd_{2-1} = 0$
1–4	adiabatic compression	compressor	$q_{1-4} = 0$; $wd_{1-4} = h_1 - h_4$
4–3	heat rejection	condenser	$q_{4-3} = T_{3-4}(s_3 - s_4)$; $wd_{4-3} = 0$

Consider first the component that provides for the expansion (process 3–2) of the R-22 gas, which we can refer to as refrigerant. Its purpose is to provide an adiabatic pressure drop that results in a corresponding temperature drop. This prepares the temperature of the refrigerant to be able to absorb Btu's from a low-temperature space. Although in the theoretical analysis this process is assumed reversible, in fact the device that supports this expansion is not nearly that complex. It is simply a valve. (Figure 17–6) The valve is almost shut so that a high pressure ahead will push a small amount of refrigerant into a region of low pressure behind. This valve might be a kink in a straight piece of tubing or a section of very small diameter. This provides a calibrated amount of

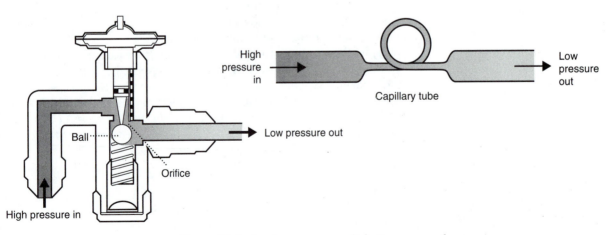

Figure 17–6. A valve causes an adiabatic pressure drop.

restriction that allows the refrigerant to expand when it flows through. Such a small tube is called a capillary tube.

How does a valve cause a pressure change? Imagine a water pump that is used to water the lawn from a pond. The purpose of the water pump is to circulate water from the pond to the lawn. Now imagine that you instantly pull the suction hose out of the water and jam the nozzle from the other end into it. The water inside the pump and hose system is caught and just circulates around and around. You have made a type of cycle. Since the nozzle is a restriction, the pump begins to build up pressure in the front of the nozzle and attempts to pull water on the back side of the nozzle. (Figure 17–7) Still water flows, but not so much. The water is under high pressure on one side of the nozzle, or restriction, and as it squirts through the nozzle, it "feels" a much lower pressure. The nozzle is just a very crude valve, but the example illustrates how a valve can cause a drop in pressure when a pump or compressor is in the circuit.

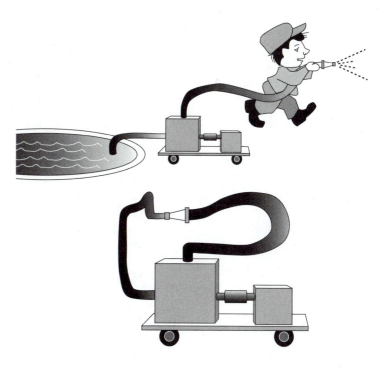

Figure 17–7. A compressor causes high pressure on one side of the valve and pulls low pressure on the other side.

The next component in the cooling cycle houses the constant temperature expansion (process 2–1). The key element here is that it is a *heat exchanger.* To absorb the required amount of heat, a very long device is anticipated—a long pipe. To conserve space, this tube is often wrapped in a serpentine configuration and fitted with aluminum fins to maximize the amount of heat exchange. The device is called an evaporator (Figure 17–8) for reasons that will be more clear in Chapter 19.

After heat has been picked up from the low-temperature source, the refrigerant carries it to process 1–4, the adiabatic compressor. Some compressors look very much like our diagram, a shell with a fan inside. They are called screw compressors. Others, called rotary-vane compressors, look like the rotary compressor of Chapter 13. But the most popular refrigeration compressors in use today are piston-and-cylinder compressors. (Figure 17–9) They are much like the units studied in Chapter 9. No matter what type of compressor is used, it provides all of the net WORK to run the cycle. As in the water pump analogy, the compressor may be thought of more as a circulating pump.

The power to run the compressor comes from an electric motor. This motor is often linked directly to the compression unit in a "direct drive compressor." One such unit has the motor welded inside of a shell that includes the refrigerant compression section. This called a hermetic compressor. It has the advantages of eliminating any possibility that refrigerant will leak out of the system and utilizing the cool refrigerant to keep the motor from overheating.

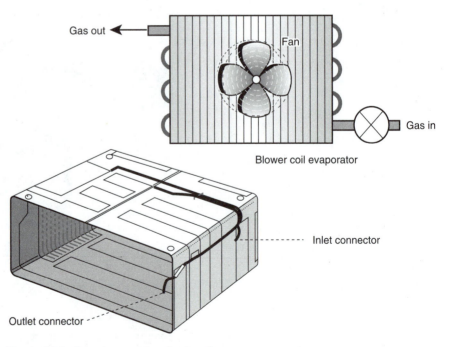

Figure 17–8. Low-temperature heat exchangers: evaporators.

Another direct drive unit is called a semi-hermetic compressor. It functions much like a hermetic unit, but it is not sealed with a weld. Instead, the shell will come apart with a bolt-and-gasket combination for easier maintenance. A less popular style is the belt-drive unit. The electric- or gas-driven motor is separate from the compression chamber, but they are coupled together by a belt-and-pulley arrangement. The advantage is that the motor can be replaced without tearing into the compression section and loosing the refrigerant gas.

Whichever style of compressor is used, the function of the compressor in the system is twofold. First, it supplies the sum total of the WORK to the system. This implies that it accounts for the circulation of the refrigerant. But

(A) Hermetic

(B) Semi-hermetic

(C) Belt drive

Figure 17–9. Refrigeration compressors.

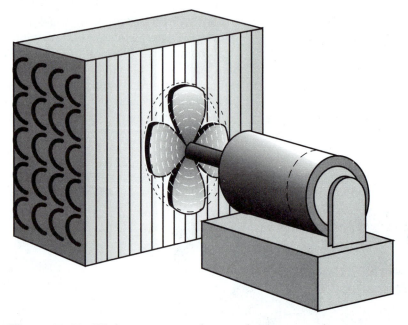

Figure 17–10. High-temperature heat exchangers: condensers.

more critical for the thermodynamics of the cycle, it compresses the heat-laden, low-temperature gas to high temperatures. Once at a high temperature, the gas is ready to move on to the next device for the final process in the cycle.

All that remains in the cycle is to reject the heat from the refrigerant so that it can start the refrigerating process over (process 4–3). Since the gas is at such a high temperature once it is discharged from the compressor, the task is easily accomplished by running it through a constant temperature heat exchanger. (Figure 17–10) This heat exchanger is cooled by a fan that forces atmospheric air across it.

Figure 17–11 is a pictorial schematic of a refrigeration system showing all four components. The analysis of the reverse Carnot cycle in this section is not precisely the way a refrigeration system operates. That will be covered in Chapter 19. Before we go on to that chapter, however, there is more that can be learned using the Carnot model.

17.6 THERMODYNAMIC ANALYSIS OF THE REFRIGERATION CYCLE

The refrigeration cycle is simply a thermodynamic cycle similar to ones that we have analyzed so many times before. In Sample Problems 17–2 and 17–3, all the calculations have been made to analyze it. But what are the performance indicators of this reversed cycle, and what is the capacity of the cycle? Power cycles have their efficiency and horsepower. What are the similar indicators

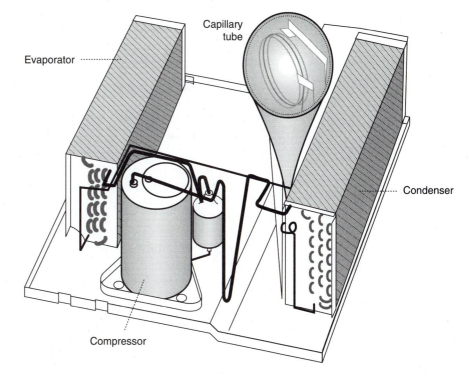

Figure 17–11. A complete refrigeration system.

for the reverse cycle? To analyze efficiency, we should not attempt to relate it to the efficiency of a power cycle but should, instead, start over with the definition of *efficiency*: useful output divided by costly input.

The costly input to the cycle is the WORK needed to drive the system—the net work. This must be supplied by an outside source (usually electricity) and will be costly. The useful output is the heat that is absorbed from the low-temperature reservoir, q_{2-1}. Therefore, the efficiency for this case will have the form

$$\eta = \frac{q_{2-1}}{|net\ work|}$$

The signs | | in the denominator indicate that if the net work is negative, as it will be for the cooling cycle, the sign should be dropped for this calculation. Notice that this formula for efficiency is different from that for a power cycle because the output and input are different.

SAMPLE PROBLEMS 17–5

What is the efficiency of the refrigeration cycle of Sample Problem 17–3?

SOLUTION:

$$\eta = \frac{q_{2-1}}{|net\ WORK|} = \frac{6.5}{2.6} = 2.5 = 250\%$$

The result of Sample Problem 17–5 is quite embarrassing for the unaware thermodynamicist. It appears that we have devised a cycle with an efficiency greater than one hundred percent. It appears that we are getting more out of this cycle than we put into it!

But remember, this is not a power cycle. Carnot stated that when you convert heat into power, it always takes more heat to drive the cycle than that which comes out as WORK. In the case of the refrigeration process, WORK is converted into *absorbed heat*. Our solution indicates that just a little WORK will power the cycle sufficiently to absorb a great deal of heat.

In fact, in Sample Problem 17–3, we see that for every Btu that is invested as WORK, there are 2.5 Btu's absorbed from the cold space. So as not to confuse the issue, thermodynamicists prefer to call the efficiency of a reversed cycle the *coefficient of performance* (COP).

Remember that the cooling cycle of this section is none other than the Carnot cycle. For the power cycle, the efficiency was calculated as

$$\eta = \frac{Net\ work}{q_{input}} = \frac{T_s - T_r}{T_s}$$

In the reverse Carnot cycle, the COP is calculated with the net work in the denominator, and the heat absorbed is the numerator, somewhat the inverse of the formula above. It is little wonder, then, that for the Carnot cooling cycle, the COP can be calculated by

$$COP = \frac{T_s}{(T_s - T_r)}$$

SAMPLE PROBLEM 17–6

Use the Carnot COP formula to calculate the efficiency of the cycle in Sample Problem 17–3.

SOLUTION:

$$\eta = \frac{T_s}{T_s - T_r} = \frac{(120 + 460)}{(120 + 460) - (33 + 460)} = \frac{580}{87} = 6.67$$

The capacity of a refrigeration system is an indication of how much useful output it will provide in a given amount of time. Specifically, it states the number of Btu's that the system will absorb in a minute, hour, or day. It is not the same as q_{2-1}, since this quantity is based on Btu's absorbed per lb_m of refrigerant flowing. To find capacity, multiply q_{2-1} by the number of lb_m of refrigerant that flows per hour:

$$Capacity = q_{2-1} \times \frac{lb_{mref}}{hr}$$

SAMPLE PROBLEM 17–7

If the amount of refrigerant flowing in Sample Problem 17–3 is 5550 lb/hr, what is the capacity of the system?

SOLUTION:

$$Capacity = |q_{2-1}| \times \frac{lb_{mref}}{hr} = 6.5 \times 5550 = 36,000 \frac{Btu}{hr}$$

The calculations demonstrated in this chapter provide the basis for a complete analysis of heat pumps.

COMPRESSOR DETAILER-DESIGNER

Almost everyone in the engineering department of Tecumseh Products in Tecumseh, Michigan, is gone today. They are at the annual convention of the American Society of Heating, Refrigerating, and Air-Conditioning Engineers (ASHRAE) in Chicago, Illinois. The engineers and technicians of this company represent just a few of the experts in the United States in the field of hermetic compressors. Although hermetic compressors are used on almost eighty percent of the new refrigerators and air conditioners in this country, much of the production of such compressors is done in countries in South America and the Far East. Hermetic compressors are unique in that the motor driver section is internal to the piping of the system and, therefore, is surrounded by pressurized refrigerant gases. This means that the driver shaft does not have to cross the boundaries of the refrigerant piping, with the possibility of leaking through shaft seals. It also means that the motor section, which consists of wound copper wire with varnish insulation, can contaminate the refrigerant. This is particularly true if the motor overheats and causes the varnishes to bubble and give off contaminating gases.

Tecumseh Products builds many models of hermetic compressors, each individually designed for a specific refrigerant, a specific capacity (Btu/hr), and for a specific application: high–back-pressure (HBP) compressors for air-conditioning, medium–back-pressure (MBP) compressors for food storage, and low–back-pressure (LBP) compressors for freezers.

**SAMPLE
PROBLEM 17–8**

Back pressure is a term used for the inlet pressure of the compressor. It is also called suction pressure. In Sample Problem 17–2, this pressure is $P_1 = 40$ psia, which is considered MBP. HBP is for systems with evaporator temperatures of 50°F or higher. Determine the corresponding back pressure for the system of Sample Problem 17–2 (120°F condensing temperature, same state 4, and the same amount of heat absorbed, q_{2-1}) but with an evaporator temperature of 50°F. Also determine the pressurization ratio.

SOLUTION:

Create a state table and write in state 4 from Sample Problem 17–2. Also write in the temperature for all states and the entropy for state 1 (same as 4). These quantities are in bold on the table on page 413.

Since $q_{2-1} = 6.5$, the entropy of state 2 can be found from

$$q_{2-1} = T(s_1 - s_2)$$

HIGH BACK PRESSURE				
State	**Temperature**	**Pressure**	**h**	**s**
1	50	50	111.95	**.2390**
2	50	90	110.6	**.2261**
3	120	170	118.7	**.2261**
4	120	110	120.8	**.2390**

For

$$T = 510°\text{R}, s_2 = -\frac{6.5}{510} + .2390 = .2261$$

Enter this number on the table in bold. This is enough information to go to Appendix II to fill in the remaining information. Interpolation to get the value is more difficult this time. From the table, the suction pressure for HBP systems will be 50 psia (10 psi higher than MBP), and the pressurization ratio is $110/50 = 2.2$ compared to $110/40 = 2.75$. Because the pressurization ratio is different for each type of compressor, their bore and stroke dimensions will be different.

The Tecumseh technicians go to the ASHRAE convention to compare notes on compressor design with other technicians from all over the world. These are busy times for compressor manufacturers, since hermetic compressors are individually designed for use with different refrigerants. A few years ago there were five popular refrigerants, which meant that Tecumseh had five complete lines of hermetic compressors. Today, three of those refrigerants have been found to have detrimental effects (ozone depletion) when released into the atmosphere. Production of these refrigerants has stopped, and new refrigerants (primarily R-134a) have been developed to replace them. This requires Tecumseh to scrap over half of its existing lines of products and replace them with redesigned compressors.

The engineers and technicians of Tecumseh are primarily detailer-designers. This means that each is in charge of the design of one component or the assembly of the final product. They make their own design calculations, build prototypes, supervise testing, and prepare the final production drawings on CAD. A few technicians work in the final testing and specification of the complete final product. One thing they look for is how close the compression process is to being adiabatic. To do this, they install the compressor on a complete refrigeration system in the lab, then measure suction and discharge pressures and temperatures. With this data, the polytropic index of the process can be calculated (see Problem 10 in Chapter 10) and compared against the value of k for the refrigerant. You can duplicate this technique in your lab if you have an operational air-conditioning or refrigeration system (preferably R-22) using a manifold pressure gauge set and an electronic thermometer (see Figure 17–12).

Figure 17–12. Lab setup to determine the polytropic index of a refrigeration compressor.

CHAPTER **17**

PROBLEMS

1. If R-22 gas undergoes an adiabatic compression from $P_1 = 25$ psig to $P_2 = 155$ psig with $T_1 = 20$ °F, find T_2:
    a. using adiabatic formulas
    b. using Appendix II
    Compare these results. Which one is more accurate?

2. An adiabatic metering valve on a refrigeration system drops the pressure from 170 psia to 110 psia. If the original temperature was 200°F, what is the enthalpy change of the refrigerant?

3. Investigate Appendix II to describe what kind of process has a constant enthalpy, that is, what is the same about two states on the table that have the same enthalpy: v, T, P, or s.

4. Investigate Appendix II and describe what states have the same entropy: constant v, T, P, or adiabatic.

5. Use the definition of enthalpy ($h = cvT + Pv$) and the perfect gas law ($Pv = RT$) to write h in terms of T only.

6. Interpolate from Appendix II to find the entropy and enthalpy of R-22 at $T = 100$°F and $P = 140$ psia.

7. A Carnot heat pump is used to hold ice cream in a 10°F chamber ($P_1 = 40$ psia and pressurization ratio (P_3/P_1) = 4.25) and rejects that heat into a 130°F atmosphere. If the heat pump has a capacity of 20,000 Btu/hr, find its
    a. $\dfrac{\text{net WORK}}{\text{lb}_{mref}}$
    b. $\dfrac{\text{amount of heat absorbed}}{\text{lb}_{mref}}$
    c. COP
    d. the lb_{mref}/hr circulated

8. Consider the gas turbine of Sample Problem 16–1 but with the following specifications:
    The gas is R-22 (not air)
    $P_1 = 40$ psia, $T_1 = 20$°F

$P_2/P_1 = 4.25$
Combustion temperature $T_3 = 200°F$
$P_4 = 40$ psia
a.   Complete the table below

State	Temperature	Pressure	v	h	s

b.   Find wd_{12}, q_{12}, wd_{23}, q_{23}, wd_{34}, q_{34} , wd_{41}, and q_{41}
c.   Compute efficiency

9. Although reversed cycles in this chapter have been constructed from the Carnot cycle, they also could be designed using the Brayton cycle (similar to the one in Problem 8 above). If the cycle of Problem 8 were reversed:
   a.   Find wd_{21}, q_{21}, wd_{32}, q_{32}, wd_{43}, q_{43}, wd_{14}, and q_{14}
   b.   Determine the net work of the cycle
   c.   In what process(es) is heat absorbed? How much total heat is absorbed?
   d.   Determine the COP of the cycle

10. The cycle in Problem 9 rejects heat at up to a temperature of 200°F, which is very hot—hot enough to heat a house.
   a.   In what process(es) is heat rejected? How much total heat is rejected?
   b.   If the heat pump is to be used for heating, what is its COP?
   c.   Where did the heat come from that was used to heat the house?

CHAPTER
18

The Thermodynamics of Material Science

PREVIEW: Most of our thermodynamics in previous chapters has involved gases. In technology, especially for manufacturing applications, the thermodynamics of liquids and solids is important, too, in particular the description of materials that are part liquid and part solid or materials that are in transition from liquid to gas or solid to liquid. As in Chapters 16 and 17, materials in transition are analyzed by using tables and graphs rather than calculations to make things both simpler and more accurate. Reading the information-rich graphs and tables of thermodynamics requires experience and patience before rewards can be appreciated.

OBJECTIVES:

❑ Describe the properties of solids, liquids, and gases in tabular and graphic form over ranges of temperature.

❑ Analyze the phase change from liquid to gas and vice versa in graphic and tabular form.

❑ Develop a strategy for reading and understanding sophisticated thermodynamic graphs and become proficient in utilizing the concentrated data they provide.

18.1 THE SCIENCE OF MATERIALS IN A PHASE CHANGE

The study of the physical properties of a substance—whether brittle or ductile, smooth or rough, and so on—is called material science. It is obvious that thermodynamics plays a role in material science because materials at high temperatures have different properties than the same materials at low temperatures. Even more important, materials in a solid form have much different characteristics than the same materials in a liquid form. The difference between the phases of the material is a matter of thermodynamics. In this chapter, we will look at the thermodynamics of material science, paying particular attention to what happens to a material that goes through a change in phase—from solid to liquid or liquid to gas, and the reverse.

Consider the material butter. Butter straight from the refrigerator is a solid. It is not a solid like steel, which has a lattice structure with strong bonds between molecules. It is not a fiber solid like wood, which has a nonhomogeneous nature of strong striations with softer material filling in between. Butter is more like plastic. Solid plastic, like butter, can be cut with a knife.

If we take a stick of butter from the refrigerator ($T = 40°F$) and place it in a saucepan in an oven and heat it slowly, the butter initially is a solid and has a specific volume of .0165 ft³/lbm. (Table 18–1) As it warms several degrees, it remains a solid, and its density and specific volume change little, with a coefficient of thermal expansion of .17 × 10⁻⁴%/°R. A P-v diagram of the material (Figure 18–2) indicates that specific weight of a solid is constant no matter how much pressure is put on it (vertical line starting at v_{solid}). The fact that there is a series of vertical lines closely spaced from each other indicates that there is a slight change in specific volume due to rising temperature. At a constant 15 psia, increasing the temperature of the butter moves the state of the material to consecutive vertical lines from left to right.

Table 18–1

DENSITY AND SPECIFIC VOLUME OF SOLIDS AND LIQUIDS

	ρ lbm/ft³	v ft³/lbm
Solids		
Nonmetallic		
Glass	156.25	.0064
Ice	57.80	.0173
Hard rubber	74.62	.0134
Butter	60.50	.0165
Wood	40.00	.0250
Metallic		
Aluminum	169.49	.0059
Copper	555.55	.0018
Steel and iron	487.80	.0020
Liquids		
Ammonia	37.73	.0265
Oil (automobile)	55.55	.0180
Ethyl glycol	69.93	.0143
Water	62.4	.0160
Butter	58.50	.0171
Gases		

The specific volume and density of gases depend on their state.

At some point in the heating process, the expansion rate is increased significantly. The butter becomes more molten and begins turning into a liquid. For a time, the butter keeps its rectangular shape, though somewhat less defined—edges are rounded and corners start to fall. If you poke the butter with your finger, you can penetrate the form as if it were liquid, yet the form remains partially rigid as if it were a solid. (Figure 18–1) Is it a solid or a liquid?

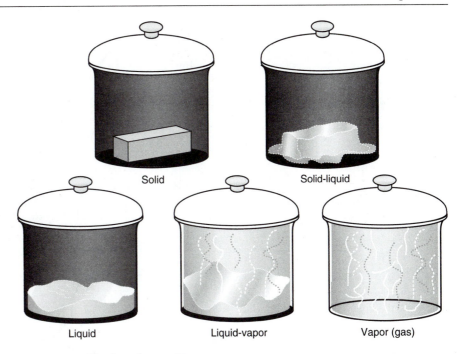

Figure 18–1. The five phases of butter.

It is neither. It is both. It is in a special solid-liquid state. Some finished goods are made of materials are in their solid-liquid state. Silly Putty and Gak Slime are examples of such products.

Figure 18–2 shows the liquid-solid state in the area where the vertical lines become horizontal, indicating a rapid expansion to the liquid specific volume. Notice that at the liquid specific volume, the lines again go vertical indicating that the specific volume of a liquid is essentially constant.

When the butter becomes a total liquid, it melts into a pool in the bottom of the saucepan. As heating continues, the liquid becomes misty. The level of the liquid rises; it is expanding again. If you stick your finger into the mist now (it's hot, wear a glove), the butter barely clings to the glove. The mist is like a gas, but rotating the pan shows that it flows like a liquid. It is in another in-between state, between liquid and gas.

In thermodynamics, this state is called the supercritical state. Some manufacturing processes require that a substance be in the supercritical state. Supercritical carbon dioxide is used to decaffeinate coffee. Supercritical carbon dioxide also is used to make plastic from chemical raw materials called polymers. By melting them in the presence of carbon dioxide heated to a supercritical state, a vapor almost as dense as a liquid, the polymers grow into long chains and become plastic. This new process eliminates much of the toxic waste by-products of plastic manufacture.

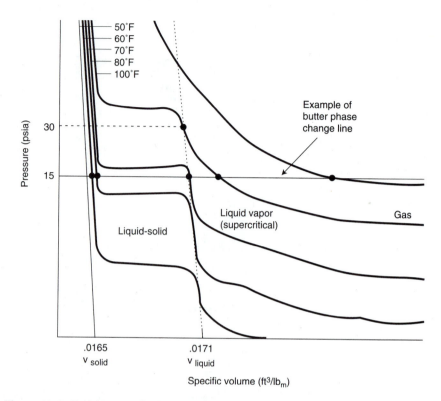

Figure 18–2. P-v diagram for butter phase change.

❖ **KEY TERM:**   **Supercritical Phase:** A phase in which the material has properties of both a liquid and a gas.

Back in the saucepan, the gaseous mist has now grown and threatens to overflow the pan. The butter becomes a true gas and continues to expand as its temperature rises. Soon its specific volume can be predicted by the perfect gas law (see Figure 8–6 for comparison). In Figure 18–2, the curves to the far right are familiar to us as those predicted by the perfect gas law.

SAMPLE PROBLEM 18–1

Butter is at a state of 100°F and P = 30 psia. What phase is it in, and what is its specific volume?

SOLUTION:

By finding the point $T = 100°F$ and $P = 30$ psia in Figure 18–2, the butter is in the liquid phase, and its density is 58.55 lb_m/ft_3 ($v = .0171$).

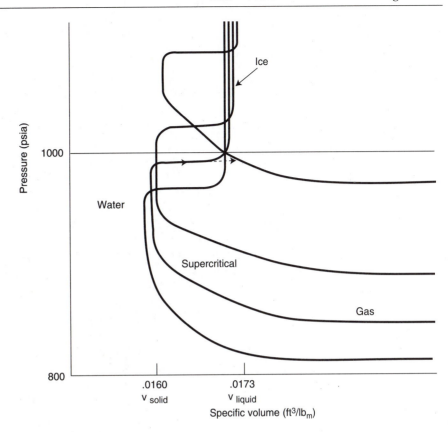

Figure 18–3. Phase changes of water at 1000 psia.

18.2 WATER AND ITS PHASE CHANGE

If we place an ice cube at 20°F into the butter saucepan of Section 18–1, then place the lid on and pressurize it to 4000 psia, the ice cube will melt and vaporize in the same manner as the butter. The solid cube of ice gets mushy in transferring to the liquid state. The icy liquid gets soft, with its edges becoming rounded, and you could put your finger into it and deform the ice and shape it something like snow. But as the solid turns to a mushy solid-liquid, the cube actually shrinks in size. The specific volume gets smaller, as the liquid is more dense than the solid. Some materials show this property, and we know it must be true for ice because ice floats on water. Solid steel does not float in liquid steel, however, but falls to the bottom. Figure 18–3 illustrates a P-v diagram for a material whose liquid specific volume is smaller than its solid specific volume.

The mushy material turns more and more to a liquid and soon fills the bottom of the pan with what is clearly recognizable as water (v_{liquid} on Figure 18–3). Further heating takes the liquid water into the misty, light, expanded phase where the fog is hard to describe as a liquid or a vapor. It rises from the bottom of the pan in an increasing specific volume, but if the pan is stirred, it swirls like a liquid. This is the supercritical state. (Figure 18–3) Finally the water fills the pan as vapor and expands rapidly as a gas in a clearly perfect-gas–like manner.

If we watch the heating of the ice cube in the same pan but under less pressure, say atmospheric pressure rather than the 1000 psia, a much different melting process occurs. We can see immediately when the ice begins to turn to a liquid. The ice does not turn mushy in the liquid-solid phase. Rather, beads of liquid fall off the ice. There is simultaneously both liquid and solid water in the pan, and as the heating goes on, the amount of solid ice decreases and the amount of liquid increases. Instead of turning to a liquid through a homogeneous mush phase, a continually increasing percentage of the ice block turns immediately to water. This means that instead of having a constantly changing specific volume, the mixture separates into two parts (Figure 18–4), the solid portion keeping the specific volume of the solid and the liquid portion assuming the density of the liquid. The specific volume of this phase, then, has two distinctly different values.

Figure 18–5 shows the distinctly different specific volumes on each side of the region called solid (A)–liquid (B), with an arrow running between them indicating that the material is jumping from the higher specific volume of the solid (A) to the lower specific volume of the liquid (B) as the ice melts, both specific volumes being for the same temperature, 32°F.

Once the water is completely in the liquid phase (B–C), very little expansion takes place until the temperature reaches 212°F (C). Then a strange occurrence takes place. Instead of the water getting lighter and foggy in a mist of growing vapor, bubbles appear in the bottom of the pan and rise to the top to be released as steam. Again there is no continuous change of phase, but instead, an ever-increasing percentage of the water goes directly to bubbles of

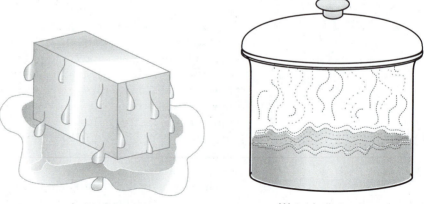

Ice melts to water
in two distinct phases

Water boils to steam
in two distinct phases

Figure 18–4. At 15 psia, water melts and boils in two distinct phases.

steam (C–D). There is an immediate and instantaneous jump in specific volume for a portion of the mixture to the steam phase. There are two distinct phases present.

In Figure 18–5, the specific volume line called the saturation liquid line is the specific volume of the water during boiling, and the line marked saturation vapor is the specific volume of the percentage that evaporates to steam. The line with the arrow going between the saturated vapor and the saturated liquid lines indicates that more and more liquid turns to steam. It is interesting to note that during this process, there is no material which has a specific volume between the two lines.

If Figure 18–5 seems confusing, that's because it is. It takes time to digest the many concepts that the graph describes. Another graph and another way of looking at the phase change process may help. Figure 18–6 illustrates the temperature versus the specific volume during phase changes. On the graph, the ice begins heating at 20°F, begins to melt at 32°F(A), is completely melted at 32°F(B), and warms the water until it boils at 212°F(C). As boiling makes more and more steam, the material is at a dual state of C and D, with more and more material going to state D. Once all the water is converted to steam, increased heating makes hotter, less dense steam (E). In this diagram, there is no supercritical phase where liquid expands and becomes foggy and begins

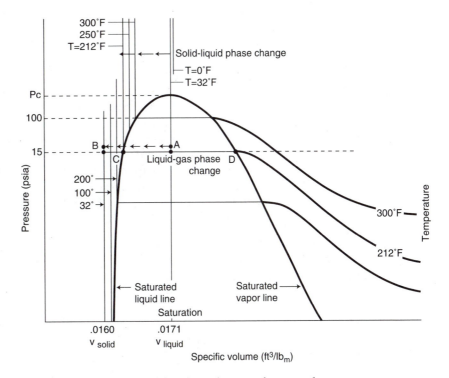

Figure 18–5. P-v diagram of the phase changes of water at low pressures.

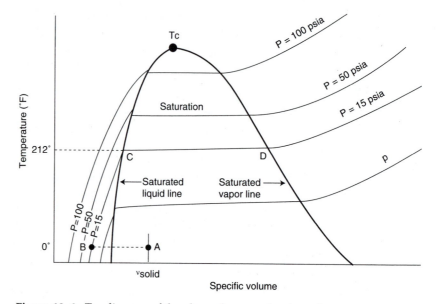

Figure 18–6. T-v diagram of the phase changes of water at low pressures.

to act like gas. The region from A–B is called melting, from B–C is called liquid, from C–D is called saturated (two phases), and from D–E is called vapor, or superheated vapor. The state at C is called saturated liquid and the phase at D is called saturated vapor.

We have described two ways that water changes phase. Which one is the correct one? Both are correct. Under high pressures and corresponding high temperatures, water goes through the solid-liquid mushy phase change and the supercritical liquid-gas phase change. Under low pressures and corresponding lower temperatures, water changes phase by going through the saturated phases. The temperature and pressure that separate these two completely different phase change mechanisms are called the critical pressure and critical temperature. For water, $T_c = 705.4°F$ and $P_c = 3206.2$ psia.

SAMPLE PROBLEM 18–2

For the states of water given below, use Figure 18–7 or 18–8 to find the specific volume and the phase of the material.
a. $P = 15$ psia, $T = 25°F$
b. $P = 15$ psia, $T = 200°F$
c. $P = 15$ psia, $T = 212°F$
d. $P = 15$ psia, $T = 300°F$

SOLUTION:

    a. solid, $v = .0171$

    b. liquid, $v = .0160$

    c. in saturation, average specific volume is unknown

    d. vapor, $v = 30$

SAMPLE PROBLEM 18-3

From Figure 18–5, find the temperature at which water begins to boil when under a pressure of 100 psia.

SOLUTION:

In the figure, locate 100 psia and follow it until it intersects the saturated liquid line. The constant temperature line that intersects this point is $T = 330°F$ (above 300°F).

18.3 REAL GASES VERSUS THE PERFECT GAS LAW

In Chapters 4 through 11, we studied processes using the perfect gas law. Pressure-volume diagrams using the perfect gas law looked like that in Figure 18–7(a). In Section 18–3, P-v diagrams with phase changes that looked like those in Figure 18–7(b) and (c) were developed. The three diagrams in Figure 18–7 look completely different, yet they all represent the same thing.

To see that the three diagrams are, indeed, the same, Figure 18–8 shows all of them assembled into one. Notice that all the graphs are valid, but in

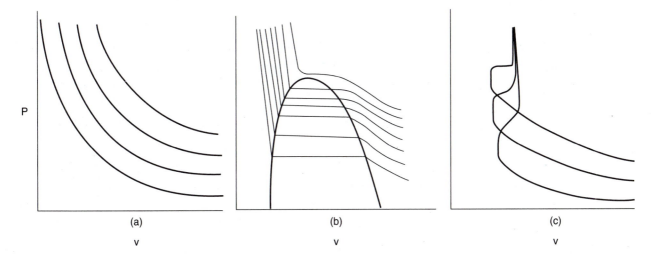

Figure 18–7. P-v diagrams for water.

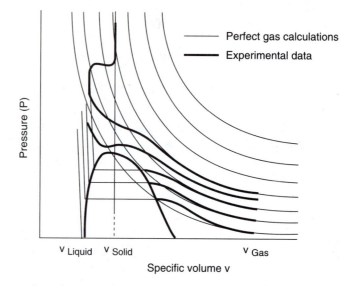

Figure 18–8. Complete P-v diagram for water.

different regions of the diagram. In particular, the perfect gas law is valid in most of the gas region, but it does not predict good results for the supercritical region, the saturated region, or the liquid and solid phases. Its prediction has varying degrees of accuracy in the gas region near transition, which may best be called the vapor region.

In general, when working a problem using the perfect gas law, the idealized results should be checked against a graph like this to see whether they are valid.

SAMPLE PROBLEM 18–4

In Sample Problem 12–1, the automobile engine was analyzed by using perfect gas equations. Show on a P-v diagram whether the perfect gas law approximation was accurate for the gas (air) at the states involved in this problem.

SOLUTION:

A real gas P-v diagram for air is shown in Figure 18–9. The states of the Otto cycle from Sample Problem 12–1 are plotted on the graph. It is clear that these states are well within the region of accuracy for the perfect gas.

18.4 STEAM TABLES

Although Figure 18–9 is useful in determining the states of water in all phases, it is difficult to be precise in reading this graph. More precise tables that represent the same data are presented in Appendix III (a and b). The first table

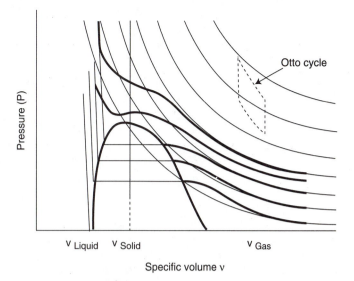

Figure 18–9. Otto cycle on a real gas P-v diagram.

represents the saturated phase, and the second table represents the gas phase, called the superheat region. The second table is easily recognizable, as it is identical to that in Appendix II for R-22. This table is used in the same manner.

In Appendix III(a), the first two columns represent the pressure and corresponding temperature that will cause steam to saturate, that is, to begin to change phase. For a given pressure, if the temperature of the water is less than the saturated temperature, then the material will be liquid water, while if the temperature of the water is greater than the saturated value, the moisture will be steam. The specific volume, enthalpy, and entropy are given for the saturated liquid, the left side of the saturation zone and the saturated vapor, the right boundary of the saturation zone.

18.5 MOLLIER DIAGRAMS

Figures such as P-v diagrams are helpful in describing phase changes, but other diagrams are needed for a complete description. In our initial investigations of water and butter, we paid no attention to the amount of heat required to raise the temperature of the materials, nor to the amount of heat necessary to fuel the phase change operation. To do this, we would have to track the enthalpy of the material, since for a constant pressure process, Chapter 16 has shown that

$$q = h_2 - h_1$$

The determination of the enthalpy of a material at all state points requires a Mollier diagram. (Figure 18–10) In the Mollier diagram for water, pressure is shown on the vertical axis, and enthalpy (h) is on the horizontal axis. Lines

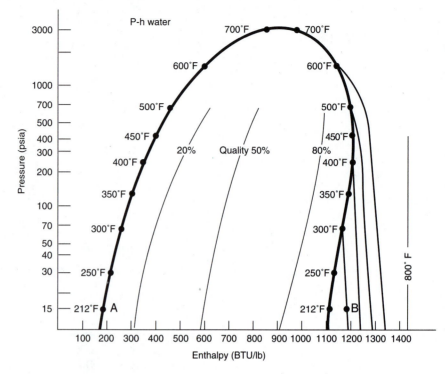

Figure 18–10. Mollier diagram for water.

of constant temperature are shown in the graph. Horizontal lines have been placed on the graph corresponding to the two constant pressure phase change experiments we did in the last section for water. To find the amount of heat required to heat one pound of water from 212°F (point A in Figure 18–10) to steam at 300°F (at 15 psia, point B), find the enthalpy of these two points—h_2 and h_1—then the difference between these two numbers:

$$h_2 - h_1 = 1190 - 185 = 1005 \frac{\text{Btu}}{\text{lb}}$$

This graph also tells us other important information. For example, the difference in enthalpy from the saturated liquid line to the saturated vapor line is the amount of heat necessary to totally vaporize one pound of water. In the terminology of Chapter 5, this is the latent heat, L. Although the original definition of enthalpy did not include latent heat ($h = c_v T + Pv/J$, the enthalpy of this diagram is defined as $c_v T + Pv/J + m_L L$, which makes the fundamental equations dealing with enthalpy in Table 16–2 true even for processes that involve phase changes.

SAMPLE PROBLEM 18–5

From the Mollier diagram,
a. Find the latent heat of water at 15 psia
b. Compare it against the value we have previously used, L = 970 Btu/lb$_m$
c. Find the latent heat of water at 100 psia and compare it against both a and b

SOLUTION:

The enthalpy of saturated liquid at 15 psi is 185, and for saturated vapor is 1140. Differencing these gives a latent heat of 955, close to the value we have previously used for L (970). The enthalpy of saturated liquid at 100 psi is 278, and for saturated vapor is 1140, yielding a latent heat of 922. This is different from the previous cases indicating that the value of L changes, depending on pressure. Our original concept of using the term $m_L L$ would have created some error because L is not constant. Our current procedure of incorporating the latent heat into enthalpy eliminates such error.

SAMPLE PROBLEM 18–6

Repeat Sample Problem 18–5 using Appendix III(a).

SOLUTION:

From that table, for T = 212°F, the enthalpy of saturated liquid is 180.07, and for saturated vapor is 1150.4, with a difference of 970.3. For P = 100 psia (corresponding to a saturation temperature of about 328°F), we extrapolate the following values from the table:

h_{liquid} = 299, h_{vapor} = 1187

for a difference of 887.

The region between the saturated liquid and saturated vapor line is called the saturation region. Although the material exists only at either the liquid or the vapor points and never in the dome, the enthalpy needed to evaporate different amounts of the one pound does exist in between. Therefore, in the saturated dome, lines are drawn indicating how much of the material has changed phase. These are called quality lines. The .2 quality line means that 20% of the water has been turned to steam. At a pressure of 15 psia and enthalpy of 930 Btu/lb, the quality of the steam is 80 (80% vapor).

SAMPLE PROBLEM 18–7

From the Mollier diagram in Figure 18–10, find the phase, enthalpy, specific volume, and quality (if appropriate) for water at the following states:
a. P = 500 psia, T = 400°F

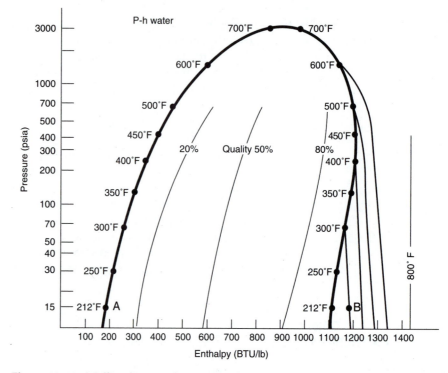

Figure 18–10. Mollier diagram for water.

b. $P = 400$ psia, $T = 445°F$, $h = 405 \dfrac{\text{Btu}}{\text{lb}_m}$

c. $P = 300$ psia, $T = 425°F$, $h = 760 \dfrac{\text{Btu}}{\text{lb}_m}$

d. $P = 15$ psia, $T = 300°F$

SOLUTION:

   a.   liquid, $h = 300$
   b.   saturated liquid
   c.   saturation, quality $= 50\%$
   d.   vapor, $h = 1250$

SAMPLE PROBLEM 18-8

A piston-cylinder container contains 1 lb_m of saturated steam-water at 300°F and 80% quality. When the material is cooled and the piston adjusted so that the water becomes a saturated liquid at 300°F, how much heat is rejected and how much WORK is done?

SOLUTION:

$$q_{reject} = h_2 - h_1 = 260 - 1025 = 765 \frac{Btu}{lb_m}$$

$Wd = 0$ (constant P process)

STEEL MILL PLANT MANAGER

You don't usually think of Texas when the subject of steel mills comes up. Steel mills are in Pennsylvania. But just outside Mesquite, Texas, a short drive from Dallas, is the Nucor Mini-Mill. Steel from this mill is shipped north for automobile manufacture. So why a steel mill in Texas? One of the main reasons comes from the nature of steel itself.

Steel is the material of preference for most consumer products, even with the strong growth of plastic and aluminum technologies. Steel, which is iron processed by thermodynamics and chemistry, can be refined to meet almost any specification: machinability, formability, hardness, brittleness, and so on. Steel is not just one material; rather, there are many formulas for the composition of steel. From stainless steel so tough you can't drill through it to strip steel you can form with your hands, steel is so versatile it can be anything the user wants it to be. Iron is the basic ingredient for steel, but carbon additives from .1 to 5% change the nature of the product from extremely malleable to very brittle. Nickel and tin also are also trace elements that greatly vary the nature of the product.

Steel is a material that has three basic phases: solid, liquid, and gas, although it is never processed in the gas phase. As liquid steel is poured from the mill ladle at 4000°F, it solidifies at 2800°F into a crystalline lattice called ferrite. Upon further cooling (2540°F), the crystalline lattice of ferrite suddenly changes, collapsing into a different lattice, with a corresponding change in density or specific volume. This phase is called austenite. At 1675°F, the lattice again falls apart and reforms a newer, more dense lattice. This lattice appears similar to ferrite, and this phase is called alpha-ferrite, whereas the original ferrite phase is called delta-ferrite or delta-iron. At 1420°F, the lattice changes slightly to what is known as magnetic alpha-ferrite. Therefore, steel has the amazing property of having four distinct phases as a solid. Figure 18–11 is a diagram refining technicians use that shows these solid phases.

The plant manager at Nucor understands well the four phases of solid steel because he worked in the mills of Pittsburgh for fifteen years. Originally hired as an engineering technician, he has worked in almost every department of the mill and knows the metallurgy of steel well. Last year, he was selected to be plant manager at this new facility.

This small mill is in Texas because of another feature of steel; it is recyclable. Steel from tin cans, scrap automobiles, and other discards is not only acceptable to be recast, but preferable. Typically a batch of new steel will contain fifty percent recycled product. The mill is in Texas because it uses scrap collected in the region, and locating the mill there saves shipping the scrap long distances.

The plant manager/technician's responsibilities in this highly automated plant are varied. He must plan each batch of steel in detail because the scrap

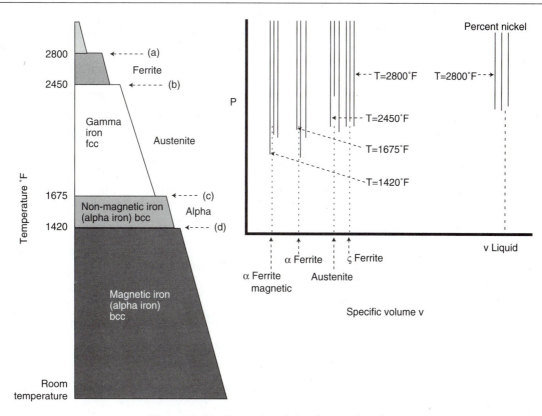

Figure 18–11. Properties of the phases of steel.

that is used has a wide variety of carbon content, along with varying amounts of tin, nickel, and so on. He must select the right mix of scrap to achieve the proper percentage of finished product. The plant manager also has primary responsibility for the maintenance of the equipment, although he has a staff of technicians who perform the work. The equipment in his control includes furnaces, metal transport equipment, hydraulic roller mills that form the molten product into long thin strips, a galvanizing line that puts a protective coating on the strips for rust protection, and finally a coiling machine that wraps the strips into coils for shipment. The galvanizing line in the mill is unusual and is there only because the primary customer for this mill's output is the automotive industry. When these coils are received at the manufacturer, they will be uncoiled, cut into flat sections, and fed into hydraulic presses that will stamp them into door panels, hoods, fenders, and trucks. One of the reasons that steel has maintained its advantage in automobile construction, over plastic and aluminum, is the recent advances in the galvanizing process. Cars today are basically rustproof.

On this day, the plant manager is faced with a problem that his repair technicians have not been able to solve. The galvanizing solution is not adhering to the strip steel as it goes through the solution vat. In fact, the sheets are galvanizing in streaks, and this will be unacceptable to the customer. The technicians have checked the galvanizing solution, and it is the proper mix. They have checked that the tank is at the proper temperature. The strip steel also is the correct composition, and the line feed rate has been checked. The technicians are baffled and need a solution.

SAMPLE PROBLEM 18-9

The plant manager pulls the drawings of the galvanizing line (Figure 18–12). He knows that the galvanizing material is heated to 2500°F while the strip steel goes into the tank at 200°F. The galvanizing material must heat the steel into the austenite phase before the proper chemical reaction will take place at the surface of the strip steel. Tables indicate that the steel picks up heat at a rate of 100 Btu/hr/in². How long must the steel be immersed in galvanizing solution to reach the austenite phase? Assume the carbon content of the steel is 1%, the density is 487.8 lb_m/ft^3, the specific heat is .1 $Btu/lb_m/°R$, and the thickness of the strip is .125 in.

SOLUTION:

The equation for the heating of the steel is

$$Q = cm\,(T_2 - T_1)$$

Considering a square inch of the strip steel,

$$m = \rho \times \frac{1}{1728} \times .125 = .035\ lb_m$$

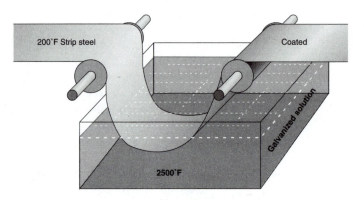

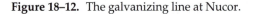

Figure 18–12. The galvanizing line at Nucor.

Checking Figure 18–11, the temperature of austenite phase change with 1% carbon is 1675°F, but with a little safety factor, the plant manager uses 1800°F. Therefore

$$Q = .1 \times .035 \times (1800 - 200) = 5.5 \text{ Btu}$$

Since 100 Btu are absorbed per hour, the heating process should take $5.5/100 = .055$ hr, or 3.2 min.

The plant manager took his information down to the idle production line where he cut a notch in the strip steel at a point before it was immersed in the vat. Starting the line, he measured the time from when the notch was immersed until it emerged as 1.95 minutes. This was the answer; the steel was not being given enough immersion time to get the temperature up to the point at which the galvanizing bonding could properly take place. This was the answer, but what was the solution? The feed rate was the same as it had always been. The length of the vat had not changed. What could have changed the immersion time? As he studied the diagram of the galvanizing vat (Figure 18–12) one more time, it suddenly came to him. The strip was too taut in the vat; there wasn't enough loop. The amount of time the strip steel was in contact with the solution depended on how much loop was there. Somehow during the starting and stopping or reloading of the strip steel, the roll had been strung too tight. By stopping the outlet roller and indexing the inlet roller, the loop was increased to get the proper immersion time.

As he left the production floor, the manager/technician felt good about having solved another problem. Now, to keep the problem from happening again, he would write up a work order and have the line supervisor add to the operating manual the procedure for calibrating the loop every time the line was restarted.

CHAPTER **18**

PROBLEMS

1. Use Figure 18–2 to determine for the following states of butter whether the butter will be
   a. solid
   c. gas
   b. liquid
   d. other

State	T (°F)	P (psia)
a	500	30
b	100	30
c	100	15
d	80	30
e	80	15
f	70	30
g	70	15
h	70	10
i	70	2

2. Use Figure 18–5 to determine for the following states of water whether it will be
   a. ice
   c. saturation
   b. liquid
   d. steam

State	T (°F)	P (psia)	$v \left(\dfrac{ft^3}{lb_m} \right)$
a	300	100	
b	300	15	
c	212	100	
d	212	15	
e	212	10	
f	100	100	
g	32		.0173
h	32		.0160

3. Use Figure 18–10 to determine for the following states of saturated water-steam (water vapor) the percent of the mixture that is liquid.

State	$T(°F)$	$h\left(\dfrac{\text{Btu}}{\text{lb}}\right)$
a	400	400
b	400	1000
c	350	400
d	350	300
e	250	1150
f	212	610

4. In Sample Problems 4–5 and 4–6, we determined for the chef at the Old Salt Seafood Restaurant some conditions for the steam in his kettles. In that analysis, we used the perfect gas law, but the PGL is not valid if the steam is in a saturated state (some liquid, some gas). Below are the states of steam for those examples. Use Figure 18–5 to determine if these states are vapor (gas) or saturated states.

State	$T(°F)$	P (psia)
a	260	14.7
b	350	16.5
c	350	14.7

5. Repeat problem 4 but use Figure 18–6.

6. Repeat problem 4 but use Figure 18–10.

CHAPTER
19

The Power of Latent Heat: Refrigeration

PREVIEW: Cycles that absorb heat using a low temperature gas as the working medium (Chapter 17) are limited by the amount of internal energy this gas can store. If the heat is absorbed by changing the phase of the absorbing fluid, then the absorption is due to the latent heat of the working medium and the amount of heat that can be absorbed is greatly increased.

OBJECTIVES:

❑ Demonstrate that a liquid absorbs heat by evaporation and the temperature at which it evaporates depends on the pressure imposed on the liquid.
❑ Create a cycle that will continuously evaporate a working fluid and recondense it in a manner by which heat is absorbed at low temperature and rejected elsewhere into a higher temperature environment.
❑ Reduce all calculations of vapor-liquid refrigeration cycles to a graph of working medium (refrigerant) properties.

19.1 INTRODUCING LATENT HEAT REFRIGERATION

The refrigeration system of Sample Problems 17–3, 17–4, and 17–5 has an excellent efficiency and achieves temperatures low enough to easily air condition a house. There is one flaw to the system that makes it impractical, however, and that is demonstrated in Sample Problem 17–7. There it was calculated that for 36,000 Btu/hr of air-conditioning (a typical household requirement), the refrigerant must be circulated at a rate 5550 lb$_m$/hr. This is a tremendous amount of gas and suggests that the compressor would have to be bigger than all the air fans in the Houston Astrodome, or the air conditioner for a house would have to be about as big as the house itself. The equipment needed to operate the cycle is gigantic.

The reason for the great mass flow rate required is that the refrigerant passing through the evaporator picks up only 6.5 Btu/lb$_m$ (14 kJ/kg) of heat by increasing its internal energy, and this isn't very much heat. There is a technique by which each pound of refrigerant can become more powerful in its ability to pick up heat. It can be discovered by recalling Equation 5–1, the fundamental energy equation:

$$Q = c_v m (T_2 - T_1) + m_L L + WORK$$

One interpretation of this equation is that if a substance (m) absorbs heat (Q) out of its surroundings, that heat will increase the temperature (increase in internal energy) of the substance, or the absorbed heat will be put to WORK, or that heat will change the phase of the substance. The latent heat of refrigerant R-22 (L) is about 70 Btu/lb$_m$ (162 kJ/kg), and comparing this against the internal energy increase per pound of Sample Problem 17–4 (6.5 Btu/lb$_m$ or 14 kJ/kg), this is a large amount. If latent heat would have been used in this sample problem, five times more heat could have been absorbed per pound of R-22.

Imagine a refrigeration system that utilizes latent heat. (Figure 19–1) The refrigerant would have to enter the evaporator (the heat absorption process) as a liquid and change phase there. It would be pulled into the compressor as a gas and enter the high-temperature heat exchanger as a gas where cooling would cause it to condense. Finally, it would go through a pressure drop expansion as a liquid as it entered the evaporator again.

Let us look in detail at Figure 19–1 to compare this *vapor-liquid* cycle, formally called the vapor compression cycle, against the *all-gas* cycle of Chap-

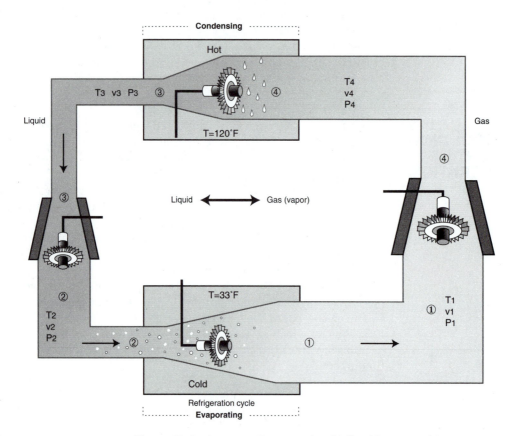

Figure 19–1. A reverse Carnot cycle with liquid-vapor refrigerant.

ter 17. Start at the entrance to the low-temperature heat absorber, process 2–1. In the gas cycle, the R-22 is expanded at a constant temperature to absorb heat, but in the vapor-liquid cycle, liquid is injected and evaporated to a gas to absorb heat by the mechanicism of latent heat. No wonder this component is called an evaporator—liquid enters and is evaporated to a gas (vapor).

The next process, 1–4, in the gas cycle is an adiabatic compression. In the vapor-liquid cycle, the gas enters the compressor and is adiabatically compressed to a higher temperature gas. The process is exactly the same in both.

In the high-temperature heat rejection process, the gas cycle requires further compression to squeeze out the heat. But in the vapor-liquid cycle, the gas is condensed to a liquid by taking out heat. The component that does this is aptly named a condenser (4–3).

The final process is an adiabatic expansion in the gas cycle (3–2) where the pressure and temperature drop as necessary to pick up heat in the evaporator. In the vapor-liquid cycle, neither the incoming nor the outgoing refrigerant is a gas, but rather a liquid. Therefore, this is a pressure drop on a pure liquid and does not follow the perfect gas law or the WORK formulas. This process will be evaluated in the next section. The device, however, is still a valve, as it was in the gas cycle.

To see the benefit of the vapor-liquid cycle, solve Sample Problem 17–7 again utilizing latent heat.

SAMPLE PROBLEM 19–1

How many lb_m/hr of refrigerant must be circulated in the air conditioner of Sample Problem 17–3 ($P_2 = 60$ psia) to achieve 36,000 Btu/hr of cooling at 33°F if the system utilizes latent heat and the refrigerant has a latent heat of 70 Btu/lb_m? Also find the COP.

SOLUTION:

We will solve this problem here very simplistically, then work on making the analysis more complete in the remainder of the chapter. We can take the results of Sample Problem 17–4 and add latent heat where it is appropriate.

$$Wd_{1-4} = -10.9 \ \frac{Btu}{lb_m}$$

$$Wd_{4-3} = -7.7 \ \frac{Btu}{lb_m}$$

$$q_{4-3} = (-7.7 - 70) = 77.7 \ \frac{Btu}{lb_m}$$

$$Wd_{3-2} = 9.9 \ \frac{Btu}{lb_m}$$

$$Wd_{2-1} = 6.5 \ \frac{Btu}{lb_m}$$

$$q_{2-1} = (6.5 + 70) = 76.5 \frac{\text{Btu}}{\text{lb}_\text{m}}$$

The heat absorbed in the evaporator is $Q_{4-3} = 76.5$ Btu/lb$_\text{m}$, and to achieve 36,000 Btu/hr of cooling requires about 444 lb$_\text{m}$/hr of refrigerant compared with 5500 lb$_\text{m}$/hr for Sample Problem 17–7.

The COP of the system is

$$COP = \frac{q_{4-3}}{net\ WORK} = \frac{77.76}{2.5} = 26 = 2600\%$$

The power of latent heat is obvious. Much less refrigerant must be circulated to achieve the same cooling results, so the equipment can be much smaller and is more efficient.

The analysis of Sample Problem 19–1 is *very* simplistic. Controlling latent heat so that it is triggered in the evaporator requires an understanding of how change of phase takes place.

19.2 THE MECHANICS OF LATENT HEAT

The Otto cycle, represented by the automobile engine, is an effective way to generate mechanical power from a fuel. One of the reasons that this small engine develops so much horsepower is the speed at which the cycles can take place, approximately 1500 cycles per minute (3000 RPM). The ability to inject Btu's into the cycle at a speed of 1/1500 minute calls for a heat generation process that is *extremely* fast. Combustion is a perfect process to create the necessary Btu's in a split second.

Is there a similar phenomonon that will absorb heat at a similarly fast speed? To some extent, the answer is yes. The phenomenon is the process of evaporation, which can take place at relatively rapid rates. If evaporation can take place almost instantaneously, the latent heat of the substance will be absorbed from its surroundings almost instantly. Materials that evaporate rapidly have low evaporation or boiling points, so low that if they were left under atmospheric conditions the materials would be in a vapor rather than a liquid state. They can be maintained in a liquid state only by keeping them under extreme pressure.

Even air, which we always consider a gas, can be kept in the liquid state if the pressure exerted on it is great enough. In fact, cylinders of compressed oxygen usually contain oxygen in the liquid state. Once the valve is open, however, the liquid oxygen explodes into evaporation, providing oxygen gas and immediately colder temperatures inside the cylinder.

Table 19–1 presents data for the popular chemical called R-22, in particular, the temperature at which it will evaporate corresponding to the pressure that is exerted on it. If a canister of this chemical were opened to the atmosphere and allowed to evaporate as rapidly as possible, its temperature and everything around it would drop to –40°F, because that is the temperature at which R-22 evaporates with zero pressure (gauge). If the canister were throttled by a valve so that some higher pressure could be maintained and the evaporation rate reduced, the evaporating temperature would be correspondingly higher.

For Sample Problem 17–3, if we want an evaporating temperature of 33°F, we must establish an evaporating pressure of 73.48 psia. Note in Sample Problem 17–3, the pressure at the inlet of the evaporator was 60 psia. Therefore, we have a major difference in the two cycles.

Table 19–1

TEMPERATURES AND PRESSURES FOR R-22 CANISTERS

Temperature		Pressure	
°F	°C	psia	kPa
–40	–40	14.70	2.2
–20	–28.9	24.84	3.6
0	–17.7	38.66	5.6
20	–6.6	57.72	8.4
33	.5	73.48	10.7
40	4.4	83.20	12.1
50	10.4	98.92	14.3
60	15.5	116.31	16.8
80	26.7	158.33	23.0
100	37.8	210.60	30.5
120	49.0	274.60	39.8

SAMPLE PROBLEM 19–2

A freezer is to maintain 8000 Btu/hr of cooling at a temperature of 0°F with R-22 refrigerant. What should be the evaporating pressure? What should be the temperature of the condenser? How high should be the pressure of the compressor pumps?

SOLUTION:

To find the evaporating pressure of the R-22, look on Table 19–1 to find that the pressure that corresponds to 0°F evaporating temperature is 38.66 psia. If the storage cylinder is to reject heat into the room in which the freezer is located, then the gas must be compressed to a temperature warmer than the room, say 80°F. Table 19–1 indicates that this requires a pressure in excess of 158.33 psia, which is the pressure required by the compressor.

19.3
P-H DIAGRAMS

One of the most powerful tools for analyzing the evaporation process and a latent heat refrigerant cycle is the Mollier diagram that was introduced for steam-water in Figure 18–10. Figure 19–2 is the Mollier diagram for R-22 (see Appendix IV for more detail), which looks almost exactly like that for water, except for the relative temperatures, pressures, and enthalpies. On that diagram is drawn the cycle of Sample Problem 17–3 ($T_1 = 33°F$, $T_3 = 120°F$).To locate state 1 (the exit to the evaporator) on Figure 19–2, look on the saturated vapor line at a temperature of 33°F (pressure of 78.48 psia). State 1 is directly to the right of state 2 on the saturated vapor line because the evaporator process continues from left to right until all the liquid is gone. The Mollier

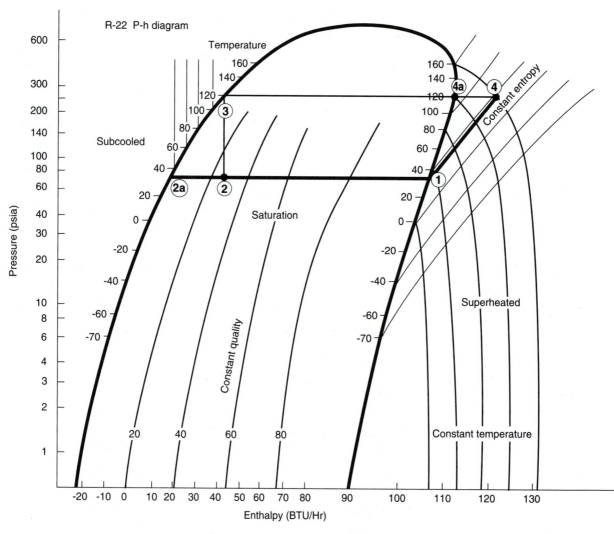

Figure 19–2. The refrigerant system on a P-h diagram.

diagram for this process shows a subtle feature about the evaporator process—not only is it constant temperature, but it also occurs at constant pressure. Evaporation that takes place at *constant temperature* also takes place at *constant pressure*.

The refrigerant gas enters the compressor at state 1 and is compressed to a higher pressure adiabatically and reversibly (state 4). Therefore, this process also can be called one that occurs at a constant entropy. Lines of constant entropy on the Mollier diagram slope steeply up from left to right. The compressor gas discharges (state 4) at a pressure that corresponds to the condensing pressure (the pressure that corresponds to 120°F from Table 19–1, 274.6 psia).

At this point (state 4), the gas enters the condenser and the process continues at constant pressure until the gas is completely condensed (saturated liquid line). The Mollier diagram indicates that this process is more complex than we expected. In particular, the gas exits the compressor in a superheated state—it is not ready to condense. As heat is taken away, the gas proceeds to state 4a (saturated vapor line) where its temperature has been reduced to its saturation temperature (120°F) and it begins to condense. To determine the superheated temperature of state 4, note that the lines of constant temperature slope sharply up and down from left to right, all starting at the saturated vapor line where the corresponding temperature is recorded. Locate the sloping line that crosses state 4 and follow it up to the saturated vapor line where the temperature is recorded. Notice that as you follow the line, this is no process that you are following. We perform this maneuver only to find what the temperature is at state 4. The resulting temperature is about 160°F. The consequence of the process line 4–3 is that the refrigerant first enters the condenser as a superheated vapor, and the first bit of cooling in the condenser desuperheats the gas to saturation. Only the remainder of the condenser actually has liquid condensing in it.

From state 3, the liquid goes to the metering device where the pressure is rapidly reduced, but the refrigerant exits again as a liquid. How do we describe this line on the diagram? The process is adiabatic, but the working medium is a liquid, not a gas. Furthermore, the valve does not have a turbine in it to catch the WORK available from an adiabatic process. Therefore, the process is irreversible—a free expansion (Chapter 11). An adiabatic free expansion is not only constant entropy, but also constant enthalpy:

$$h_3 = h_2$$

Constant enthalpy processes on the P-h diagram are represented by vertical lines. Therefore, process 3–2 is a vertical line from state 3 straight down to the constant pressure line that represents process 2–1.

Figure 19–2 demonstrates that the end of process 3–2 does not meet the state point 2a as we have drawn it. This implies that the original state point 2a was positioned wrong. We assumed that the liquid entering the evaporator was saturated liquid, but now we see that, in fact, it is partially evaporated liquid; it is a vapor-liquid combination in saturation. In fact, on this diagram, lines of constant quality are shown that represent the percentage of vapor that has been evaporated away as the process goes from the saturated liquid line to the saturated vapor line. Along the original evaporator process line going from left to right, these lines enable you to pinpoint when twenty percent of the liquid is evaporated, and so forth. The metering device process lines indicate that the refrigerant enters the evaporator at a quality of twenty-five percent, and the actual evaporator process is 2–1, not 2a–1.

How is it that some refrigerant is evaporated before entering the evaporator? The energy equation for process 3–2 is

$$q = c_v(T_2 - T_3) + m_L L = 0$$

When the pressure drops on the saturated liquid as it goes through the valve, the refrigerant begins to evaporate in the valve. The only way that the equation above can continue to equal zero when $m_L L$ becomes a positive number is for $c_v(T_2 - T_3)$ to become negative, implying that the refrigerant temperature drops. Said another way, as the refrigerant is forced to evaporate because the pressure is dropping, the only place for the refrigerant to find energy to evaporate is from its own internal energy. Some refrigerant evaporates; the

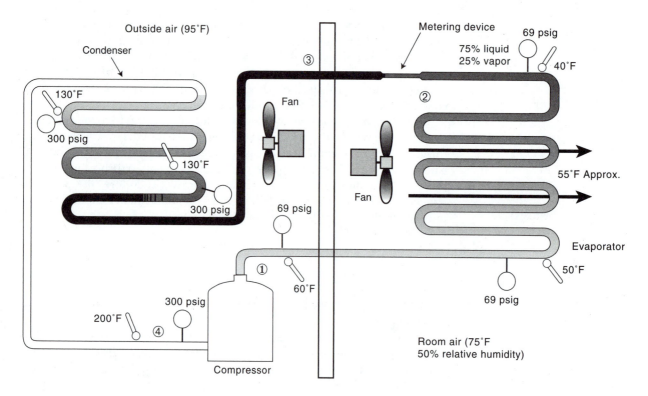

Figure 19–3. Cutaway diagram of a vapor-liquid refrigeration cycle.

remaining liquid loses internal energy to sponsor that evaporation. The purpose of the metering device is not only to drop the pressure but also to make the temperature of the refrigerant compatible to pick up heat in a cold space. Therefore, the drop in temperature through the metering device is essential to prepare the refrigerant for its function in the evaporator. The vapor created in the metering device is referred to as *flash gas*.

SAMPLE PROBLEM 19–3

How much flash gas is generated through the metering device of Figure 19–3?

SOLUTION:

Figure 19–3 shows a more precise view of a vapor-liquid refrigeration (air-conditioning) system and the changes of phase of the refrigerant. Check it against Figure 19–2. Point 2 in the figure is at a quality of 25%, or .25 lb_m flash gas/lb_m refrig . Point 3 is saturated liquid at 300 psig.

19.4 PERFORMANCE OF A REFRIGERATION SYSTEM

From Figure 19–2, the performance characteristics of the refrigerating system are easily found, including COP and capacity (Btu/hr). The enthalpy difference between state points 2 and 1 represents the heat absorbed in the evaporator, called the refrigerant effect. For the example of Figure 19–3, the refrigerant effect is $h_1 - h_2$.

❖ **KEY TERM:**   **Refrigerant Effect:** The amount of heat removed in the evaporator (Btu/lb_m or kJ/kg).

The WORK done during the adiabatic compression is the difference between the suction enthalpy and the discharge enthalpy, $h_4 - h_1$. The COP of the device is the ratio of refrigerant effect to compression WORK.

SAMPLE PROBLEM 19–4

What is the compressor work (Btu/lb) and COP of the R-22 reverse cycle that operates between 33°F and 120°F?

SOLUTION:

$$q_{2-1} = h_1 - h_2 = 105 - 45 = 60 \frac{Btu}{lb_m}$$

$$Wd_{1-4} = h_4 - h_1 = 120 - 105 = 15 \frac{Btu}{lb_m}$$

Therefore

$$COP = \frac{60}{15} = 4.00 = 400\%$$

Notice that although this is the same system as in Sample Problem 19–1, this efficiency is somewhat lower than was computed in that sample problem because the total latent heat was taken as the refrigerant effect in that problem, but we now know that this is not the case.

19.5 EFFECT OF SUPERHEAT AND SUBCOOLING ON REFRIGERATION SYSTEM EFFICIENCY

While Figure 19–2 illustrates a basic refrigeration cycle, Figure 19–4 shows a more typical system. Notice that state 1 is not on the saturated vapor line but, instead, in the superheated vapor region. Suppose the refrigerant goes through the evaporator and is completely evaporated before it reaches the end of the evaporator. As the vapor completes its traverse of the evaporator, it is at a low temperature and, therefore, will continue to pick up heat. Since there is no longer any latent heat action, however, the additional heat is stored in the refrigerant as internal energy, with a corresponding increase in temperature. When the temperature of a vapor is increased above its saturation point, it is called superheated, and the superheat is described as the actual temperature of the refrigerant exiting the evaporator in Figure 19–4 minus the temperature of *saturation* of the gas. Notice that this increases slightly the refrigerant effect.

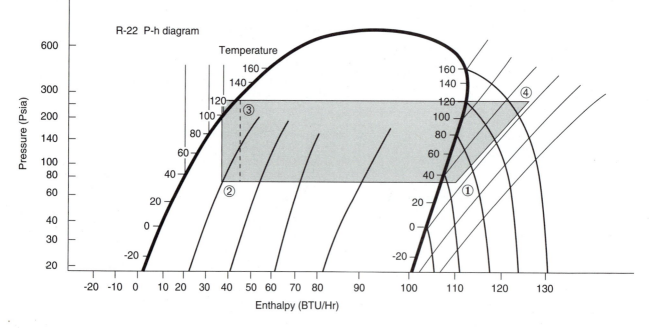

Figure 19–4. Refrigeration system with superheat and subcooling.

SAMPLE PROBLEM 19–5

Find the refrigerant effect, compressor discharge temperature, and COP of the refrigerating system of Sample Problem 19–4 that has 20°F evaporator superheat (suction superheat).

SOLUTION:

$$q = h_1 - h_2 = 110 - 45 = 65 \ \frac{\text{Btu}}{\text{lb}} \ \text{(refrigerant effect)}$$

$$T_1 = 53°F \ (20° \ \text{superheat from } 33° \ \text{saturation})$$

$$Wd = 126 - 110 = 16 \ \frac{\text{Btu}}{\text{lb}}$$

$$COP = \frac{65}{16} = 4.05 = 405\%$$

Notice from Sample Problem 19–5 that the effect of suction superheat is to increase the superheat of the compressor discharge and to increase the compressor WORK. Therefore, the minor increase in the refrigerant effect comes at some expense.

The interest in suction superheat is not that it improves COP, but rather that it ensures that the refrigerant gas exiting the evaporator is totally a gas and contains no liquid. If the refrigerant were to exit the evaporator before all the liquid refrigerant was evaporated, it could enter the compressor as a partial liquid, and the compression pump is not designed to handle liquid compression. This could cause piston rods to fail or exhaust valves to break.

Figure 19–4 also differs from Figure 19–2 at state point 3. Notice that state 3 in Figure 19–4 is to the left of the saturated liquid line. What happens if the refrigerant is completely condensed in the condenser before the refrigerant exits the condenser? The saturated liquid is hot and continues to give up heat to the surroundings. But the heat does not come from latent heat; instead, the internal energy of the refrigerant is reduced—the liquid temperature drops. This implies that the refrigerant is subcooled and that the liquid refrigerant that enters the metering device is subcooled. To locate the subcooled state 3, follow the condenser constant pressure line to the left of the saturated liquid line. Notice that vertical lines on the P-h diagram are drawn to represent lower liquid temperatures. Stop the process line at the temperature of subcooling of the refrigerant liquid. On the figure, the temperature of state 3 is 100°F, which represents 20° of subcooling.

❖ **KEY TERM:**

Subcooled: A liquid is said to be subcooled if its temperature is below its saturation temperature, and the "degree of subcooling" is this temperature difference.

Notice that moving state point 3 to the left with subcooling also moves state point 2 to the left, thereby increasing the refrigerant effect and COP.

Subcooling is a beneficial feature of a refrigeration cycle and is accomplished by oversizing the condenser.

SAMPLE PROBLEM 19–6

Find the efficiency of the system in Sample Problem 19–5 with 20°F subcooling in the condenser.

SOLUTION:

$$\text{Refrigerant effect} = h_1 - h_2 = 107 - 35 = 72 \ \frac{\text{Btu}}{\text{lb}_m}$$

$$Wd = h_4 - h_3 = 16 \ \frac{\text{Btu}}{\text{lb}_m}$$

$$COP = \frac{75}{16} = 4.9 = 490\%$$

Sample Problem 19–6 proves that subcooling in the condenser increases the refrigerant effect in the evaporator and does not increase the WORK of compression; therefore, it improves efficiency. It illustrates that the more heat that can be rejected from the condenser, the more heat the refrigerant can absorb in the evaporator.

19.6 EFFECT OF CONDENSING AND EVAPORATING PRESSURE ON SYSTEM EFFICIENCY

What determines the pressure in the condenser? Is it the designer of the system? Is it the compression capability—the compression ratio of the compressor? Is it the temperature of the cooling medium (air or water)?

A *yes* to all three of these questions is partially correct, but the one of overwhelming importance is the last one—the temperature of the cooling medium. This temperature determines the pressure of condensation, and this pressure is imposed on the complete high-pressure side of the system. The rule of thumb is that the condensing temperature is 30°F greater than the air cooling medium or 5°F greater than the water cooling medium. This determines the operating high side pressure. On a hot day, the condensing pressure will be higher than on a cool day, particularly in an air-cooled system.

What determines the evaporating pressure? Can the designer just pick one, or does the compressor or the temperature of the space to be cooled determine it? Primarily, it is determined by the fact that the designer must make a system that will achieve temperatures low enough for the application. If the system is for air-conditioning, the evaporating temperatures must be about 50°F. If the system is for maintaining fresh foods or beverages, temperatures must be above freezing, about 36°F. If it is a freezer, the evaporating temperatures must be 10°F or below. Overall, the evaporating pressure is determined by the application of the system, and a freezer will have a lower evaporating pressure than an air conditioner, assuming both use the same refrigerant.

Sample Problem 19–7 illustrates the effect of lower evaporating pressures on the efficiency of refrigerating systems.

SAMPLE PROBLEM 19–7

The evaporator temperature of the system illustrated in Sample Problem 19–5 is consistent with a medium-temperature system, 33°F. If a freezer were to be designed with the same specifications as this system, but with a desired evaporating temperature of 0°F, what would be the refrigerant effect, the WORK of compression, the efficiency, and the lb_m/hr of refrigerant circulating?

SOLUTION:

From Figure 19–5:

$$Refrigerant\ effect = 104 - 44 = 60\ \frac{Btu}{lb_m}$$

$$Wd = 127 - 104 = 23\ \frac{Btu}{lb_m}$$

$$COP = \frac{60}{23} = 2.61 = 261\%$$

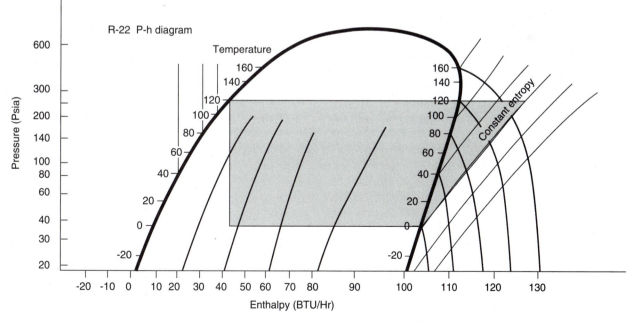

Figure 19–5. Sample Problem 19–7.

Sample Problem 19–7 clearly shows that the efficiency of a freezer is sufficiently less than the efficiency of a similar medium-temperature unit. This should not come as a surprise, since Carnot predicted that the ultimate efficiency of a cycle depends on the difference between the temperature of the source (evaporator) and the temperature of the sink (condenser). This temperature difference is greatest for a freezer application, and, therefore, Carnot would predict it has the poorest efficiency.

This result shows up in the calculation for the lb_{mref}/hr. The flow rate is much higher to achieve the same 36,000 Btu/hr of cooling for a freezer than for a food cooler. This implies that the capacity of the compressor and the horsepower must be greater for a freezer that puts out the same amount of cooling as a medium-temperature or high-temperature application.

This fact is illustrated on the manufacturers' data for compressors. Table 19–2 demonstrates that if a model of a compressor is to be utilized in system applications of different evaporator temperatures, the capacity of the system in which it is used will decrease as the design evaporator temperature decreases. The capacity ratings on the table are in tons of cooling, equivalent to 12,000 Btu/hr of cooling.

Table 19–2

COMPRESSOR CAPACITIES (TONS)							
Suction Temperature	Discharge Temperature	R-500			R-22		
		Model A	Model B	Model C	Model A	Model B	Model C
10°F	90°F	25.0	33.8	49.9	35.2	45.8	65.8
	100°F	22.8	30.5	45.3	32.4	42.0	60.8
	110°F	21.6	28.9	43.1	31.0	39.2	58.4
30°F	90°F	40.9	54.6	81.7	56.3	77.1	117.8
	100°F	38.0	50.6	75.9	52.6	72.2	108.6
	120°F	35.1	48.6	73.0	50.7	65.7	98.2

REFRIGERATION MECHANIC

All-Seasons Service, Inc., employs only one engineering technician. That is because there is only one employee of the company for now. This one person is mechanic, salesman, billing clerk, and owner. Receptionist he is not, since he has an answering service. On his way to his first call of the day, he is thinking that he has the greatest career in the world. It seems that all refrigeration and air-conditioning mechanics truly enjoy their work. Where else can you find a job that guarantees you will get five to ten problems thrown at you each day—problems you have never seen before, but that you have full confidence you will solve?

This technician didn't always work for himself. After earning his engineering technology degree, he worked for a large service contractor in town and fulfilled his requirements to get a license to become an independent contractor. Two years later, he started his own service organization.

Today he arrives at his first call, a residence in which he is to repair a split-system air conditioner (the condenser and compressor are outside the house and lines pipe refrigerant to the evaporator in the air duct system in the house). As he opens the cover that houses the condenser and compressor (the condensing unit), he notes that the compressor is running. Even before he gets his tools and instruments from the truck, he lets his sense of touch be his troubleshooting tool. He feels the suction line as it enters the compressor. It should be cool, but it is not. Maybe there is no refrigerant inside, he thinks. Maybe there is a leak in the system. Gingerly he touches the top of the compressor. It should be hot if there is a compression process going on inside. It is hot; therefore, there is refrigerant in the system and a leak seems less likely.

SAMPLE PROBLEM 19–8

If the air-conditioning system were the system of Sample Problem 19–5 (it is not, since that example is for a medium-temperature application while this one is a high-temperature application), what temperature should the technician feel at the suction line? What temperature should he feel at the discharge line?

SOLUTION:

Even with a 20°F superheat, the suction temperature should be 53°F (state 1). The discharge temperature should be 170°F (state 4). The mechanic had better be careful when he touches this line—it could burn him.

Now the technician gets his instruments from the truck, a pressure measuring set (manifold gauge set) and an electric thermometer. He connects his pressure gauges to the suction line and the discharge line at the compressor. The suction pressure reads 20 psig (35 psia). This is too low for an air-conditioning application (even for the medium-temperature application in Figure 19–2 in which the suction pressure was 73.48 psia). The discharge pressure is 200 psig (215 psia). This is too high for the 95°F ambient conditions.

What could be making the suction pressure too low and the condensing pressure too high? The technician reviews the cycle diagram (Figure 19–4) in his mind and realizes that the metering valve must be partially plugged. The compressor is pushing against the valve restriction, creating high discharge pressures, and pulling refrigerant from an evaporator that is getting very little refrigerant.

Now he goes into the basement where the evaporator and metering device are and, by inspection, confirms his analysis. The metering device has frost all over it. The compressor is pulling such a low pressure because there is little refrigerant, and the refrigerant that is coming through evaporates immediately at a very low pressure, freezing the moisture in the air around it. Unfortunately, by the time the refrigerant enters the evaporator, there is no liquid left and, therefore, no potential to cool the air.

Two hours later, the system is running perfectly. The technician made a trip to the supply store to get a new valve, performed a refrigerant recovery procedure to remove the refrigerant from the system, made a quick changeout of the valve, and then recharged the system. The first problem of the day has been solved. When he gets back to the truck, his beeper is on. The answering service tells him that the blood bank at South Memorial Hospital is having a problem with their main cooler. Can he come right away? He pencils this call into his log book ahead of his next scheduled call, then turns his truck south.

CHAPTER **19**

PROBLEMS

1. An air-conditioning plant uses R-22 and has evaporating and condensing temperatures of 32°F and 130°F. What will be the percent of liquid refrigerant entering the evaporator?

2. List the four components of a refrigerating system. Identify each component by its state numbers from Figure 19–2.

3. Consider the following states of R-22 refrigerant by noting whether the same represents the refrigerant in saturation, subcooling, or superheat.
   a. $T = 40°F$, $P = 100$ psig
   b. $T = 40°F$, $P = 80$ psig
   c. $T = 40°F$, $P = 65$ psig
   d. $T = 40°F$, $P = 40$ psig
   e. $T = 40°F$, $P = 20$ psig

4. For a refrigeration system using R-22 with condensing at 110°F and evaporating at 20°F and no superheating or subcooling, find the refrigerant enthalpies
   a. at compressor suction
   b. at compressor discharge
   c. at end of condenser

5. The vaporizing and condensing temperatures of an air-cooled R-22 reach-in cooler are 40°F and 110°F, respectively. Assuming no superheating or subcooling, determine:
   a. the refrigerant effect (Btu/lb)
   b. the mass of refrigerant circulated for 12,000 Btu/hr ("one ton of cooling")
   c. Wd (Btu/lb)
   d. the amount of heat rejected in the condenser
   e. the efficiency (COP)

6. On the drawing of the P-h diagram in Figure 19–6, label each of the four process lines that represent the refrigeration cycle with the device that supports the process: evaporator, compressor, condenser, and metering device.

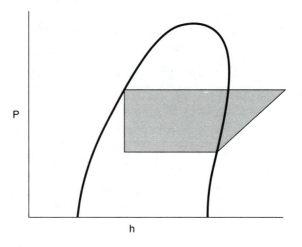

Figure 19–6. Problem 6.

7. In Problem 5, if the surface of the condensing heat exchanger became plugged with leaves or dirt and the condenser became very inefficient, the condenser would react by increasing in temperature. Assuming the condensing temperature increased to 140°F but the evaporating temperature in Problem 5 remained the same, repeat a through e in that problem.

8. Determine the proper evaporating and condensing pressures for the following system:
   a. air-cooled R-22 air conditioner in Arizona (ambient temperature 100°F).
   b. air-cooled R-22 air conditioner in Ohio (ambient temperature 85°F)
   c. water-cooled R-22 air conditioner (800°F)
   d. water-cooled R-22 freezer (–100°F)
   e. air-cooled R-22 freezer in Ohio (–100°F)

9. An air-cooled R-22 air-conditioning system in Arizona (100°F ambient) has a measured condensing pressure of 350 psig. Does this unit appear to be malfunctioning? If so, what could possibly be wrong?

10. By analyzing Figure 19–3, how would you expect the compressor to react to a system that did not have a sufficient charge—that had so little that only a few drops of refrigerant were in the evaporator at all times? In particular, what would happen to the suction pressure being pulled by the compressor, and how would this affect the corresponding evaporating temperature of the system?

11. A system is to be built for a freezer application using R-22. The system will be air-cooled, with a typical ambient air temperature of 70°F. If the system must have a capacity of 40 tons, which compressor should be used from Table 19–2?

12. Make a copy of Appendix IV and draw on it the refrigeration cycle with the following characteristics:

Refrigerant	R-22
Condensing temperature	86°F
Evaporating temperature	25°F
Suction superheat	20°F
Condenser subcooling	5°F

13. An R-22 air-conditioning system yielded the following specifications in the test lab:

Evaporating pressure	120 psia
Condensing temperature	250 psia
Actual air-conditioning effect	$16,000 \frac{Btu}{hr}$
Compressor input kilowatts	3.0

Find the COP (3.4 Btu/hr = 1 watt).

14. An R-22 freezer system condenses refrigerant at 120°F with no subcooling. Evaporation is at –10°F, with dry-saturated refrigerant entering the compressor. Find the refrigerant flow rate through the metering device in lb_m/hr, and find the CFM at the compressor inlet.

15. In the freezer of Problem 14, consider that the liquid line from the condenser is soldered to the suction line from the evaporator to form a perfect heat exchanger, and that by doing this, the condenser user is subcooled by 10°F.
    a. What is the relationship between the enthalpy drop of the condenser liquid and the enthalpy rise of the suction vapor?
    b. This implies that the resulting suction refrigerant will be superheated by how many °F?

CHAPTER

20

The Power of Latent Heat: Steam Heat and Steam Power

PREVIEW:

It is not just refrigeration cycles that can benefit from latent heat. Power-generating cycles can benefit as well. Here we study steam and water as a combined-phase working medium for both heating systems and power-generating systems. The analysis is done primarily by table and graph, as in previous chapters, but some new graphic methods are introduced.

OBJECTIVES:

- ❑ Compare a hot water heating system to a steam heating system.
- ❑ Perform design calculation for steam heating systems.
- ❑ Distinguish the differences and similarities of a steam heating system and a steam power generator.
- ❑ Use T-s diagrams for making easy calculations for a steam power cycle.
- ❑ Use steam tables for analyzing Rankine cycles.
- ❑ Use computerized tables for analyzing Rankine cycles.

20.1 STEAM HEAT

Latent heat was a powerful tool in making the refrigeration cycle commercially feasible. It can be applied to other areas of thermodynamics as well, for example, heating systems.

Heat distribution from a centralized boiler was studied in Section 6–6. Water was heated in a boiler to approximately 180°F, then pumped to individual heat exchangers throughout a building. The water was returned to the boiler at about 130°F after delivering its heat. With a 50°F temperature drop, each pound of water delivered 50 Btu/lb$_m$ to the space.

Contrast that amount of heat to the latent heat of vaporization of water, about 970 Btu/lb$_m$. Latent heat can deliver twenty times as much heat per pound as was delivered through internal energy in the hot water heating system of Section 6–6.

Using latent heat implies *steam heat* and Figure 20–1 illustrates the components of a steam system. Steam is generated in the boiler, and to do this, the burners must supply the full latent heat of 970 Btu/lb. The steam then travels to the radiator where it condenses back to liquid, referred to as condensate, which requires the full latent heat to be given off into the room. Although the system looks similar to a hot water system, there are several major differences, one of which is illustrated in Sample Problem 20–1.

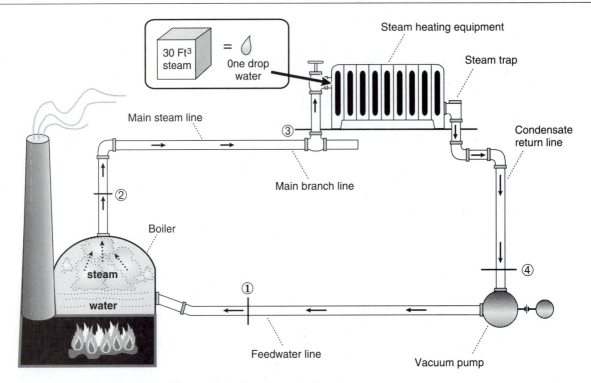

Figure 20–1. Components of a steam system.

In Sample Problem 6–4, a hot water system was designed that delivered 18,000
Btu/hr with 108.5 lb_m/hr of water flowing. If this would have been a steam
system, how much steam would circulate (lb_m/hr)?

SOLUTION:

We can first solve this problem very simplistically, then investigate its
details. If the water is turned to steam at atmospheric pressure, it boils at
212°F, and it condenses at 212°F in the radiators. In the radiator, there is
no temperature change and no WORK done. Therefore

$$q_{radiator} = m_L L$$

Then

$$q_{radiator} = 18,000 \ \frac{Btu}{hr} = m_L \times 970$$

or

$$m_L = \frac{18,000}{970} = 18.55 \ \frac{lb_m}{hr}$$

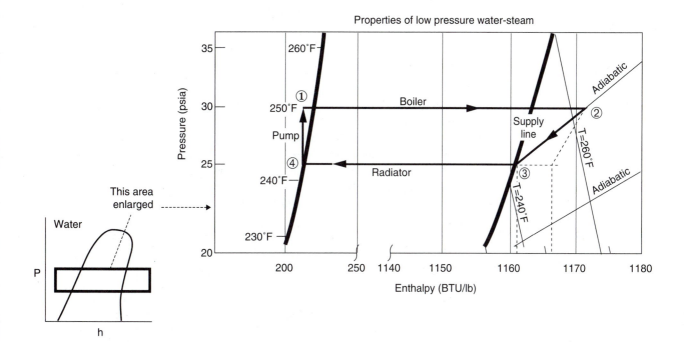

Figure 20–2. Mollier diagram of water showing steam system thermodynamics.

Sample Problem 20–1 clearly shows that the steam system with latent heat reduces the circulation rate substantially over the hydronic system (108.5 lb_m/hr to 18.55 lb_m/hr). There is another benefit, however, that is not so clearly seen. The release of the latent heat is performed while the steam condenses in the radiator, and this condensation converts large volumes of vapor into very small volumes of liquid (see Figure 20–1). In fact, the specific volume of steam at atmospheric pressure and 212°F is about 30 ft^3/lb_m, but when it is converted into liquid, it occupies less than .1 ft^3/lb_m. As the steam "collapses" into liquid, something must take the place of the volume it occupied, and this something is more steam. Steam is *pulled* into the radiator from the boiler without the need of a pump, but rather by the natural action of condensation. More precisely, the condensation causes lower pressures in the radiator, and the steam flows from high pressure to low pressure. The condensing of the steam acts like the suction of a pump.

The properties of steam in a steam heating system can best be described using a P-h Mollier diagram such as that in Figure 20–2. The diagram shows the state changes for the steam system of Figure 20–1. The dark horizontal line marked "boiler" exhibits the making of steam in the boiler, with a boiler pressure of 30 psia. This is more realistic than the sample problem where steam was made at atmospheric pressure. Boilers of this type are called medium-pressure boilers, and some in the high-pressure boiler category make steam up to 125 psia. The temperature of the steam from the diagram is 250°F, which is the saturation temperature of steam at this pressure.

The boiler line on the Mollier diagram starts on the left (1) in the subcooled portion of the diagram, demonstrating that water enters the boiler in an unsaturated (subcooled) state. As heating continues, the liquid turns to steam, and the diagram shows that heating continues into the superheated region (2), indicating that the steam usually picks up more heat than is needed to evaporate all the water.

The lower dark line exhibits the condensing of steam in the radiators or room heat exchangers. This line is at a pressure of 25 psia, indicating that there is a 5-psi pressure drop from boiler to radiator. From the Mollier diagram, steam condenses in the radiator at 243°F. Notice that sufficient heat is given off to condense the steam and generate 100% liquid, called condensate.

❖ **KEY TERM:**   **Condensate:**   The (liquid) water that is created from steam in radiators. It must be returned to the boiler to generate more steam.

Since the return water is 4 psi lower than the supply water, some device must be placed in the return line to increase the pressure back to the boiler pressure. We might think of this as a pump, a condensate return pump. The thermodynamic action of the pump is represented by the vertical line in Figure 20–2 that connects the radiator to the boiler (process 4–1). Actually this pump may not be needed, since this pressure rise can be accomplished naturally by the condensate water itself in a manner called *static head*.

To illustrate static head, Figure 20–3 shows water stacked higher in the condensate return line than in the boiler because the boiler pressure is higher than the radiator pressure. The height of the water in the return line in excess of that in the boiler can be predicted by the fluid mechanics equation

$$H = \frac{\Delta P}{\rho_{\text{water}}}$$

where

$$\rho_{\text{water}} = 62.4 \, \frac{\text{lb}_m}{\text{ft}^3}$$

and ΔP is the difference in pressure in $\text{lb}_f / \text{ft}^2$ between boiler and radiator. In our example, the static head will be

$$H = 5 \, \frac{\text{lb}_f}{\text{in}^2} \times \frac{144 \, \frac{\text{in}^2}{\text{ft}^2}}{\left(62.4 \, \frac{\text{lb}_m}{\text{ft}^3} \right)} = 11.63 \text{ ft}$$

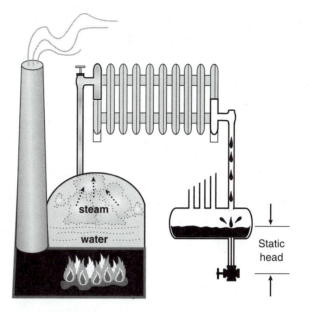

Figure 20–3. Static head eliminates the need for a water pump on some steam systems.

The higher column of water on the right forces water into the boiler, even though the boiler is under higher pressure. Therefore, the static head acts like a pump.

The final line on the P-h diagram that must be made to complete the steam heating cycle is the line that connects the boiler with the radiator (Figure 20–2, process 2–3). There is a pressure drop in this supply line that delivers the steam due to friction in the steam line. Table 20–1 presents typical values of pressure drop in steam supply lines.

Table 20–1

PRESSURE DROPS IN COMMON USE FOR SIZING STEAM PIPE (FOR CORRESPONDING INITIAL STEAM PRESSURES) (ASHRAE)	
Initial Steam Pressure psig	Pressure Drop Per 100 Feet psi
0	1/2 oz
2	2 oz
10	8 oz
15	1 psi
30	2 psi
100	2–5 psi
150	2–10 psi

For our example, the supply steam pressure is at 30 psi, and the table indicates that this implies a 2-psi/100-ft pressure drop. If we consider a 250-ft steam supply line, this results in a total pressure drop of 5 psi, which is illustrated in Figure 20–2. This pressure drop process is adiabatic, since the supply line is insulated to eliminate any unnecessary heat loss. Lines of constant entropy are drawn on the P-h diagram, and it was found in Chapter 19 that adiabatic reversible processes follow lines of constant entropy. But this process is not reversible, since no WORK is delivered and the pressure drop is sustained by friction in the supply line. The process may be better described as an adiabatic free expansion, and in this case, $h_2 = h_1$, and the line drops straight down. Even more realistically, the process is a semi-free expansion with an index of irreversibility (β) between 0 and 1. In this case, the actual point where the steam enters the radiator (point 3) can be found from the equation

$$h_3 = h_2 - \beta\,(h_2 - h_{3\text{ rev}})$$

Equation 20–1

where these enthalpies are defined in Figure 20–2.

For the steam system described above with a boiler pressure of 30 psia and a 4-psi drop in the supply line, using an index of irreversibility of .5, 7°F subcooling of condensate, and 20°F superheating of boiler steam, find
a. the amount of heat added in the boiler/$lb_{m\,\text{steam}}$
b. the temperature of the steam as it enters the radiator
c. the amount of heat released in the radiator/$lb_{m\,\text{steam}}$
d. the heating capacity of the radiator if 12,000 lb_m/min of steam is circulated
e. the static head of the condensate

SOLUTION:

Figure 20–2 traces this system perfectly, since the temperature of point 1 is 243°F, which represents 7°F subcooling, and the superheat of the supply steam is 20°F (260°F).

a. The enthalpy change of the steam-water in the boiler is (Figure 20–2: $h_2 = 1179$, $h_1 = 205$)
$$h_2 - h_1 = 1179 - 205$$
and this is all heat, since for a constant pressure process
$$q_{2-1} = h_2 - h_1 = 974\ \frac{\text{Btu}}{lb_m}$$

b. $h_3 = h_2 - \beta(h_2 - h_{3\text{rev}}) = 1179 - .5\,(1179 - 1167) = 1173\ \dfrac{\text{Btu}}{\text{lb}}$
This enthalpy represents $T_3 = 240°F$ on the $P = 25$ psia line.

c. The enthalpy change of the steam-water in the radiator is

$$h_3 - h_4 = 1173 - 205 = 968 \ \frac{\text{Btu}}{\text{lb}_m}$$

and this heat is given off by the radiator.

d.   The heating capacity

$$q = q_{2-1}m = 968 \times 12,000 \ \frac{\text{lb}_m}{\text{min}} \times 60 \ \frac{\text{min}}{\text{hr}} = 69,696,000 \ \frac{\text{Btu}}{\text{hr}}$$

e.   The static head is 9.23 ft.

Notice that the procedure followed in Sample Problem 20–2 is more realistic than the simple analysis of a steam heat system in Sample Problem 20–1. In the first problem, irreversibilities were not considered, nor were the possibilities of subcooling and superheating. Sample Problem 20–2 is the preferred analysis of steam heating systems.

20.2 STEAM POWER: THE RANKINE CYCLE

The steam system described in the previous section is unique in that it is self-powered. Steam circulates naturally by the mechanism of the drawing power of condensation and the pressurizing power of static head.

The ability of condensation to create low pressures can be utilized in power cycles as well. To best see how, reconsider the jet engine cycle of Section 16–2 (see Figure 16–2). In particular, the adiabatic turbine is where power is taken off, and it is clear how this is accomplished. High-pressure gases are poised at the inlet of the turbine, and as they are expanded through the turbine to lower pressures, WORK is taken off by the fan blades. The critical question is: What causes the low pressures at the turbine outlet? Our previous discussions have implied that the adiabatic process itself created them. But if the outlet of the turbine is capped off, the adiabatic expansion itself cannot continue to support lower pressures at the outlet. The pressures at the outlet will rise because the gases have nowhere to go, and even if the compressor section continues to run, forcing more gases into the turbine, the adiabatic process will be unable to maintain low pressures at the outlet.

In the power cycles of previous chapters, the turbine often exited to the atmosphere; therefore, it was the duty of the atmosphere to maintain low turbine-outlet pressures. However, it appears that the process of condensation also can be used, since collapsing vapors cause pressures to drop dramatically. (Figure 20–4) High pressures to drive the turbine are created by a boiler, and low pressures are maintained by a condenser or radiator. The system looks like that in Figure 20–5, similar to the steam heating system except that the supply line connecting the boiler to the radiator is now fitted with a power turbine, a steam turbine. In the heating system, the supply line supported an irreversible adiabatic pressure drop, but now the process is the same except reversible with useful WORK as the result.

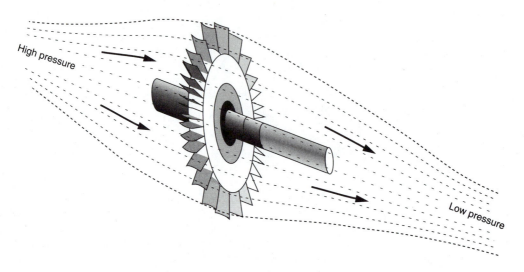

Figure 20–4. Maintaining low pressure at the outlet of a turbine is as important as having high pressure at the inlet.

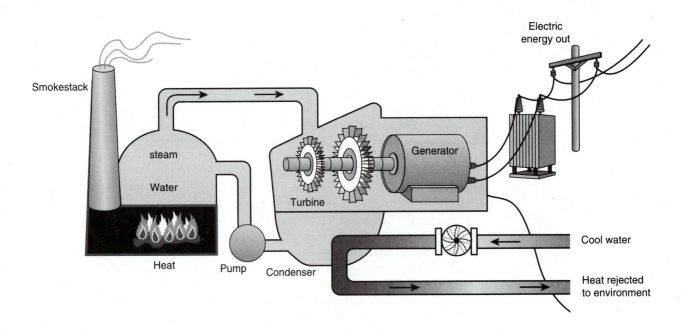

Figure 20–5. A steam power cycle.

In a steam power cycle, the boiler is called a steam generator, and the radiator is called a condenser. The pressure difference between the generator and the condenser is much greater than in the heating system, since power is the end objective. A liquid pump must be used to return condensate rather than a static head because the pressure difference is so great. Figure 20–6 illustrates the appropriate P-h diagram for a typical vapor power system operating between 2500 psia and 15 psia. In this device, steam is delivered from a generator through the supply line to a nozzle at the turbine which directs the steam against buckets attached to a turbine wheel. In turn, the turbine rotates an electrical generator. For a frictionless turbine allowing reversible adiabatic expansion of the steam, we find from Table 13–4 that

$$Wd_{turb} = h_3 - h_4$$

A condenser receives the exhaust steam, condenses this steam to a liquid, and then feeds this liquid water to the pump. In the condenser, the amount of heat rejected from the steam is

$$q_{rej} = h_1 - h_4$$

Whether the condenser coolant is air, river water, or some other fluid, it is not in direct contact with the steam inside. A variation of this design is the cooling tower, which relies on heat transfer to the surrounding air.

The steam cycle described here is formally called the Rankine cycle. It consists of the following four processes:

1. Constant pressure heat addition to vaporize steam to one-hundred-percent quality or to superheat.
2. Constant entropy (reversible adiabatic) expansion through a turbine.
3. Constant pressure condensation in a heat exchanger to maintain low turbine-outlet pressure.
4. Constant entropy pressurization in a liquid pump.

The analysis of the Rankine cycle is very similar to the analysis of the steam heating system of Sample Problem 20–2.

SAMPLE PROBLEM 20–3

Using Figure 20–6(a), a Rankine cycle is operating between 400 psia and 15 psia. Find
a. the superheat of steam leaving the boiler
b. the amount of heat absorbed in the boiler (/lb$_m$)
c. the WORK generated in the turbine
d. the efficiency of the cycle

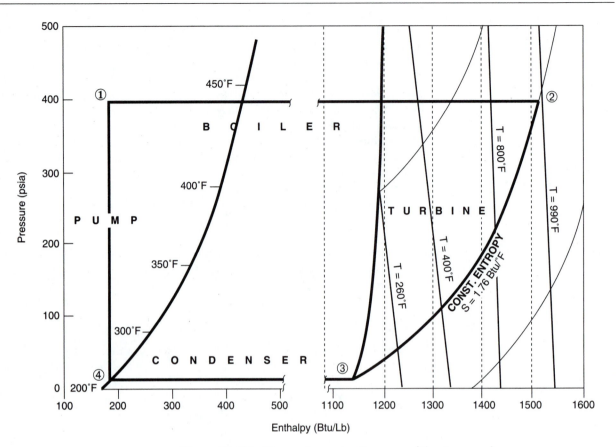

Figure 20–6(a). Thermodynamics diagrams of the steam cycle.

SOLUTION:

a. Figure 20–6(a) indicates that the saturation temperature of the steam in the boiler is 440°F. Since the temperature of the steam leaving the boiler (2) is 990°F, the superheat is 990 − 440 = 550°F. From Figure 20–6(a), make the following state table:

State	Temperature	Pressure	Enthalpy
1	215	400	190
2	990	400	1520
3	215	15	1152
4	215	15	190

b. From the table,

$$q_{boiler} = h_2 - h_1 = 1520 - 190 = 1330 \; \frac{Btu}{lb_m}$$

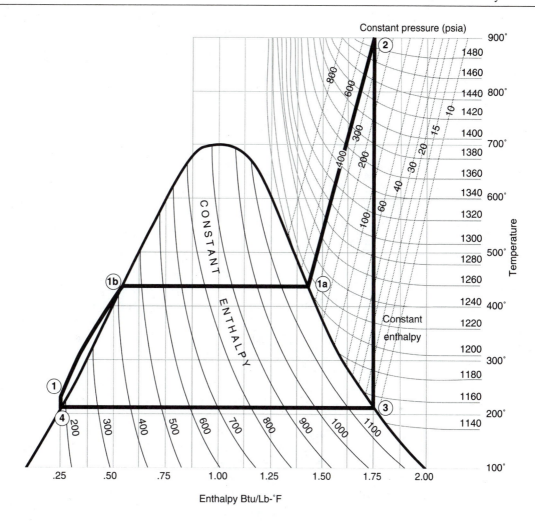

Figure 20–6(b).   Thermodynamics diagrams of the steam cycle. (*continued*)

c.   From the table,

$$Wd_{\text{turbine}} = h_2 - h_3 = 1520 - 1152 = 368 \; \frac{\text{Btu}}{\text{lb}_m}$$

d.   Efficiency:

$$\eta = \frac{368}{1330} = .27$$

Figure 20–6(b) is another diagram often used to analyze steam cycles. It contains the same information as Figure 20–6(a) but in a different format. In Figure 20–6(b), the temperature of the state is plotted on the vertical axis and the entropy of the state is plotted on the horizontal axis, with the lines of constant pressure and constant enthalpy marked. This diagram also shows the region of saturation; any state within the dome is in a partial gas-liquid state. The diagram is called a temperature-entropy chart, or T-s diagram. It is another type of Mollier diagram. One advantage of this diagram is that the area inside the cycle diagram represents the net WORK done by the cycle. Another advantage is that the heat given off or added to the steam is represented by the area under the process in which it occurs. This makes analysis of improving efficiency of the cycle easy. A further advantage is that adiabatic reversible processes (constant entropy) are vertical lines on the graph, which makes computations of adiabatic processes particularly easy.

The steam cycle of Figure 20–6(a) has been replotted in Figure 20–6(b) point for point. Note that point 1a is determined by the fact that it is on the saturation liquid line and at a pressure of 400 psia. Point 1b is found on the saturated vapor line directly across from point 1a. Point 2 is determined by following a line of constant pressure (400 psia) into the vapor region (upward and to the right). Point 3 is directly below point 2 because the turbine process is adiabatic and reversible and that means constant entropy, $s_2 = s_3$. Point 4 is the end of the condenser where all steam has been converted to liquid. The short line going vertically from 4 to 1 represents the pump. It is vertical because the pump process is constant entropy, and it is short because pumping water doesn't increase the temperature of the water much at all. Therefore, although the pump creates a great increase in pressure, it is only a short line on the T-s diagram. In Figure 20–6(a), the pump line is long and easy to read. But as Sample Problem 20–3 indicates, the thermodynamic properties of 4 and 1 are almost identical, except for the large increase in pressure. This process is liquid pumping; no temperature change occurs, and it is hardly a thermodynamic process at all. Usually for analysis, we can consider this process a fluid mechanics rather than thermodynamics process and use the fluid mechanic formula

$$WORK = \frac{(P_1 - P_4)}{\rho g} \quad (P \text{ in psf, } \rho = 1.93 \text{ for water, } g = 32.2 \, \frac{ft}{sec^2})$$

Point 1 represents the true beginning of the boiler. The line from 1 to 1a represents the heating in the boiler necessary to begin making steam.

To see how the T-s graph simplifies both the calculations and understanding of steam power systems, consider the sample situation below.

SAMPLE PROBLEM 20–4

Redo Sample Problem 20–3 using the T-s diagram of Figure 20–6(b) to analyze the Rankine cycle. Also, find the horsepower of the cycle if 120 lb$_m$/min is flowing.

SOLUTION:

Figure 20–6b has the cycle drawn on it. To establish this cycle, first find the 400 psia constant pressure line in the upper right-hand corner (the superheated zone). Follow this line down to the saturated vapor kernel. Draw a horizontal line over to the saturated liquid line. Label this point 1a. The line drawn represents most of the heating in the boiler. Next find the 15-psia constant pressure line in the superheat region and trace it down to the saturated liquid kernel (dome). This establishes point 3. Draw a horizontal line from 3 to the saturated liquid side of the dome. This point is point 4, and the line represents the condensation in the condenser. From point 3, draw a vertical line up until it reaches the constant pressure line corresponding to $P = 400$ psia. Label this point 2, and the line from 2 to 3 represents the turbine. Draw the line down the constant pressure line from 2 to the saturated vapor kernel. This represents the superheating of the steam in the boiler.

To finish the cycle, label point 4 also point 1. This double point is an approximation that the pump process (4 to 1) does not show up on the T-s diagram and that the state properties of the liquid water are almost identical before and after the pumping process except for pressure (from 15 psia to 400). In fact, since by this assumption, $h_4 = h_1$, which is only slightly inaccurate, we will not use the state table to determine $h_1 - h_4$ but, instead, use a calculation for pressurizing pumps:

$$Energy\ of\ pumping = \frac{(P_1 - P_4)}{\rho g}$$

Taking the state properties from the diagram results in

State	Temperature °F	Pressure psia	Enthalpy Btu/lb	Entropy Btu/lb –°R
1	215	400	190	.25
1a	440	400	430	.60
2	990	400	1520	1.76
3	215	15	1152	1.76
4	215	15	190	.25

Once this state table is set up, the analysis follows identically that in Sample Problem 20-3 except the calculation for

$$Wd_{pump} = \frac{(P_1 - P_4)}{\rho g} = (400 - 15) \times \frac{144\,\frac{in^2}{ft^2}}{1.93} \times 32.2 = 892.1\,\frac{ft-lb_f}{lb_m} = 1.15\ Btu$$

Also

$$Hp = (h_2 - h_3) \times 120 \times 778 \frac{ft-lb}{Btu} = 35{,}476{,}800 \frac{ft-lb}{min} = 1075 \text{ hp}$$

With Sample Problem 20–4, you are able to solve simple steam power problems. Now we will look at refinements to this basic cycle.

20.3 SUPERHEAT IN THE POWER CYCLE

Sample Problem 20–4 calls for a maximum steam temperature of 990°F. This is pretty hot! The boiler, the combustion process, and the heat transfer to the steam all become inefficient when trying to elevate steam to this condition. A more realistic maximum temperature for steam is 600°F. If the maximum steam temperature is limited to this value, it means that the superheat of the steam is not as great as in the simple cycle. How does limiting the superheat of the steam in the boiler affect the cycle—in its performance, and in its operating characteristics? We will investigate this in Sample Problem 20–5.

SAMPLE PROBLEM 20–5

Consider the Rankine cycle of Sample Problem 20–3 but with a limit on the superheat of 160°F ($T_3 = 600°F$). Fill in the state table for this cycle, and determine the thermal efficiency and the horsepower generated.

SOLUTION:

We begin the solution by drawing the cycle on the T-s diagram. Points 1, 1a, and the boiler line through the saturation dome are identical to the previous sample problem. Point 2 is up the superheat constant entropy line but stops at a temperature of 600°F. To find Point 3, draw a vertical line down to the condenser line ($T = 215°F$ and $P = 15$ psia). This establishes Point 3. Then complete the condenser line to the left to Point 1,4. The diagram is shown in Figure 20–7.

From the diagram, make the following state table:

State	Temperature	Pressure	Enthalpy	Entropy
1	215	400	190	.25
2	600	400	1306	1.59
3	215	15	1075	1.59
4	215	15	190	.25

$$Wd_{2-3} = h_2 - h_3 = 231 \frac{Btu}{lb_m} \qquad q_{1-2} = h_2 - h_1 = 1116 \frac{Btu}{lb_m}$$

$$Wd_{4-1} = \frac{(P_1 - P_4)}{\rho g} = 1.15 \frac{Btu}{lb_m} \qquad q_{3-4} = h_4 - h_3 = -885 \frac{Btu}{lb_m}$$

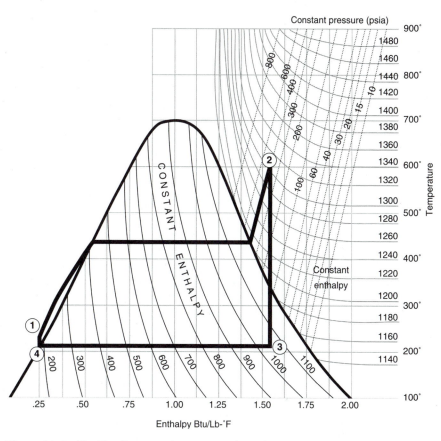

Figure 20–7. The T-s diagram of a steam cycle with $T_{max} = 600°F$.

$$Heat\ added = 1116\ \frac{Btu}{lb}$$

$$Net\ WORK = 230\ \frac{Btu}{lb}$$

$$Efficiency = \frac{230}{1116} = .20$$

$$Horsepower = (h_2 - h_3) \times 120 \times 778 = 21,566,160\ \frac{ft-lb_f}{lb_m} = 653\ hp$$

To compare this cycle against the simple cycle, notice that the efficiency did not decrease much (22% versus 27%). In fact, if we consider that the combustion process and heat transfer in the boiler itself is more efficient at the lower temperatures, the complete efficiency for the system actually may improve for the lower superheat case. However, the solutions show that the horsepower is off 30% with the lower superheat. This means that although the steam power system is just as fuel-efficient with lower superheat, the boiler, turbine, and all components must be much larger, definitely an undesireable characteristic, since it adversely affects the cost of building power plants.

Also of interest is that Figure 20–7 clearly shows that the steam that goes through the turbine begins to condense. Notice that line 2–3 shows that the steam exiting the turbine has a quality of 90%, indicating that the turbine blades are dripping with condensate. The steam exiting the turbine would be called *wet steam*. Wet steam and condensate cause havoc in the turbine. Lubrication of the turbine blades can be washed away, blades may break when hit with liquid, and condensate removal is a problem.

Therefore, there are benefits and drawbacks to reducing boiler superheat. Another modification to the cycle may eliminate the problems and maintain the benefits of reducing superheat.

20.4 REHEAT IN THE POWER CYCLE

A modification to the steam power cycle yields the benefits of the high-super-heat effect without actually having to attain the high temperatures. It is called *reheat*. With this adaption, steam does not expand in the turbine in one stage but, instead, expands through a preliminary turbine called the first stage, then

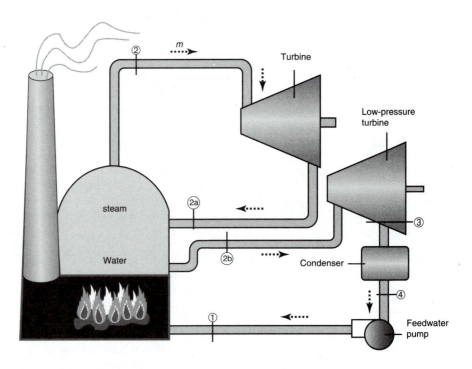

Figure 20–8. The steam cycle with reheat.

exits the turbine to be heated again in the steam generator before being ducted to a second-stage turbine for final expansion. Figure 20–8 shows the Rankine cycle with reheat and demonstrates how wet steam is avoided by using moderate amounts of superheat and superheating the steam twice instead of once. Sample Problem 20–6 investigates the effect of using reheat.

**SAMPLE
PROBLEM 20–6**

Consider a Rankine power generator with pressure limits of 400 psia and 15 psia, with a stage of reheat at 100 psia, and 600°F before a second stage of the turbine. Determine the heat added, net WORK, thermal efficiency, and horsepower if 120 lb$_m$/min of steam is generated. (Figure 20–9)

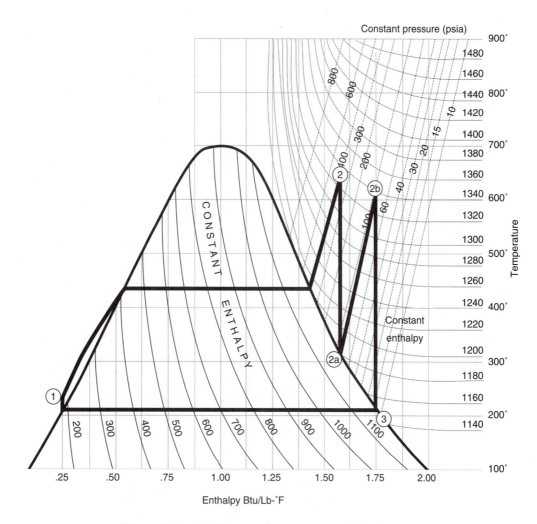

Figure 20–9. T-s diagram of steam power with reheat.

SOLUTION:

STATE	Temperature °F	Pressure psia	Enthalpy Btu/lb	Entropy Btu/lb –°R
1	215	400	190	.25
2	600	400	1306	1.59
2a	325	100	1185	1.59
2b	600	100	1306	1.76
3	212	15	1150	1.76
4	212	15	390	.25

$$q_{1-2} = h_2 - h_1 + h_{2b} - h_{2a} = 1306 - 190 + 1306 - 1185 = 1237$$

$$Net\ WORK = h_2 - h_{2a} + h_{2b} - h_3 - W_{pump} = 121 + 156 - 1.46 = 276$$

$$Efficiency = \frac{276}{1237} = .22$$

$$Horsepower = 276 \times 120 \times 778 = 25,760,000 = 780\ hp$$

Notice for this case that some of the efficiency and over half of the horsepower were recovered as compared to Sample Problems 20–3 and 20–5.

20.5 USING STEAM TABLES FOR CYCLE ANALYSIS

The thermodynamic diagrams of P-h and T-s give visual descriptions of the operation of the Rankine cycle that provide insight into its operation. Once mastered, these diagrams eliminate the tedious calculations required to perform such analysis. Another method for generating the state values necessary to evaluate cycles is use of the steam tables in Appendix III(a) and b.

Use of the tables to complete a cycle state point table requires a different approach than that taken in Sample Problem 20–3, and it requires an understanding of the cycle that is being represented. The state points 3 and 4 are found first. Both of these points are in Appendix III(a), since both of these are saturated points. Enter the table by finding the row that represents the pressure of state points 3 and 4—15 psia for this example. The temperature of both of these state points is found from the table to be 213°F. For state 3, the saturated vapor state, the enthalpy is found in the "saturated vapor" column: 1150.8 Btu/lb. For state 4, the value is found from the "saturated liquid" column: 181.11 Btu/lb. The entropies for both states are noted in the same manner.

State 1 can be filled in by noting that the enthalpy and temperature of the state is the same as for state 4, a fact that is known because it is consistent with our knowledge of the cycle (a liquid pump changes neither the temperature of the fluid nor its enthalpy). The entropy of that state is not immediately known but is not necessary for the operation of the cycle.

The next data to fill in is state 2. The pressure of state 2 is 400 psia, and it is a superheated state that requires using Appendix III(b). The last row of that table represents this temperature. To find which column is proper, we again use knowledge of the cycle, this time the fact that the entropy of state 2 is the same as the entropy of state 3—in between the temperature columns of 900°F and 1000°F. To find the precise temperature of state 2, interpolate between these two temperature columns.

Since these tables have direct information of state properties with gaps that are 100°F apart, it is clear that the tables include only a small fraction of the data presented on the P-h or T-s diagrams. Interpolating between the direct information is a critical part of using the tables. Eyeballing or guessing between values is very imprecise, so calculators can be used to get more precise data with an interpolation formula. Consider that we want to interpolate the value x from between two values of that quantity, x_1 and x_2. Also, we know property Y at both state points 1 and 2 (Y_1 and Y_2), and we know the value of this property at interpolated point Y. Then to interpolate x, use

$$x = \frac{(Y - Y_1)}{(Y_2 - Y_1)} \times (x_2 - x_1) + x_1$$

SAMPLE PROBLEM 20-7

Interpolate the value from Appendix III(b) for the temperature of state 2 in Sample Problem 20–4 ($P = 400$ psia, $s = 1.76$).

SOLUTION:

From the table the following values are found (Y is entropy, s):

$Y = 1.76$ Btu/°F

$Y_1 = 1.7247$ Btu/°F $Y_2 = 1.7623$ Btu/°F

$x_1 = 900$°F $x_2 = 1000$°F

Then

$$x = \frac{(1.76 - 1.7247)}{(1.7623 - 1.7247)} \times (1000 - 900) + 900$$

$$= \frac{.035}{.0376} \times 100 = 990°F$$

To complete state 2, the enthalpy of the state also must be found by interpolation. This time

$Y = 1.76$ Btu/lb-°F

$Y_1 = 1.7247$ Btu/lb-°F $Y_2 = 1.7623$ Btu/lb-°F

$$x_1 = 1469.4 \ \frac{\text{Btu}}{\text{lb}} \qquad\qquad x_2 = 1522.4 \ \frac{\text{Btu}}{\text{lb}}$$

Then

$$Y = \frac{.035}{.0376} \times (53) + 1469.4 = 1520$$

SAMPLE PROBLEM 20-8

Redo the Rankine cycle of Sample Problem 20–4, this time using the steam tables, to determine:
a. the amount of heat added in the boiler per pound of steam
b. the amount of heat released in the condenser per pound of steam
c. the WORK done in the turbine
d. the cycle efficiency
Compare these results with the results of Sample Problem 20–4.

SOLUTION:

State	Temperature	Pressure	Enthalpy	Entropy
1	215	400	183	.316
2	990	400	1512.00	1.76 [Appendix III(b)]
3	215	15	1154	1.76 [Appendix III(a)]
4	215	15	183	.3160 [Appendix III(a)]

a. $q_{boiler} = h_2 - h_1 = 1512 - 183 = 1329 \ \dfrac{\text{Btu}}{\text{lb}_m}$

b. $q_{condenser} = h_4 - h_3 = -971$

c. $Wd_{turbine} = h_2 - h_3 = 1512 - 1154 = 358$

d. $\eta = \dfrac{358}{1329} = .27$

Steam tables have been a traditional method of Rankine cycle analysis, but they do suffer from the bother of interpolation and the inaccuracies that it entails. When computers are available and the proper software exists, steam tables can be accessed with great accuracy. Use of computerized steam tables requires a new set of skills—the knowledge of how to operate a computer. Try computerized steam tables on your computer disk, Lesson 8.

20.6 COGENERATION

Steam heating systems and steam power cycles are nearly the same. One performs a heating function and one performs a power generation function. But heat and power can be created in the same device if the turbine acts to generate the power and the heat from the condenser is put to use in a heating system. When both functions are put to use simultaneously, the system is called a total energy package or a cogeneration system.

SAMPLE PROBLEM 20-9

If the heat from the condenser of the Rankine cycle in Sample Problem 20–3 were used for heating, what is the capacity of the unit as a steam heat system (120 lb/min)? What is the temperature of the steam that would be fed to building radiators?

SOLUTION:

In that problem,

$$q_{3-4} = h_4 - h_3 = -962 \ \frac{Btu}{lb}$$

so the capacity of the heating system would be

$$\text{Capacity } q = mq_{3-4} = 120 \times 60 \ \frac{min}{hr} \times 962 = 6,926,000 \ \frac{Btu}{hr}$$

The temperature of the radiator steam is 215°F, somewhat low for a heating system. The condensing pressure might have to be increased if this system were used as a cogeneration system.

Cogeneration systems have received some degree of popularity in recent years, but they suffer from one problem. It is not often that one location needs both the level of power that a steam turbine provides and the amount of heat that it rejects. In particular, power plants that use great amounts of power usually are located in unpopulated areas where the heat is useless. On the other hand, office buildings and other commercial buildings require great amounts of heat, but not as much power as a full power plant. One application particularly strong in cogeneration is in hospitals where the need for heat is tremendous, not only for comfort but also for washing and sterilization, and the need for power is somewhat large. Even in this situation, though, a small power plant will not only generate sufficient heat for their needs but also usually have power left over.

Some cogeneration installations have been successful in selling power in the form of electricity to the local electrical utility company. In this way, they actually receive some revenues from their equipment as well as a savings of efficiency.

COGENERATION SPECIALIST

Fifteen years ago, Mercy Hospital of Queens, New York, hired their first engineering technician as a Climate Control Specialist. Until that point, their heating and air-conditioning maintenance technicians were hired through classified advertisement with no educational requirements. Their job description consisted primarily of cleaning and routine maintenance. But at that time, the control systems on the equipment were becoming quite sophisticated, and the hospital became reliant on outside contractors to fix their everyday problems. The engineering technician was hired straight out of college and sent to schools to learn from the manufacturers how to service the controls. Slowly, he began to take over the responsibilities of maintaining the total hospital system, which included an air distribution system with pneumatic and electric controls at each room outlet for proper air flow and temperature, a state-of-the-art filter system to remove airborne germs and pathogens, a large refrigerating plant for air-conditioning, and a 12,000-pph (lb_m/hr) steam boiler for the heat requirement.

Within two years, the outside repair contractors had been reduced to only occasional visits and the technician had a staff of five working around the clock to maintain the building. After five years at the hospital, the technician made a startling proposal to the hospital administration. Through his involvement with manufacturers and attendance at industry-wide meetings, he had learned about a method of generating electricity for the hospital and creating all the steam necessary for heating the hospital with one piece of equipment. The hospital was buying electricity from the local utility and had no idea that they could make their own electricity. The technician told them that because they had a large requirement for electricity, they should make their own power through a steam turbine, and because they had a large requirement for heating steam, they should use the waste heat off the turbine condenser to generate that steam. This concept is called cogeneration. The technican convinced the hospital that with cogeneration, it could recoup the cost of the equipment through savings on electricity within three years.

The equipment that was installed included a high-pressure steam boiler that was capable of generating steam at 800 psia with a superheat of 300°F. It was connected to a steam turbine with reheat that exhausted saturated steam at 30 psia to a condenser. The condenser steam is not used directly for space heat but was passed through a heat exchanger cooled by a circuit that converted water to steam at 15.0 psia and could superheat the steam by 40°F. This condenser, therefore, was a heat exchanger that had a high-pressure side for the power turbine steam and a low-pressure side for heating steam.

SAMPLE PROBLEM 20-10

Determine the state points for the Mercy Hospital steam turbine power plant using Figure 20–6(b).

a. Determine how much superheat is necessary for reheat to meet the condenser conditions at saturation.
b. Calculate the efficiency of the power cycle.
c. Determine the electrical rating of the power plant if it processes 600 lb_m/min. Find the rating in megawatt/hr (1 watt = 3.41 Btu/hr)
d. How many lb_m/min of water at 100°F can be turned into heating steam at 15 psia and 252°F from the heat off the condenser?

SOLUTION:

POWER CYCLE STATE TABLE

State	T	P	h	s
1	250	800	240	.33
2	820	800	1410	1.67
3a	315	70	1180	1.67
3b	460	70	1258	1.75
3	250	30	1168	1.75
4	250	30	240	.37

a. From the table, reheat superheat is
$$T_{3b} - T_{3a} = 145°F$$

b. $Net\ WORK = Wd_{2-3a} + Wd_{3b-3} - W_{pump}$

$$= h_2 - h_{3a} + h_{3b} - h_3 - \frac{(P_1 - P_4)144}{1.93 \times 32.2 \times J}$$

$$= 1410 - 1180 + 1258 - 1168 - \frac{730 \times 144}{62.4 \times J}$$

$$= 230 + 90 - 2.17 = 308\ Btu$$

$$\eta = \frac{308}{q_{boiler} + q_{reheat}} = \frac{308}{1410 - 240 + 1258 - 1180} = \frac{308}{1248}$$

$$= .25$$

c. $$308\frac{Btu}{lb} \times 600\frac{lb}{min} = 1,848,000\frac{Btu}{min} \times 3.3\ megawatt\frac{60\ min}{1\ hr}$$

$$= 11,088,000\frac{Btu}{hr}$$

d. $$q_{cond} = h_3 - h_4 = 1168 - 240 = 928\frac{Btu}{lb}$$

$$q_{cond}\frac{Btu}{hr} = 928 \times 600 \times 60 = 33,408,000$$

$h_{leaving\ cooling\ steam} - h_{enter\ cooling\ water} = 1170 - 180$

$$= 990 \frac{Btu}{lb}$$

$$990 \frac{Btu}{lb} \times \frac{lb}{hr} = 11,088,000$$

$$\frac{lb}{hr} = 11,200$$

Once the cogeneration system was installed and personnel were trained to operate it, the technician's career took an abrupt change in direction. He left the hospital to become a cogeneration consultant. Based on his experience, he now designs cogeneration systems for other hospitals. He knows the turbine equipment so well that he has set up a shop to perform special service. In particular, the turbine impeller blades can become worn from use over the years, and he performs the precision machining needed to bring them back to peak efficiency. His repair business is more successful than the design business, and he now has a staff of three people in his machine shop, all engineering technicians.

CHAPTER **20**

PROBLEMS

1. A hot water heating system has 200 gal per minute of water flowing at 160°F and returning to the boiler at 130°F. What is the heating capacity of the system?

2. The supply line from a steam boiler carries steam at a temperature of 240°F and a pressure of 23 psia. How many degrees of superheat are there? If the pressure drops 2 psi in the supply line, will the steam begin to condense before the radiator?

3. In Sample Problem 4–5, the chef is cooking with a pot of steam at 350°F and a pressure of 16.5 psia. Is the steam superheated, and if so, by how much? Did the solution to that problem require that the steam be superheated, or would the solution have been acceptable if the steam were in saturation?

4. Is steam at a temperature of 600°F and a pressure of 320 psia saturated?

5. Steam at 600°F and 320 psia is made in a boiler from 190°F water. What is the change in enthalpy, specific volume, and entropy of the medium?

6. For the case of Sample Problem 20–2, consider that the pressure drop in the pipes was a perfect free expansion, $\beta = 0$. Calculate the heat released in the radiator for this case and compare the results to those in the sample problem.

7. For the case of Sample Problem 20–2, consider that the pressure drop in the pipes was a reversible adiabatic process. Calculate the heat released in the radiator for this case and compare the results to those in the sample problem. What device in the supply line would have made this process a reversible adiabatic one?

8. A low-pressure steam heating system operates between 17 psia and 15 psia with no superheat and an irreversibility index of $\beta = 1.00$ (free expansion). At what temperature and with what quality does the steam enter the radiator? How much heat is given off per pound by the steam in the radiator? What is the static head?

9. Repeat Problem 7 but with 10°F of superheat.

10. Repeat Problem 7 but with 10°F of superheat and an irreversibility index of .8.

11. The Brayton cycle of Chapter 16 consisted of adiabatic processes and constant pressure processes. How does it differ from the Rankine cycle?

12. A sealed pot contains one pound of steam and water in saturation. If the temperature of the steam is 280°F, what is its pressure? If the quality of the steam is 50%, how much (lb$_m$) of the mixture is liquid? If the temperature of the pot is raised to 290°F but remains saturated, what is the pressure in the kettle? How much liquid is changed to steam when the temperature increases by 10°F? (Hint: Use Figure 20–5.)

13. Determine the Carnot efficiency of a cycle with the same operating limits of Sample Problem 20–2 and compare it to the results of the sample problem.

14. A steam turbine requires maximum inlet pressure of 800 psia and superheat of 30°F. If the turbine process is adiabatic and reversible and the exit pressure is 15 psia, what is the quality of the steam as it exits?

15. A Rankine cycle turbine receives steam at a condition of 950 psia and 760°F and exhausts it at 17 psia. What is the superheat of the entrance steam? What is the heat added, net WORK, thermal efficiency, and amount of steam flow required to achieve 5000 horsepower?

16. Consider that reheat is added to the system of Problem 15 at a point in the turbine expansion that brings the steam back to saturation, and that the reheat is such that the steam leaves the reheat stage with 100°F superheat. Determine how much reheat in Btu/lb$_m$ must be added.

17. Assume that in Problem 15, there is friction in the turbine that makes the adiabatic expansion irreversible. The irreversibility index is .9 (the change in enthalpy during that process is only .9 of its reversible value). What is the temperature and quality of the steam as it exits the condenser?

18. Find the kilowatts of power that can be generated from the Rankine cycle of Sample Problem 20–2. (1 hp = 2545 Btu/hr; 1 kW = 3413 Btu/hr).

19. A saturated mixture of water and steam is placed into a constant pressure vessel set to maintain 200 psia. The vessel is placed in an oven set for 400°F. Once the vessel reaches equilibrium, what will be the quality or superheat or subcooling of the medium?

20. Compare the heat flows in the diagram for a typical power plant in Figure 20–10. Why is the energy into the turbine and the energy into the generator not the same?

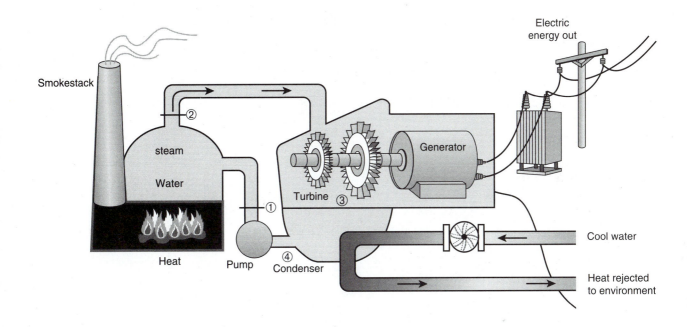

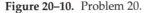

Figure 20–10. Problem 20.

21. A diagram of a steam power plant is shown in Figure 20–10. Consider that the pump (1–2) takes in 10,000 lb/hr of condenstate at 220°F and 20 psia and compresses it to 50 psia. Find:
   a. T_2
   b. Wd

22. The boiler in Figure 20–1 heats the steam at constant pressure to 1300°F. Find:
   a. T_3
   b. q
   c. Wd

23. The adiabatic-expansion (power-stroke) turbine in Figure 20–10 generates power and reduces the temperature to 300°F. Find:
   a. P_4
   b. Wd

24. Since the steam cycle of Figure 20–10 must be a "closed" cycle, what must be the last process (4–1)? Find:
    a. *Wd*
    b. *q*

25. For the steam cycle of Figure 20–10, find:
    a. *Net WORK*
    b. *Efficiency*
    c. *Hp*

CHAPTER
21

Psychrometrics: The Interaction of Moisture and Air

PREVIEW: When two materials come in contact with each other and one is in the gas phase and one is in the liquid phase, there is a slight amount of mixing between the two. How much mixing occurs is important to meteorologists, who must determine the humidity in air, and to designers of air-conditioning systems, where humidity plays an important role. This situation can be analyzed with thermodynamics and an important new graph called the psychrometric chart.

OBJECTIVES:
- ❏ Learn the vocabulary of humidification.
- ❏ Calculate the numerical indicators of humidity.
- ❏ Use the psychrometric chart to make the calculations easy.

21.1 WATER VAPOR IN AIR (HUMIDITY)

The properties of a material in a phase change were introduced in Chapter 18. There it was shown that a material can be in a solid phase, be a subcooled liquid, be in saturation, or be a gas. During saturation, both gas and liquid forms of the material exist simultaneously and in varying portions, depending on the quality of the material. As a subcooled liquid, the material exists only as a liquid, with no vapor component. Figure 21–1 shows water in a pan at 70°F. Under atmospheric pressure, the water is subcooled, and there is no vapor above it. However, the pressure is applied to the liquid by the air above

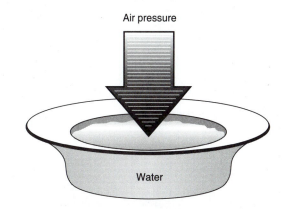

Figure 21–1. Water in a pan under atmospheric pressure.

it; therefore, the system is really one of liquid water with gaseous air above it. It is actually a two-component system—a mixture of gases and liquid.

An important thermodynamic situation that we have not studied is the thermodynamics of different liquids and gases together in the same space. Figure 21–2 demonstrates an application of this—the meeting of sky and ocean, a gas-liquid mixture. A microscopic view of a sample of air over the water reveals that air and water molecules are intermixed. The fact that water molecules exist in the vapor phase seems to be a violation of what we know about a subcooled liquid. Moisture in air is called *humidity*, but how did it get there if the liquid is subcooled?

Molecules in a liquid are vibrating, rotating, and translating, and this motion is collectively called "speed," which is measured as the temperature. Individually, the molecules do not all have the same speed, but the temperature records the average speed. In fact, Figure 21–3 shows a distribution of speeds, from slow to fast, for the molecules. Although the average speed of molecules in this graph correspond to a liquid temperature of 70°F, a few of

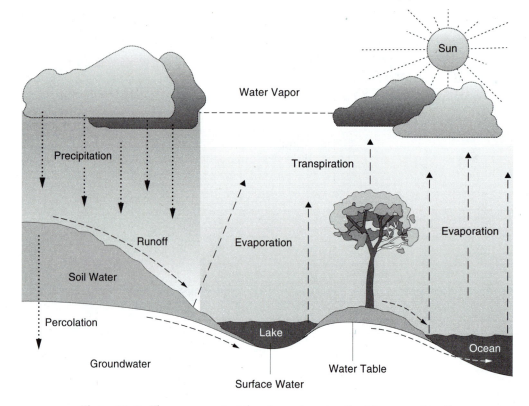

Figure 21–2. The ocean against the atmosphere is a liquid-gas combination.

the molecules have speeds greater than that corresponding to 212°F, and these molecules could escape as vapor. If these molecules are located at the bottom of the liquid, they have nowhere to which to escape. But if they rise to the top surface, they can escape. Therefore, even though a liquid is subcooled, there is a small percentage that will vaporize in a process called evaporation. Boiling, or rapid evaporation, is a phenomenon that occurs when molecules are transitioning to vapor all throughout the liquid (therefore, the bubbles in a boiling liquid), whereas evaporation is a slow vaporization at the surface of the fluid.

❖ **KEY TERM:**   **Boiling:** Rapid evaporation of molecules from within the confines of the liquid.

❖ **KEY TERM:**   **Evaporation:** The separation of faster molecules from a liquid as they reach the surface of that liquid.

A consequence of the evaporation process is that, if all the faster molecules escape the liquid, the average speed of the remaining molecules is less. This explains the well-known fact that *evaporation is a cooling process.* As the liquid temperature drops during evaporation, it seems that evaporation should stop when all of the fastest molecules have made it to the surface. But if the local temperature of the air surrounding the water heats it back up to its original temperature, molecules again gain speed, and some of them gain a sufficient speed to be a gas. The water will continue to evaporate until all of it is gone.

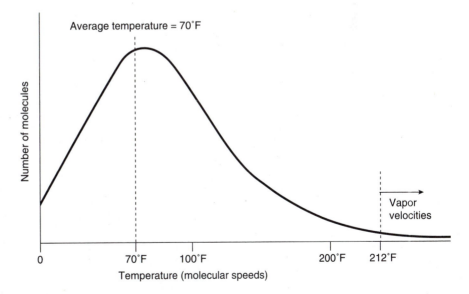

Figure 21–3. Molecules in a gas have different velocities.

This explains the difference between evaporation and boiling. Now consider something quite different. A lid is placed over the evaporating pan of water. As the liquid is evaporated away, it is not lost. It remains in the air over the water. More and more water molecules become vapor, and the air over the top contains a higher and higher percentage of water vapor. Soon so many water molecules are in suspension that some of them strike the surface of the water, become bound by the water molecules again, and become liquid molecules. Eventually there will be as many molecules going back to liquid (condensing) as there are evaporating. This is a state of equilibrium where the same amount of water vapor is held in the air as both evaporation and condensation takes place. It is called *saturation*. (Figure 21–4)

❖ **KEY TERM:**   **Saturation:** The condition in which the maximum amount of water vapor is suspended in air.

The word *saturation* is used in Chapter 18 to mean the state of a substance that is going through a liquid-vapor transition. In this chapter, we use the same word to mean the state where two different gases (air and water vapor) are mixed and the percentage of one of the gases is the highest it can be. The word is often used in its two different meanings. "Saturated steam" can mean only wet steam that is a combination of vapor and water droplets. "Saturated air" must be air saturated with moisture. The word *evaporation* also is used with

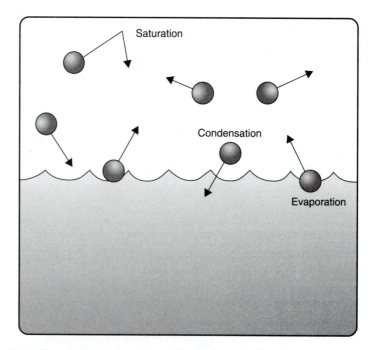

Figure 21–4. Evaporation takes place at the surface of a liquid.

more than one definition. In Chapter 19, the refrigeration system evaporator was the component in which the refrigerant was evaporating; but in fact, the refrigerant was in saturation. This evaporation was not a surface activity, but true boiling. Low temperature boiling, however, often is referred to as evaporation.

Once saturation of air is achieved, the amount of water vapor suspended in air can be measured and recorded as specific humidity (lb_{water}/lb_{air}); however, if the temperature of the water is increased at some later time, molecules begin to evaporate at a greater rate, and more of them reach suspension before a new equilibrium is reached. When the temperature of the water is increased, a new saturation is reached at a higher specific humidity. Table 21–1 represents the saturated properties of air and water vapor. The left column of that table is the temperature of the combination dry air and water vapor. The next column is the amount of water vapor that can be suspended in the air at that temperature (specific humidity). Notice that this column is specific humidity, and that the value increases as the temperature rises. Meteorologists, or weathermen, can use this table to find *relative humidity*. This term is used for humid air samples that are not at saturation, often the case for weather conditions.

Table 21–1

PROPERTIES OF MIXTURES OF AIR AND SATURATED WATER VAPOR

Temperature (°F)	Moisture Content (lb_{water}/lb_{air}) (Specific Humidity)	Enthalpy(Btu/lb) Dry Air	To Sat	Sat Air
20	.0021	4.80	2.30	7.10
25	.0027	6.00	2.93	8.93
30	.0034	7.20	3.71	10.92
35	.0042	8.41	4.60	13.01
40	.0052	9.61	5.62	15.23
45	.0063	10.81	6.84	17.65
50	.0076	12.01	8.29	20.30
55	.0092	13.21	10.02	23.23
60	.0110	14.41	12.05	26.47
65	.0132	15.61	14.45	30.07
70	.0158	16.82	17.28	34.10
75	.0188	18.02	20.59	38.615
80	.0223	19.22	24.48	43.70
85	.0264	20.42	29.02	49.44
90	.0312	21.62	34.32	55.95
95	.0367	22.83	40.51	63.34
100	.0432	24.03	47.73	71.76

Abstracted with permission from American Society of Heating, Refrigerating and Air Conditioning Engineers, ASHRAE Handbook of Fundamentals, 1968.

❖ **KEY TERM:**   **Relative Humidity (RH):** The ratio of the actual amount of specific humidity in a sample compared with the specific humidity of that sample at saturation.

As an example, at 70°F, the amount of moisture in the air at saturation is .0158 lb$_{water}$/lb$_{air}$. If an air sample tests for only .0074 lb$_{water}$/lb$_{air}$, then it can be concluded that the sample is not saturated, and, in fact, has a relative humidity of .0074/.0158 = .47 or 47%.

SAMPLE PROBLEM 21–1

How much moisture does the air hold (lb$_{water}$/lb$_{air}$) on a 90°F day with a relative humidity of 60%?

SOLUTION:

If the air were saturated, Table 21–1 indicates a specific humidity of .03118. Since the RH is 60%, the moisture content is .60 x .03118 = .01872 lb$_{water}$/lb$_{air}$.

Another important concept can be seen from the table by considering again 70°F air with a moisture content of .0074 lb$_{water}$/lb$_{air}$. If the air was cooled to a lower temperature without changing the moisture content, what would happen? If it were cooled to 62°F the total possible moisture would reduce to .0119. Comparing that against the actual .0074, the relative humidity has changed to .0074/.0119 = 62%. Continuing to cool the temperature of the air down to 50°F reduces the capability of air to hold moisture down to .0076, giving us .0074/.0076 = .96, a relative humidity of 96%.

When it is cooled down to 46°F the air can hold only .0065, which indicates that there is too much moisture in the air for its ability to hold it. Some moisture then must condense out. When the temperature is reduced to this level, moisture begins to form in the bottom of the container. This happens in nature as dew, and we call the temperature at which the moisture condenses the dew point.

What has been described is a natural phenomenon that often occurs early in the morning, when the temperatures of the day are the lowest. Air during the afternoon has acquired a certain amount of moisture and a relative humidity that is very low, but as the temperatures drop, the air can no longer hold the moisture and it condenses out as dew. Condensation does not actually have to take place to consider the dew point temperature of air. The dew point of air at 70°F and a moisture content of .0074 lb$_{water}$/lb$_{air}$ is 49°F. Since the dew point is so low compared to the actual air sample temperature, it gives an indication that the moisture content of the humid air is, in fact, small.

❖ **KEY TERM:**   **Dew Point:** The temperature at which an air sample would reach 100% RH (saturation).

SAMPLE PROBLEM 21-2

Find the dew point of 90°F of air that has a relative humidity of 60%. If the relative humidity was 70%, would the dew point be higher or lower?

SOLUTION:

The humidity of the 60% sample is 60 x .0312 = .0187 lb_{water}/lb_{air}. Searching Table 21–1, this amount of humidity corresponds to a saturated condition at approximately 75°F. Thus, the dew point of the 60% RH air is 75°F. The 70% sample has a dew point of 79°F.

Columns 3, 4 and 5 on Table 21–1 deal with the thermal energy of air: the enthalpy. Table 21–1 lists three columns of enthalpy. Column 3 is for dry air, column 4 is the additional enthalpy to saturate the dry air (at that temperature), and column 5 is the total enthalpy of the saturated air.

SAMPLE PROBLEM 21-3

How much heat is needed to raise the temperature of dry air from 50°F to 74°F?

SOLUTION:

From Table 21–1,

h_1 = 12.010 Btu/lb_{air} (at 50°F)

h_2 = 17.778 Btu/lb_{air} (at 74°F)

Therefore

$q = h_2 - h_1$ = 5.768 Btu/lb

SAMPLE PROBLEM 21-4

How much heat is needed to raise the temperature of one pound of saturated air at 50°F to saturated air at 74°F? How much water must be evaporated to keep the sample saturated?

SOLUTION:

$q = h_2 - h_1$

But now

h_1 = 20.301 Btu/lb

because it is saturated, and

$h_2 = 37.660$ Btu/lb

because it is saturated, too.

$q = 37.660 - 20.301 = 17.359$ Btu/lb

The moisture content is .00678 lb_{water}/lb_{air} before and .01890 lb_{water}/lb_{air} after. Therefore, the moisture added is .018190 − .007658 = .010532 lb.

The table gives enthalpy values only for dry air and saturated air, but the enthalpy for partially saturated air can be calculated by

$$h = h_{dry} + \frac{RH}{100} \times (h_{sat} - h_{dry})$$

Therefore, the enthalpy of 50°F air with RH = 60% is

$$h = 12.010 + 60 \times 8.291 = 16.984 \text{ Btu/lb}$$

SAMPLE PROBLEM 21–5

Air at 80°F, 70% RH is cooled to 54°F. How much heat must be taken out per lb_{air}, and how much water per lb_{air} will accumulate in the bottom at the chamber?

SOLUTION:

$$h_{dry} + \frac{70}{100} (h_{sat} - h_{dry}) = 19.7 + .7 \times 26.2 = 38.0$$

$$q = h_2 - h_1 = 22.6 - 38.0 = -15.4 \text{ Btu/lb}$$

Moisture content of 80°F, 70% RH air is

.7 x .0239 = .0167

Moisture content of 54°F saturated is .0078, and accumulated water is

$$.0167 - .0078 = .0089 \frac{lb_{water}}{lb_{air}}$$

21.2 THE PSYCHROMETRIC CHART

Table 21–1 has provided a wealth of information, helped to solve many problems, and introduced many terms appropriate to meteorology. Enthalpy has made calculations easier. There is an even simpler method of using graphic analysis, and that is the subject of this section.

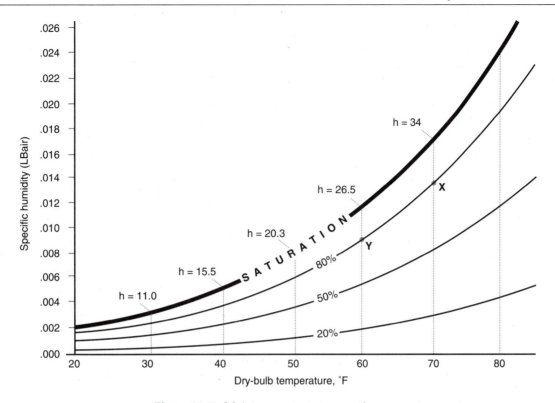

Figure 21–5. Moisture content at saturation versus temperature.

If the values from Table 21–1 are plotted on a graph, the graph will give visual representation of the change in thermodynamic quantities of saturated air at different temperatures. For example, plotting column 1 against column 2 will give a curve representing the increase of moisture content in saturated air as the temperature of the air rises. Figure 21–5 shows such a graph.

To use this graph, find the temperature of the air sample on the lower axis. Then go straight up to cross the "saturation curve," then directly to the left to find the amount of moisture content of the air (specific humidity, lb_{water}/lb_{air}). For an illustration, find that at 80°F, the specific humidity of air is .022, and for 60°F, the value is .0105.

Consider also a point 20% below the saturation line representing 20% less moisture than saturation—point X on Figure 21–5 for 70°F saturated and point Y for 60°F. Connecting these two points with a line, and then all the other points that are 20% below saturation, we get another line that represents air samples that are at 80% relative humidity. Figure 21–5 shows several lines of relative humidity drawn in.

Notice that on the saturation curve of Figure 21–5, the enthalpies have been written in for some states. These were obtained from Table 21–1. At 70°F, the saturated enthalpy is 31 Btu/lb_{air}, and at 60°F, the saturated enthalpy is

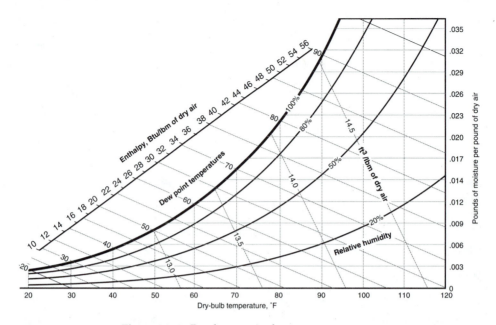

Figure 21–6. Psychrometric chart.

26.5 Btu/lb$_{air}$. Figure 21–6 extends these enthalpies to all the states on the graph and places them on the angled straight line. Notice also that the vertical axis has been moved to the right side where it is easier to read.

SAMPLE PROBLEM 21-6

Using Figure 21–6, find the enthalpy of air at 50°F with RH = 60%.

SOLUTION:

Interpolating from Figure 21–6,

$h = 16.5$ Btu/lb$_{air}$

21.3 WET BULB–DRY BULB

Specific humidity is great for problem-solving, but the specific humidity of a sample of air is very difficult to measure. To illustrate, imagine what type of instrument and what kind of sensor a meteorologist would have to use to absorb humidity and read the corresponding specific or relative humidity. Dew point temperature is also a measure of humidity, but how is dew point measured?

Figure 21–7. An adiabatic process that converts internal energy into latent heat.

A simple device to measure humidity can be introduced by investigating a rather mysterious situation. A man in a room of dry air is holding a spray bottle of water. (Figure 21–7) Suppose both the air in the room and the water bottle are 80°F. If the man begins to spray the water into the air to humidify it, what will be the resulting temperature of the air? If the air and the water are at the same temperature and no cooling or heating comes from the outside, we would expect the temperature to remain unchanged. This, however, is not what happens.

The first law of thermodynamics predicts the result properly.

$$q = c_p\,(T_2 - T_1) + m_L L \text{ (constant pressure process)}$$

The total amount of heating and cooling of the room and the water (q) is zero, and the equation seems trivial. Still, m_L is not zero; there is some evaporation of water. In fact, for 80°F, Table 21–1 indicates that .022 lb of moisture must be absorbed to reach saturation and a latent heat of 21.3 Btu/lb$_{air}$ absorbed by the water. Where does this energy come from?

The energy equation above predicts that the energy will come from the internal energy of the air with a resulting drop in temperature. Specifically

$$0 = cp(T_2 - T_1) + 21.3 \times m_L$$

$$T_2 - T_1 = -21.3 \times \frac{.022}{.24} = -2°F \quad (c_p = .24 \text{ Btu/lb} - °F \text{ for air})$$

By changing the amount of moisture vapor in the air in the room, the temperature of the room air has dropped.

This is a classic adiabatic ($q = 0$) process in meteorology or air-conditioning. It is an adiabatic process that involves the exchange of humidity and air temperature. The process can be represented on a psychrometric chart by realizing that since $q = 0$ and $q = h_2 - h_1$, then $h_2 = h_1$ or this adiabatic process is also a constant enthalpy process. Figure 21–8 demonstrates a process of constant enthalpy adiabatic cooling on the psychrometric chart. The process starts with dry air at 80°F (point 1) that is then humidified without heat addition (constant enthalpy) until the RH reaches 50% where its temperature

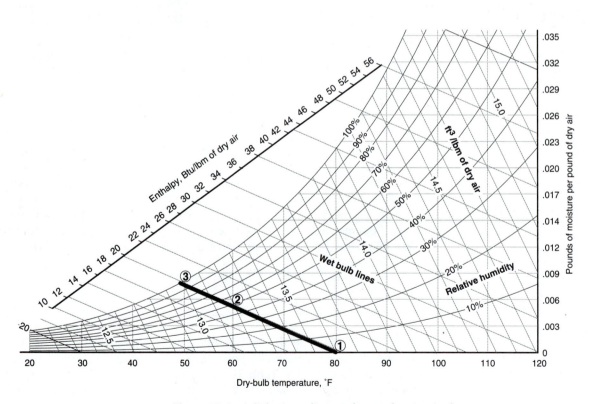

Figure 21–8. Adiabatic cooling on the psychrometric chart.

has fallen to 60°F (point 2). When it reaches saturation, the temperature will have dropped to 48°F (point 3).

The little man in the room with a spray bottle represents a technique for cooling air that on a large scale is called *evaporative cooling*. It works well in hot, dry climates.

SAMPLE PROBLEM 21-7

On a day in Albuquerque, New Mexico, the temperature reaches 105°F with RH 20%. Water is sprayed into the air to cool it. At what temperature will the air come out of the spray cooler if the relative humidity increases to 50%?

SOLUTION:

Use Figure 21–8 to find that the enthalpy of 105°F, 20% RH air is 36 Btu/lb. Follow this constant enthalpy line up to 50% RH, then drop straight down. A final temperature of 85°F is indicted.

Now we will relate this situation to a simple tool for measuring humidity. Consider a thermometer hanging on a wall. If a drop of water is placed on the thermometer or a wet cloth is placed around it, moisture begins to evaporate, and the thermometer records a decrease in temperature. (Figure 21–9) How fast the moisture evaporates depends on how rapidly the air can absorb the moisture. Drier air absorbs moisture faster so water will evaporate more rapidly. The cloth will get colder in dry air.

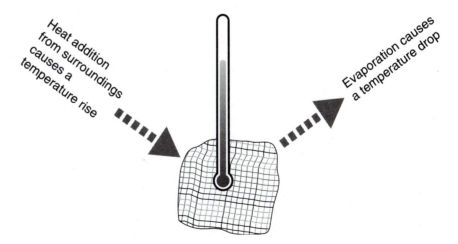

Figure 21–9. A cloth is cooled by evaporation and heated by the air around it.

Does the temperature keep dropping as evaporation continues? Perhaps the air around the cloth will saturate and cut the rate of evaporation. Assume that a small fan blows air away so that the cloth is constantly hit with fresh air. Will the temperature of the cloth continue to drop now? The answer is no, because the cloth is also heated. As the temperature of the cloth drops, it is heated by the air around it. The lower the temperature of the cloth, the more the air provides heat to the cloth. A point is reached where the heating by the air around the cloth matches the loss of heat by evaporation. Then the temperature of the cloth stops dropping.

The resulting thermometer reading is determined by the dryness and the temperature of the air. This temperature is different from the usual temperature reading and is called the *wet bulb temperature* (WB) to distinguish it from a temperature measured without a wet cloth, which is called a *dry bulb temperature* (DB).

Wet bulb temperature of air is always lower than dry bulb temperature of air. Imagine what would happen if the wet bulb thermometer were placed in a room with saturated air. Water would not evaporate from the cloth—it would have no place to go if the air were saturated. Since it would not evaporate, the temperature would not drop. Therefore, the wet bulb temperature reading would be the same as the dry bulb temperature for a saturated room. The conclusion is that wet bulb temperature reading equals dry bulb temperature in a room saturated with moisture. For drier air, the wet bulb temperature is lower than the dry bulb temperature.

An amazing thing happens when we put the wet bulb thermometer in the room with the little man who is squirting his spray bottle. As he humidifies the room, we watch the wet bulb temperature. *It never changes.* Initially, the wet bulb temperature will be low because the room is dry. As the room becomes more moist, the evaporation rate decreases, but the wet bulb temperature does not rise, since the actual temperature (dry bulb) of the room drops and less heating of the wet bulb is done. As the room air approaches saturation, the wet bulb temperature equals the dry bulb temperature. The conclusion of this illustration is that wet bulb temperature is an indication of humidity level in air samples and that all air samples that have the same wet bulb temperatures have the same enthalpy because the aspirator process is constant enthalpy.

This can be represented on Figure 21–8, since the lines of constant total enthalpy have already been drawn. These lines also represent the same wet bulb temperature. To determine what wet bulb temperature is represented by which enthalpy lines, remember that at saturation the wet bulb temperature equals the dry bulb temperature. Therefore, for 80°F saturated air, the corresponding wet bulb temperature is also 80°F. Starting at 80°F, the constant enthalpy line sloping downward and to the right also represents the air samples with a wet bulb temperature of 80°F. This means that at a dry bulb temperature of 100°F, the air sample that has a wet bulb temperature of 80°F also has a relative humidity of about 44%.

SAMPLE PROBLEM 21-8

What is the wet bulb temperature of air at 65°F dry bulb and 50% relative humidity?

SOLUTION:

Locate the point on the psychrometric chart and notice that it is just below the line corresponding to 55°F wet bulb.

Two more simple concepts complete the total power of the psychrometric chart. The first one is that the specific volume (v_{air}) of any air sample can be plotted on the psychrometric chart. It will then be easy to determine the specific volume of any air sample identified on Figure 21–6.

SAMPLE PROBLEM 21-9

What is the approximate specific volume of 65°F dry bulb air at 50% relative humidity?

SOLUTION:

$v_{air} = 13.4 \text{ ft}^3/\text{lb}$

Dew point also can be found by Figure 21–6 (see also Appendix V). Pick a point at 80°F dry bulb, 50% relative humidity. The dew point of this sample can be found by reducing the temperature of that sample until condensation is just beginning on the psychrometric chart. Start at the 80°F DB and 50% RH and go horizontally to the left, all the way to the saturated curve. Then dropping straight down to the dry bulb temperature line will pick up the temperature to which this sample has been reduced. This temperature is the dew point.

SAMPLE PROBLEM 21-10

Find the dew point of the air sample at 65°F DB and 50% RH.

SOLUTION:

The dew point is 47°F from Figure 21–8.

The graph of Appendix V is called the psychrometric chart. It is fairly complicated, but it gives so much information that it will become an easy-to-use and useful tool for working with air and moisture content. In general, there are seven properties of air and moisture that describe an air sample: dry bulb temperature, wet bulb temperature, dew point temperature, specific volume, total enthalpy, relative humidity, and specific humidity. If two of these characteristics are known, then the psychrometric chart will be able to determine the remaining five properties.

SAMPLE
PROBLEM 21–11

An air sample has a dew point of 70°F and a wet bulb temperature of 78°F. Find all the remaining psychrometric quantities.

SOLUTION:

$DB = 97.5°F$

$RH = 40\%$

Specific humidity = $.0155$ lb$_{water}$/lb$_{air}$

$v = 14.4$ ft^3/lb$_{air}$

Enthalpy = 41 Btu/lb$_{air}$

CHAPTER **21**

PROBLEMS

1. On the psychrometric chart, the saturation line is
   a. the line representing a saturated air-water-vapor mixture
   b. the line representing air with no moisture content
   c. the line where all moisture is condensed to water
   d. the 50% RH line

2. Air at 70°F DB, 60% RH has a wet bulb of
   a. 61°F
   b. 61%
   c. 55°F
   d. 55%

3. Air at 80°F DB, 50% RH has a dew point of
   a. 60°F
   b. 60%
   c. 80°F
   d. 80%

4. Air at 30°F DB, 100% RH has a dew point of
   a. below zero
   b. 30°F
   c. 10
   d. none

5. Air at 95°F DB, 40% RH has a specific humidity (lb_{water}/lb_{air}) of
   a. .013
   b. .017
   c. .023
   d. .026

6. Air at 65°F WB has an enthalpy of
   a. 84
   b. 26
   c. 30
   d. 100

7. The units of enthalpy are
   a. $\dfrac{Btu}{hr}$

   b. °F

   c. Btu/lb – °F

   d. Btu/lb

8. Air at 65° DB, 50% RH is heated to 90°F DB, 30% RH. What is the change in enthalpy of the air?

   a. 300 Btu/hr

   b. 300 Btu/lb

   c. 10 Btu/hr

   d. 10 Btu/lb

9. Air at 50°F WB is heated and humidified until it is at 60°F WB. What is the change in enthalpy?

   a. 60 Btu/hr

   b. 60 Btu/lb

   c. 6 Btu/lb

   d. 6 Btu/hr

10. Air at 85°F DB, 70°F WB has a specific volume of

   a. 62

   b. 14.1

   c. 13.5

   d. 105

11. This chapter has shown that when a liquid and gas come in contact, the liquid will evaporate to partially saturate the gas. The opposite also happens. The gas dissolves into the liquid and can even saturate the liquid. The solubility of nitrogen in liquid iron at 2500°F and 14.7 psia is .045% by weight. Calculate the volume of nitrogen dissolved in 1 ton (2000 lb$_m$) liquid iron at this temperature.

CHAPTER
22

Air-Conditioning
Processes

PREVIEW: Designers of air comfort systems use the psychrometric chart to match the proper equipment to a specific job. Many different types of conditioning processes for air are easily explained by plotting them on the psychrometric chart.

OBJECTIVES:
❑ Draw air process lines on the psychrometric chart.
❑ Convert psychrometric chart information into equipment sizing and specification details.

22.1
HUMAN THERMAL
COMFORT

The psychrometric chart is a great help for the meteorologists and designers of conditioning systems for comfort air. Comfort involves not only temperature of the surroundings but also the humidity level. Graphically, researchers have defined comfort in a chart that relates human comfort to psychrometric

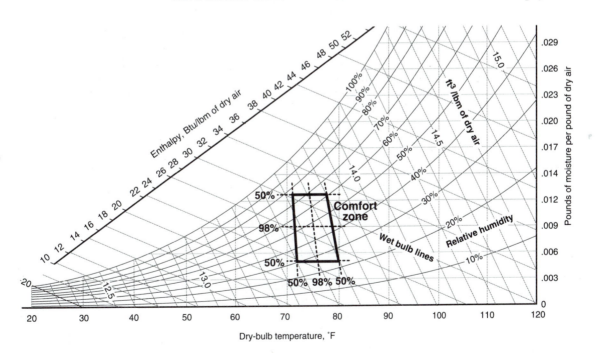

Figure 22–1. Comfort chart.

quantities: primarily temperature and humidity. (Figure 22–1) In the center of this chart is the comfort zone that implies relative degrees of comfort for an individual. In summer, 50% of the people will be comfortable indoors if the temperature is 72°F and the relative humidity is less than 75%. That same 50% will be happy even if the temperature rises to 80°F, if at the same time the humidity is dropped to 25%. The center zone of the chart covers the majority of people: 98% will be comfortable in the summer with 75°F in a room if the relative humidity of the room is 47%. However, if the room is drier, at 40% relative humidity, then the temperature in the room can rise to 78°F and still maintain comfort for 98% of the people. Humidity and temperature combined are necessary for comfort.

The process of changing the thermodynamic properties of air in order to maintain comfort in a building or room is called air-conditioning. This may involve heating to get higher temperatures, cooling the air to get lower temperatures, filtering the air to reduce contaminants, or humidifying or dehumidifying the air in order to get proper humidity levels. The air-conditioning process attempts to change the condition of air from one point on the psychrometric chart to another more comfortable point. The number of processes that can do this and the number of situations that can be involved are many.

To begin investigating different air-conditioning processes and how they relate to the psychrometric chart, consider a layout of a typical large air-conditioning system for a high-rise office building or a college classroom building. (Figure 22–2) In the upper left corner is a duct that pulls air from the rooms that are being conditioned. In that duct is a fan that provides the power to move the air. This duct is returning air to the air-conditioning process and, therefore, is called the return air duct.

In the center of the figure, there is a connecting duct from top to bottom. This section moves air to the lower right section, where it goes through several types of thermodynamic processes. The first one is a filter that can be as little as a replaceable filter in a residential furnace or as sophisticated as a "bag house" in a modern office building air-conditioning system. The second one is a cooling coil—the evaporating section of a refrigeration system or a coil through which cold water circulates. The third one is a heating coil. It might be the heat exchanger in a household furnace where combustion gases supply the heat. In larger systems, this coil might be filled with hot water or steam heated in a boiler. The final process is the humidifier coil, which can be a spray humidifier where air circulates through a water spray or electric elements that evaporate water into steam.

In the upper left of Figure 22–2 is a duct that exhausts air to the atmosphere so that fresh air, or outside air, can be added in the duct at the lower left. This fresh air comes in the exact quantity as the exhaust air and mixes with air that is returned through the center section to provide the air-conditioning coils a mixture of room air and outside air.

Now we shall consider one-by-one the air-conditioning processes that occur in the lower right of Figure 22–2.

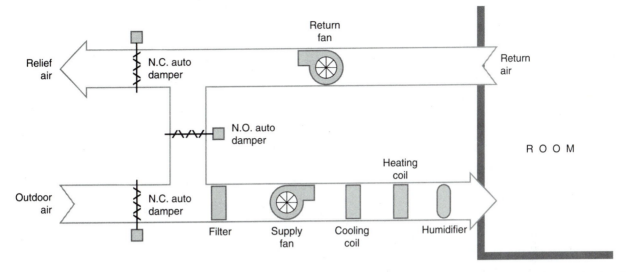

Figure 22–2. Schematic of an air-conditioning system for a commercial or institutional building.

22.2 HEATING AIR

Air that goes through a heating coil increases its temperature but has no effect on the amount of humidity in the air: none is added or condensed out. This process is called *heating without humidification*. Since no moisture is added or removed, this process on the psychrometric chart occurs along a horizontal line (constant specific humidity). With this information, we may learn something about the amount of heat needed to accomplish the heating.

SAMPLE PROBLEM 22–1

How much heat is required to change 70°F, 50% RH air to 105°F in a process without humidification? What is the final RH of the air?

SOLUTION:

Locate the initial air state on the psychrometric chart, then move horizontally (no moisture addition) to the right to a dry bulb temperature of 105°F (point 1 to 2 in Figure 22–3). The change of enthalpy is

$$h_2 = 34 \, \frac{\text{Btu}}{\text{lb}_{\text{air}}} \qquad h_1 = 26 \, \frac{\text{Btu}}{\text{lb}_{\text{air}}}$$

$$q = h_2 - h_1 = 8 \, \frac{\text{Btu}}{\text{lb}_{\text{air}}}$$

The RH of state 2 is 15%.

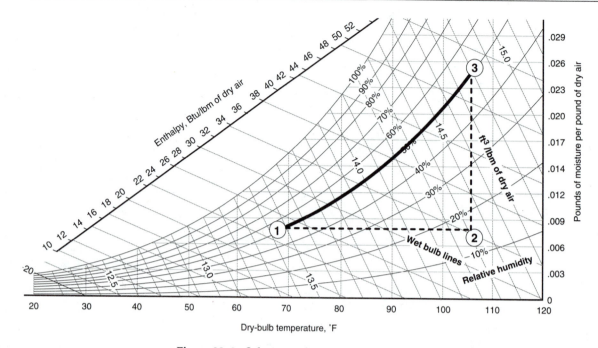

Figure 22–3. Solution to Sample Problems 22–1 through 22–4.

For the sake of discussion, we call the conditions of air going into the heating coil (1) the "on-the-coil" conditions, and we call the properties of the air leaving the system (2) the "off-the-coil" conditions.

The results of Sample Problem 22–1 are easy to calculate but do not provide the precise information that a furnace or heating coil designer needs to size the equipment necessary to do this conditioning. In particular, they do not indicate how big the heating system must be in Btu/hr. One additional step from the psychrometric chart is needed to put this information into the form needed by the designer.

This additional calculation involves an equation that relates Btu/hr and Btu/lb$_m$. This equation is

$$\frac{\text{Btu}}{\text{hr}} = \frac{\text{Btu}}{\text{lb}_m} \times \frac{\text{lb}_m}{\text{hr}}$$

Usually the lb$_m$/hr that circulates through a heating system is not known; instead, the size of the fan will relate to the CFM (cubic feet per minute) of air that the fan will push. The cubic feet per minute can be related to lb$_m$/min by the equation

$$\frac{\text{lb}_m}{\text{hr}} = CFM \times \frac{60}{v_{\text{air}}}$$

The overall designer's equation that we will use often, then, is:

$$\frac{Btu}{hr} = \frac{Btu}{lb_m} \times CFM \times \frac{60}{v_{air}}$$

Equation 22–1

This is one of the equations that links the psychrometric chart to the designer's specification of air-conditioning equipment.

SAMPLE PROBLEM 22–2

Find the size of the heating system for the situation of Sample Problem 22–1 if 2500 CFM is circulated by the fan located before the heating coil.

SOLUTION:

$$\frac{Btu}{hr} = 8\frac{Btu}{lb} \times 2500\ CFM \times \frac{60}{v_{air}}$$

What value of v_{air} should be used? Should it be the "on-the-coil" value or the "off-the-coil" value? Logically, if the CFM is specified for the fan and the fan is located before the heating coil, then the v_{air} should be for the "on-the-coil" condition, or

$$v_{air} = 13.5\frac{ft^3}{lb_m}$$

Therefore

$$\frac{Btu}{hr} = 8 \times 2500 \times \frac{60}{13.5} = 88,880\ \frac{Btu}{hr}$$

22.3 HEATING WITH HUMIDIFICATION

The results of Sample Problem 22–2 show that off-the-coil air comes out at a high temperature but a low relative humidity, too low for proper comfort during the heating season. Some humidification may be required, possibly to maintain a constant relative humidity. On the psychrometric chart, this means the heating and humidification processes follow a line of constant relative humidity.

SAMPLE PROBLEM 22–3

Air is returned to a coil of a central heating system at a condition of 70°F DB and 50% RH and heated and humidified to a condition of 105°F and 50% RH. How much heat is required per hour if a 2500 CFM fan is used?

SOLUTION:

$$q\left(\frac{\text{Btu}}{\text{lb}_m}\right) = h_3 - h_1 = 52 - 26 = 26.0 \frac{\text{Btu}}{\text{lb}_m} \text{ (see Figure 22–3)}$$

$$q\left(\frac{\text{Btu}}{\text{hr}}\right) = 26 \times 2500 \times \frac{60}{13.5} = 288{,}800 \frac{\text{Btu}}{\text{hr}}$$

Although Figure 22–3 suggests that the heating and humidification processes occur simultaneously to maintain constant relative humidity, Figure 22–2 clearly shows that the equipment required for heating precedes that required for the humidification process. The actual processes occur along the dashed lines of Figure 22–3. Breaking the processes up into their component parts is essential because the designer is actually specifying the heating coil and the humidification equipment; therefore, the total Btu/hr value from Sample Problem 22–3 has little meaning, since this is for the two processes combined. Figure 22–3 clearly shows that the enthalpy change for the air in the heating coil is $h_2 - h_1 = 8$ Btu/lb$_m$ (Sample Problem 22–1). That psychrometric chart also indicates the amount of water that must be added in the humidifier as the difference in specific humidity:

$$.024 - .008 = .016 \frac{\text{lb}_{\text{water}}}{\text{lb}_{\text{air}}}$$

More important than this value is the lb$_{\text{water}}$/hr that must be supplied to the humidifier. To convert the psychrometric information into designer's specifications, use the relationship

$$\frac{\text{lb}_{\text{water}}}{\text{hr}} = \frac{\text{lb}_{\text{water}}}{\text{lb}_{\text{air}}} \times \frac{\text{lb}_{\text{air}}}{\text{hr}} = \frac{\text{lb}_{\text{water}}}{\text{lb}_{\text{air}}} \times CFM \times \frac{60}{v_{\text{air}}}$$

Equation 22–2

SAMPLE PROBLEM 22–4

For the case of Sample Problem 22–3, determine the size of the heating system or coil (Btu/hr), the size of the humidifier heater (Btu/hr), and the amount of water that must be supplied to the humidifier (lb$_{\text{water}}$/hr).

SOLUTION:

Heating coil:

$$Q\left(\frac{\text{Btu}}{\text{hr}}\right) = 8 \times 2500 \times \frac{60}{13.5} = 88{,}800$$

Humidifier:

$$Q\left(\frac{\text{Btu}}{\text{hr}}\right) = 18 \times 2500 \times \frac{60}{13.5} = 200,000$$

$$\frac{\text{lb}_{\text{water}}}{\text{hr}} = .016 \times 2500 \times \frac{60}{13.5} = 178 \text{ (use Equation 22-2)}$$

22.4 REFRIGERATING AIR

Passing air through the cooling coil is similar to passing it through the heating coil except the temperature of the air is reduced. The humidity of the air is unchanged (dehumidification is considered in the next section); therefore, the process on the psychrometric chart occurs on a horizontal line, from right to left (see Figure 22-4, process 1-2).

SAMPLE PROBLEM 22-5

Air at 80°F and a 40% RH is to be cooled to 60°F. If 2500 CFM is circulated, how much refrigeration (Btu/hr) is required (process 1-2 in Figure 22-4)?

SOLUTION:

$$q = h_2 - h_1 = 24 - 29 = -5 \frac{\text{Btu}}{\text{lb}_{\text{m}}}$$

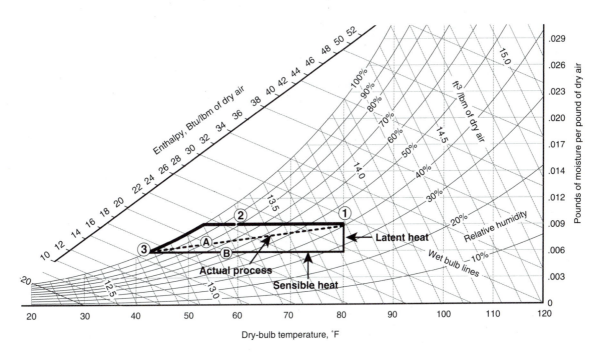

Figure 22-4. Solution to Sample Problems 22-5 and 22-6.

$$Q\left(\frac{\text{Btu}}{\text{hr}}\right) = -5 \times 2500 \times \frac{60}{13.5} = -57{,}602 \, \frac{\text{Btu}}{\text{hr}}$$

22.5 COOLING WITH DEHUMIDIFICATION

Suppose in the previous case we wish to cool the air going into the cooling coil down to 45°F. The psychrometric chart indicates that 45°F is below the dew point of the air; therefore, the process will reduce the temperature of the air to the point that some moisture will condense out. This might be beneficial if we want to remove moisture from a muggy, interior atmosphere.

The air cools down to the dew point on a horizontal line and begins to condense, moving down the saturation line until a final temperature is achieved, the "off-the-coil" temperature. On the psychrometric chart, this takes place in two steps and is easy to analyze.

SAMPLE PROBLEM 22–6

Air at 80°F and 40% RH enters a cooling coil set at a temperature of 45°F.
a. Will dehumidification of the air be possible?
b. If the air comes off the coil at the coil temperature, what will be its relative humidity?
c. What percent of the total heat removed will be latent heat?
d. What is the total Btu/hr of cooling if the fan is 2500 CFM?

SOLUTION:
a. Yes, 45°F is below the dew point of the air (dew point is 54°F)
b. 100% RH
c. On Figure 22–4, the process line (1–2–3) proceeds to the left to the saturation line, then curves down the saturation line. The overall change in enthalpy is

$$h_2 - h_1 = 17 - 29 = -12 \frac{\text{Btu}}{\text{lb}}$$

The figure shows the process broken into hypothetical horizontal and vertical sections. The vertical section represents an enthalpy change of –4 (25 – 29), and this is the latent heat component (specific humidity is dropping). Therefore,

$$\frac{-4}{-12} = .33 \text{ or } 33\%$$

of the load is latent heat.

d. $Q\left(\dfrac{\text{Btu}}{\text{hr}}\right) = 12 \times 2500 \times \dfrac{60}{13.8} = 133{,}300$

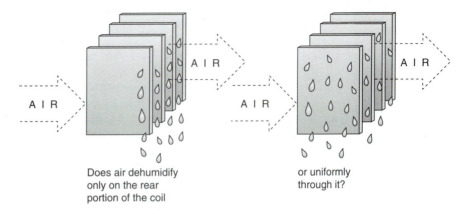

Does air dehumidify
only on the rear
portion of the coil

or uniformly
through it?

Figure 22–5. Condensation on a coil.

Sample Problem 22–6 reduces cooling and dehumidification into two steps, but is this in fact what happens? For example, is the temperature of the air reduced down to the saturation point in the first part of the cooling coil with no dehumidification taking place? If this were the case, water would not condense out on the front part of a cooling coil but start at some interior point. (Figure 22–5) But this is not what happens. Figure 22–6 illustrates this with a close-up look at the fins of such a cooling coil. These fins are extensions of the cooling medium. The air molecules that go between must come in contact with the fins before their temperature is dropped. When this happens, the temperature of that parcel is lowered immediately to the fin temperature. This means that some air parcels hit the fin upon entry into the cooling coil, the temperature is dropped, and these parcels immediately give up moisture to the coil.

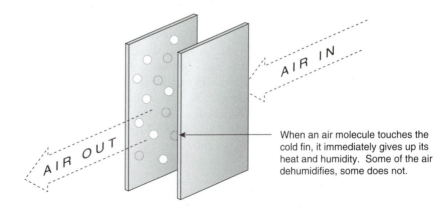

AIR IN

AIR OUT

When an air molecule touches the
cold fin, it immediately gives up its
heat and humidity. Some of the air
dehumidifies, some does not.

Figure 22–6. Some the air dehumidifiers, some does not.

Other air parcels never touch a fin and go through the coil without temperature change.

Water drips off the coil from beginning to end, and the air is being dehumidified from beginning to end. The line that represents this on the psychrometric chart is a diagonal line that directly connects the incoming air point with the leaving air point. Air in the coil goes through every condition on that line (see Figure 22–4, dotted lines).

The calculations of Sample Problem 22–5 do not have to be redone. They are correct no matter what the specific path on the psychrometric chart. Still, the line on the chart connecting initial and final air conditions is important for later analysis of cooling coils. Heat that is removed by this coil is both latent and sensible heat, and it may be important for the designer to know how many Btu/hr of sensible heat are removed by the coil and how much of it is latent heat. This is easily done by dividing the process into its sensible and latent heat portions, as was done in Sample Problem 22–6.

22.6 COOLING COIL TEMPERATURE

The cooling and dehumidification processes were investigated in the previous section. It was found that the air in a cooling coil is simultaneously dehumidified and cooled. The line in Figure 22–4 of the air sample as it passes through the coil is more or less a straight one from the initial point to the leaving air point. We have noted that at every point in the coil, some dehumidification and cooling has been accomplished. Now we ask the question whether the air leaving the cooling coil is really saturated or whether it is possible the air can leave the coil at some point along that line before saturation occurs.

We can imagine a cooling coil being cut off so that it is only half as deep. In this case, only half the cooling and half the dehumidification would be done. In such a case, we could stop along the conditioning line at a point short of the saturation line. What this implies is that some of the air can get through any cooling coil without touching the surface. For example, if 90% of the air going through the coil touches the coil and 10% does not, then air will leave the coil on the psychrometric chart at a point that is 90% along the process line but 10% short of saturation. Locating this temperature on the process line indicates that the air does not come off the coil saturated.

If some of the air in a cooling coil goes through unaffected, it can be referred to as "bypassed," and a bypass factor can be defined to indicate what percentage of the air did not touch the coil surface. The formal calculation for bypass factor introduces the terms "on-the-coil" temperature, T_{on}, "off-the-coil" temperature, T_{off}, and "coil temperature," T_{adp}. Then:

$$Bypass\ factor = \frac{\left(T_{off} - T_{adp}\right)}{\left(T_{on} - T_{adp}\right)}$$

SAMPLE PROBLEM 22-7

For the conditions of Sample Problem 22–6, consider that 90% of the air touches the coil and 10% does not.
a. What is the temperature of the off-the-coil air?
b. What is the bypass factor for this case?
c. If the bypass factor is 20%, what will be the off-the-coil temperature?

SOLUTION:

a. From Figure 22–6, measure 10% of the distance between points 1 and 3 (see point A) and find $T_{off} = 48°F$

b. $\textit{Bypass factor} = \dfrac{48-45}{78-45} = \dfrac{3}{33} = .1$ or 10%

c. $\textit{Bypass factor} = .20 = \dfrac{T_{off}-45}{78-45}$

or

$T_{off} = 51.60°F$ (point B, Figure 22–4)

22.7 DESIGN BY SENSIBLE HEAT FACTOR

In the Sample Problem 22–6, it is implied that the designer knows what the coil temperature should be for this cooling situation (45°F, in this case). In fact, the designer's responsibility is to choose the temperature for the coil that will give proper results. What determines the temperature at which a cooling coil should be set? The immediate answer is that the more Btu/hr that must be taken out, the lower should be the coil temperature. For the designer, however, this is absolutely not true. Equation 22–1 indicates that the designer has two variables that can be manipulated to generate the Btu/hr output required: enthalpy change and the fan CFM. By driving more air across the coil, more Btu/hr will be taken out. Sample Problem 22–8 illustrates the trade-off that a designer can make.

SAMPLE PROBLEM 22-8

In Sample Problem 22–6, the designer selected a coil temperature of 45°F and a cooling capability of 133,300 Btu/hr resulted with a CFM of 2500. Consider that the designer wanted the same Btu/hr cooling but with a cooling coil temperature set at 54°F. What CFM is required? What percent of the total heat removed is latent heat ?

SOLUTION:

Figure 22–7 indicates the process line for this case (1–3). The enthalpy change is

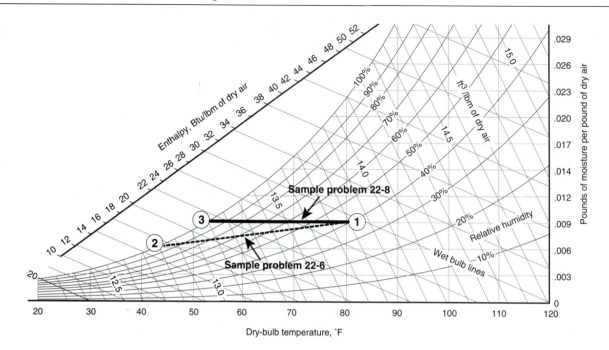

Figure 22–7. Sample Problem 22–8.

$$h_1 - h_3 = 29 - 22 = 7 \frac{\text{Btu}}{\text{lb}_m}$$

To achieve 133,300 Btu/hr of cooling (Sample Problem 22–5), the following CFM will be required:

$$CFM = 133,300 \times \frac{13.5}{(7 \times 60)} = 4284$$

Contrasting this to the original design, the required amount of Btu/hr can be removed with the warmer cooling coil but with a correspondingly higher CFM. Note, however, that the amount of latent heat removal in this design is zero.

Sample Problem 22–8 clearly illustrates that the coil temperature must be selected not on the basis of how many Btu/hr are required, but instead on the basis of how much dehumidification and CFM is necessary. This information must come from information about the building. Studies of the building will show whether large amounts or relatively low amounts of moisture need to be taken out of the air. If the designer picks a coil temperature to be above the dew point, then no moisture will be removed from the building. If the coil

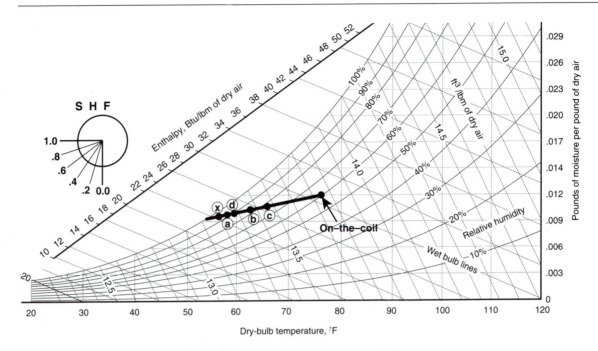

Figure 22–8. Psychrometric chart with SHF.

temperature selected is very low, below the dew point, then a great amount of moisture will be removed.

The designer must pick a cooling process line that corresponds to the percentage of latent and sensible heat that must be removed. In fact, that percentage represents the slope of the process line that must be followed. To do this formally, the designer creates a number called the sensible heat factor (SHF) that is equal to sensible heat load divided by total heat load.

$$SHF = \frac{sensible\ heat}{total\ load}$$

This is simply the percentage of the total load required of the building that is sensible heat. To determine how that slope affects the psychrometric process line, the sensible heat factor is presented on the psychrometric chart in the upper left-hand corner (see Figure 22–8). To determine the proper slope, the designer finds the sensible heat factor and uses the slope it represents. Once the proper slope in cooling is found, a straight edge is slid down the chart at the same slope until it crosses the on-the-coil temperature on the psychromet-ric chart. The line from that point with the proper slope may be drawn all the way down until it hits the saturated line. This will be the desired coil tempera-ture and is referred to as the "apparatus dew point."

SAMPLE PROBLEM 22–9

From the drawings of a 20,000-square-foot new addition to a school (one story, 10-foot ceiling), a designer finds that the building cooling load breaks down as follows:

$$800,000 \; \frac{Btu}{hr} \; \text{sensible;} \; 200,000 \; \frac{Btu}{hr} \; \text{latent}$$

The air-conditioning system will be mounted on the roof of the addition. The interior summer condition to be maintained is 76°F, 60% RH. What is the proper coil temperature of the air-conditioning system?

SOLUTION:

The sensible heat factor is

$$\frac{800,000}{1,000,000} = .8$$

and this slope is found on Figure 22–8. Taking this slope down so that it crosses through the point 76°F DB and 60% RH gives a line that represents the required dehumidification process. Any point along this straight line will satisfy the dehumidification and cooling requirements of the room in question. At each off-the-coil point, however, the amount of CFM to accomplish the total cooling will be different, as shown in Sample Problem 22–10.

SAMPLE PROBLEM 22–10

Find the CFM required and the off-the-coil temperature for the school addition cited above using each of the following criteria:
a. Consider the temperature drop across the coil to be 20°F (off-the-coil = 56°F)
b. Assume the off-the-coil (OTC) air (supply air) achieves a RH of 90%
c. Require that the building achieve six air exchanges per hour
d. Use a coil with BPF = .20

SOLUTION:

The process line on the psychrometric chart corresponding to SHR = .8 is shown in Figure 22–8. The resultant ADP is 54°F. All the calculations below use this process line, but each corresponds to a different point on the line:

a. For $\Delta T = 20°F$, $T_{off} = 56°F$ and uses point a on the diagram. Thus,

$$CFM = \frac{Btu}{hr} \times \frac{v_{air}}{(h_1 - h_a)} 60 = 1,000,000 \times \frac{13.75}{(31-23)} 60 = 28,645$$

b. For OTC RH = 80%, find point b on the diagram. Thus,

$$CFM = 1,000,000 \times \frac{13.75}{(31-26)} \, 60 = 45,833$$

c.   For six air changes/hr with a volume of $20,000 \times 10 = 200,000$, then

$$CFM = 200,000 \times \frac{6}{60} = 20,000 \text{ CFM}$$

Thus

$$h_c - h_1 = \frac{\text{Btu}}{\text{hr}} \times \frac{13.75}{CFM} \times 60 = 11.08$$

or

$$h_c = 23$$

This identifies point c on Figure 22–8. Therefore, $T_{\text{off}} = 65.6°F$.

d.   For BPF = .20, off-the-coil conditions are 20% up the process line from ADP. Measuring this amount yields point d with

$$T_{\text{off}} = 58.5°F$$

and

$$h_d - h_1 = 7$$

$$CFM = 1,000,000 \times \frac{13.75}{7} \times 60 = 32,738$$

22.8 MIXING AIR STREAMS

One thermodynamic process that is easy to relate using the psychrometric chart is the mixing of two different airstreams. In Chapter 6, we "mixed" horseshoes and water in a quenching process to determine the final temperature. When two airstreams with different thermodynamic conditions and in different proportions are mixed, not only the final air temperature but also the final relative humidity is unknown.

Pinpointing the points on the psychrometric chart that represent the two original airstreams, A and B, allows a line to be drawn between them. The final air condition is on this line. The final dry bulb temperature can be found from the equation

$$T_{\text{mixed}} = \frac{\left[\left(\dfrac{CFM_a}{v_a}\right)T_a + \left(\dfrac{CFM_b}{v_b}\right)T_b\right]}{\left(\dfrac{CFM_a}{v_a} + \dfrac{CFM_b}{v_b}\right)}$$

Often this equation is simplified by noting that if the conditions a and b are not too different, then

$$v_a = v_b$$

and

$$T_{mixed} = \frac{\left(CFM_a T_a + CFM_b T_b\right)}{\left(CFM_a + CFM_b\right)}$$

Equation 22–3

Once T_{mixed} is known, all the mixed air conditions can be identified directly from the psychrometric chart.

SAMPLE PROBLEM 22–11

If 1000 CFM of 72°F, 50% RH air (A on Figure 22–9) is mixed with 200 CFM of 100°F, 10% RH air, what is the resulting air DB temperature and RH?

SOLUTION:

$$T_{mixed} = \frac{\left(1000 \times 72 + 200 \times 100\right)}{\left(1000 + 200\right)} = 76.66°F$$

From the psychrometric chart, the RH is found to be 42%.

AIR-CONDITIONING DESIGNER

There was a light on at the offices of Blinkerson, Moore, and Crevits (BM&C), even though it was 7:30 in the evening. Inside an engineering technician stared once at the computer and then over to a laminated card on the desk. BM&C is an engineering consulting firm that specializes in the design of schools. They are contracted by school boards and architects all over the country to design the mechanical systems for school buildings—heating and cooling equipment, plumbing, and sometimes electrical systems. This technician is the head of the heating, ventilating, and air-conditioning (HVAC) department of the firm and is now finalizing the school addition project outlined in Sample Problems 22–9, 22–10, and 22–11. The calculations have been made by one of the new engineers in the technician's section, and tomorrow they will meet with the architect to finalize the design. Going over the work, she sees that an error of omission has been made.

Twelve years ago, she might have made such an error herself, as she had hired into this firm with an associate degree. After receiving her bachelor's degree in engineering technology, she was promoted from the drafting board to designer and performed these same calculations. Eight years later, last year, she sat for her professional engineer's (PE) license and, after receiving it, was promoted to head of the department.

The technician likes the responsibility. One of her obligations is to be sure that the designs from the department are accurate. Her experience has helped her find that this design is missing an important factor.

The school addition will be built with energy conservation in mind—with windows that won't open, doors that seal tightly, and exhaust fans that are miserly in their operation. Yet interior air picks up contaminants of smoke, chemicals, and carbon dioxide (a natural by-product of human breathing). To keep air fresh, a certain amount of interior air must be vented, and the same amount of fresh air from the outside must be "made up" to take its place.

A typical standard for make-up air in a school is 15 CFM of fresh air for every occupant of the building. Since the outside air is usually hot and moist during the cooling season, this puts additional load on the air-conditioning equipment, and the designer must give it close consideration. In the psychrometric process, this outside air changes the on-the-coil air so that it is no longer return air but a mixture of return air and outside air.

To calculate solutions of this type, the designer first solves the problem without outside air considerations. The new engineer has left on her computer the solution for off-the-coil conditions, the apparatus dew point, and the CFM required. Now the technician adds the portion that is missing by determining the amount of outside air needed and locating the outside air conditions on the laminated psychrometric chart that she always carries in her briefcase. The mixed air, or on-the-coil, condition she calculates by Equation 22–3, the equation that determines the mixed air dry bulb temperature.

The technician draws the actual coil process line on her laminated psychrometric chart from the mixed air point through the off-the-coil point previously found. This supply air temperature and humidity and CFM is sufficient to maintain the building, and it is still valid even though the incoming air has been changed.

This new process line intersects the saturation line at a new "apparatus dew point." There now appears to be two apparatus dew points; therefore, the second one is called the "*coil* apparatus dew point," for it truly represents the required temperature of the coil. The original is called the *room* ADP because it is the condition that will satisfy the building or room without outside air considerations.

Now the design is complete. The total CFM is identical to the original CFM. The supply air conditions remain the same, but a new ADP has been found.

Only one final calculation needs to be made, then the technician can go home for the night, knowing that the meeting will go well the next day. The enthalpy change across the cooling coil has been modified from the original problem, so a new Btu/hr requirement must be found from Equation 22–1. Sample Problem 22–12 shows how she completes the design.

SAMPLE PROBLEM 22–12

Using the conditions of Sample Problem 22–9 and an off-the-coil DB = 58.5°F (condition d in Sample Problem 22–10), but with make-up air requirements for 500 occupants in the building and outside air conditions of 90°F DB and 65% RH, find

a. total CFM
b. supply air conditions
c. room ADP
d. coil ADP
e. total sensible and latent cooling requirements

SOLUTION:

a., b., c. These conditions do not change from the original solution, since they are necessary conditions to satisfy the room load.

d. *Outside air CFM* = 500 × 15 = 7500 CFM

$$T_{mixed} = \frac{(32{,}738 \times 76 + 7500 \times 90)}{(32{,}738 + 7500)} = 78.7°F$$

From Figure 22–9 the new ADP is 49°F (coil ADP)

e. Total sensible = $(29 - 24) CFM \times \frac{60}{13.7} = 881{,}124 \frac{Btu}{hr}$

Total latent = $(34 - 29) CFM \times \frac{60}{13.7} = 881{,}124 \frac{Btu}{hr}$

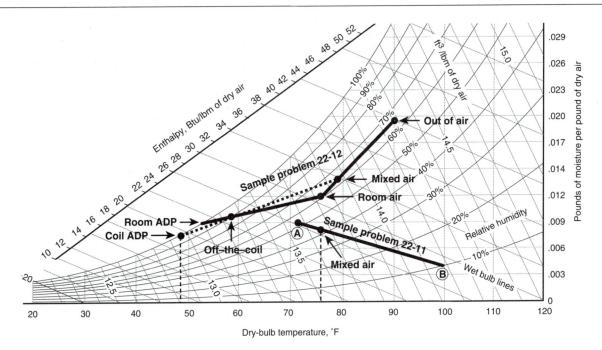

Figure 22–9. Sample problem 22–12.

Notice that with outside air (ventilation) added, the total cooling load is 1,762,248 Btu/hr, whereas the building cooling load is just 1,000,000. Also, the building load is 80% sensible heat, while the load with ventilation considered is 50% sensible, 50% latent.

CHAPTER **22**

PROBLEMS

1. Which of the following conditions of air is within the comfort zone?
   a. 70°FDB, 50% RH
   b. 60°FWB, 40% RH
   c. 74°FDB, h = 30 Btu/lb

2. If the enthalpy of air at 60°FDB, 40% RH is increased by 10 Btu/lb without humidification, what is the resulting RH?

3. If the enthalpy of 80°FDB, 30% RH is increased by 10 Btu/lb without changing the DB temperature, what is the resulting % RH?

4. If air at 80°FDB, 30% RH is humidified without adding heat to a RH of 70%, what is the resulting DB? By how much does the enthalpy change?

5. In Problem 2, assume that 2000 CFM is circulating. How many Btu/hr must be added to the air?

6. If 60°FDB, 40% RH air is heated and humidified to 80°FDB without changing the RH,
   a. what is the total enthalpy change of the air?
   b. how much enthalpy change is required for heating only?
   c. how much enthalpy change is required for humidification only?
   d. how many lb_{water}/lb_{air} must be added?

7. If the fan circulated 200 CFM in Problem 6, what is the total amount of heat that must be added (Btu/hr)? What is the amount of water that must be evaporated (lb_{water}/hr)?

8. 200 CFM of air at 60°F, 30% RH is mixed with 800 CFM air at 80°F, 80% RH. Find the resulting air temperature using
   a. the first equation in section 22.8 on page 519
   b. Equation 22–3

APPENDIX

Properties of Air at Low Pressure

T R	h Btu/lb	pr	vr	T R	h Btu/lb	pr	vr
200	47.67	0.04320	1714.9	1200	291.30	24.01	18.514
220	52.46	0.06026	1352.5	1220	296.41	25.53	17.700
240	57.23	0.08165	1088.8	1240	301.52	27.13	16.932
260	62.03	0.10797	892.0	1260	306.65	28.80	16.205
280	66.82	0.13986	741.6	1280	311.79	30.55	15.518
300	71.61	0.17795	624.5	1300	316.94	32.39	14.868
320	76.40	0.22290	531.8	1320	322.11	34.31	14.253
340	81.18	0.27545	457.2	1340	327.20	36.31	13.670
360	85.97	0.3363	396.6	1360	332.48	38.41	13.118
380	90.75	0.4061	346.6	1380	337.68	40.59	12.593
400	95.53	0.4858	305.0	1400	342.90	42.88	12.095
420	100.32	0.5760	270.1	1420	348.14	45.26	11.622
440	105.11	0.6776	240.6	1440	353.37	47.75	11.172
460	109.90	0.7913	215.33	1460	358.63	50.34	10.743
480	114.69	0.9182	193.65	1480	363.89	53.04	10.336
500	119.48	1.0500	174.90	1500	369.17	55.86	9.948
520	124.27	1.2147	158.58	1520	374.47	58.78	9.578
540	129.06	1.3860	144.32	1540	379.77	61.83	9.226
560	133.86	1.5742	131.78	1560	385.08	65.00	8.890
580	138.66	1.7800	120.70	1580	390.40	68.30	8.569
600	143.47	2.005	110.88	1600	395.74	71.73	8.263
620	148.28	2.249	102.12	1620	401.09	75.29	7.971
640	153.09	2.514	94.30	1640	406.45	78.99	7.691
660	157.92	2.801	87.27	1660	411.82	82.83	7.424
680	162.73	3.111	80.96	1680	417.20	86.82	7.168
700	167.56	3.446	75.25	1700	422.59	90.95	6.924
720	172.39	3.806	70.07	1720	428.00	95.24	6.690
740	177.23	4.193	65.38	1740	433.41	99.69	6.465
760	182.08	4.607	61.10	1760	438.83	104.30	6.251
780	186.94	5.051	57.20	1780	444.26	109.08	6.045
800	191.81	5.526	53.63	1800	449.71	114.03	5.847
820	196.09	6.033	50.35	1820	455.17	119.16	5.658
840	201.56	6.573	47.34	1840	460.63	124.47	5.476
860	206.46	7.149	44.57	1860	466.12	129.95	5.302
880	211.35	7.761	42.01	1880	471.60	135.64	5.134
900	216.26	8.411	39.64	1900	477.09	141.51	4.974
920	221.18	9.102	37.44	1920	482.60	147.59	4.819
940	226.11	9.834	35.41	1940	488.12	153.87	4.670
960	231.06	10.610	33.52	1960	493.64	160.37	4.527
980	236.02	11.430	31.76	1980	499.17	167.07	4.390
1000	240.98	12.298	30.12	2000	504.71	174.00	4.258
1020	245.97	13.215	28.59	2020	510.26	181.16	4.130
1040	250.95	14.182	27.17	2040	515.82	188.54	4.008
1060	255.96	15.203	25.82	2060	521.39	196.16	3.890
1080	260.97	16.278	24.58	2080	526.97	204.02	3.777
1100	265.99	17.413	23.40	2100	532.55	212.1	3.667
1120	271.03	18.604	22.30	2120	538.15	220.5	3.561
1140	276.08	19.858	21.27	2140	543.74	229.1	3.460
1160	281.14	21.18	20.293	2160	549.35	238.0	3.362
1180	286.21	22.56	19.371	2180	554.97	247.2	3.267

APPENDIX

II Properties of R-22 Vapor (Superheated)

Temp °F (sat pressure, psia)		Pressure psia								
		20	40	60	80	110	130	150	170	190
0 (39)	v =	2.75	...	...	...	...	...	...	...	...
	h =	105.89	...	...	...	...	...	...	...	...
	s =	.2457	...	...	...	...	...	...	...	...
10 (47)	v =	2.82	1.36	...	...	...	...	...	...	...
	h =	107.39	106.06	...	...	...	...	...	...	...
	s =	.2490	.2310	...	...	...	...	...	...	...
20 (58)	v =	2.88	1.39	...	...	...	...	...	...	...
	h =	108.91	107.67	...	...	...	...	LIQUID	...	...
	s =	.2522	.2343	...	...	...	...		...	...
30 (70)	v =	2.95	1.43	.92	...	...	...	...	...	...
	h =	110.43	109.27	108.00	...	...	...	...	...	...
	s =	.2553	.2376	.2244	...	...	...	...	...	...
33 (74)	v =	2.97	1.44	.93	...	...	...	...	...	...
	h =	110.97	109.90	109.10	...	...	...	...	...	...
	s =	.2563	.2390	.2258	...	...	...	...	...	...
40 (84)	v =	3.02	1.47	.95	.69	...	...	...	...	...
	h =	111.97	110.88	109.70	108.42	...	...	...	...	...
	s =	.2584	.2409	.2298	.2241	...	...	...	...	...
60 (116)	v =	3.15	1.53	1.00	.73	.51	...	...	...	...
	h =	115.08	114.10	113.07	111.97	110.24	...	...	...	...
	s =	.2645	.2472	.2365	.2283	.2184	...	...	...	...
80 (160)	v =			1.05	.76	.53	.44	.37	...	...
	h =			116.44	115.48	114.34	112.78	111.7	...	...
	s =			.2428	.2349	.2266	.2201	.2151	...	...
100 (223)	v =			.80	.57	.46	.40	.34	.29	
	h =			118.97	117.63	116.64	115.63	114.51	113.35	
	s =			.2412	.2326	.2270	.2223	.2196	.2138	
120 (275)	v =			.84	.58	.50	.42	.36	.32	
	h =		NOT OFTEN USED		122.45	120.92	120.39	119.51	118.69	117.61	
	s =			.2474	.2383	.2348	.2292	.2257	.2202	
140 (350)	v =				.62	.53	.44	.38	.34	
	h =				124.87	124.12	123.34	122.21	121.60	
	s =				.2440	.2390	.2357	.2317	.2277	
160 (440)	v =				.65	.54	.46	.40	.35	
	h =				128.51	127.83	127.14	126.34	125.67	
	s =				.2496	.2461	.2420	.2383	.2347	

APPENDIX

IIIA   Thermodynamic Properties of Steam

DRY SATURATED STEAM: TEMPERATURE TABLE[a]

Temp., °F T	Abs. press., psia P	Specific volume Sat. vapor v_g	Enthalpy Sat. liquid h_f	Evap. h_{fg}	Sat. vapor h_g	Entropy Sat. liquid s_f	Evap. s_{fg}	Sat. vapor s_g
32	0.08854	3306	0.00	1075.8	1075.8	0.0000	2.1877	2.1877
40	0.12170	2444	8.05	1071.3	1079.3	0.0162	2.1435	2.1597
50	0.17811	1703.2	18.07	1065.6	1083.7	0.0361	2.0903	2.1264
60	0.2563	1206.7	28.06	1059.9	1088.0	0.0555	2.0393	2.0948
80	0.5069	633.1	48.02	1048.6	1096.6	0.0932	1.9428	2.0360
100	0.9492	350.4	67.97	1037.2	1105.2	0.1295	1.8531	1.9826
150	3.718	97.07	117.89	1008.2	1126.1	0.2149	1.6537	1.8685
200	11.526	33.64	167.99	977.9	1145.9	0.3120	1.4824	1.7762
212	14.696	26.80	180.07	970.3	1150.4	0.3675	1.4446	1.7566
250	29.825	13.821	216.48	945.5	1164.0	0.3675	1.3323	1.6998
300	67.013	6.466	269.59	910.1	1179.7	0.4369	1.1980	1.6350
350	134.63	3.342	321.63	870.7	1192.3	0.5029	1.0754	1.5783
400	247.31	1.8633	374.97	826.0	1201.0	0.5664	0.9608	1.5272
450	422.6	1.0993	430.1	774.5	1204.6	0.6280	0.8513	1.4793
500	680.8	0.6749	487.8	713.9	1201.7	0.6887	0.7438	1.4325
600	1542.9	0.2668	617.0	548.5	1165.5	0.8131	0.5176	1.3307
700	3093.7	0.0761	823.3	172.1	995.4	0.9905	0.1484	1.1389
705.4	3206.2	0.0503	902.7	0	902.7	1.0580	0	1.0580

[a] Abridged from "Thermodynamic Properties of Steam," by Joseph H. Keenan and Frederick G. Keyes, John Wiley & Sons, Inc., New York, 1937.

APPENDIX

III B — Thermodynamic Properties of Steam (Superheated)

Temp °F (sat pressure, psia)		Pressure psia 10	14.7	20	60	100	150	200	250	300	400
200 (11.52)	v=	38.85	...	...	...	...	...	...	...	...	...
	h=	1146.6	...	...	...	...	...	...	...	...	...
	s=	1.7927	...	...	...	...	...	...	...	...	...
300 (67.0)	v=	45.00	30.53	22.36	7.25	...	...	...	**LIQUID**	...	...
	h=	1193.9	1192.8	1191.6	1181.6	...	...	...		...	...
	s=	1.8595	1.8160	1.7808	1.6492	...	...	...		...	...
400 (247)	v=		34.68	25.43	8.35	4.93	3.24	2.36	...	...	...
	h=		1239.9	1239.2	1233.6	1227.6	1219.4	1210.3	...	...	...
	s=		1.8743	1.8396	1.7135	1.6518	1.5996	1.5594	...	...	...
500 (680)	v=					5.58	3.69	2.72	2.18	1.77	1.28
	h=					1279.1	1274.1	1288.9	1263.4	1257.6	1245.1
	s=					1.7085	1.6601	1.6240	1.5950	1.5701	1.5281
600 (1543)	v=					6.22	4.12	3.06	2.43	2.00	1.48
	h=					1329.1	1325.7	1322.1	1318.4	1314.7	1306.9
	s=					1.7581	1.7831	1.6767	1.6497	1.6268	1.5894
700 (3093.7)	v=							3.38	2.69	2.23	1.65
	h=							1373.6	1370.9	1368.3	1362.7
	s=							1.7232	1.6970	1.6751	1.6398
800 (**)	v=								2.95	2.44	1.82
	h=								1422.8	1420.6	1416.4
	s=								1.7397	1.7184	1.6842
900 (**)	v=			NOT OFTEN USED					3.20	2.65	1.98
	h=								1474.5	1472.8	1469.4
	s=								1.7793	1.7582	1.7247
1000 (**)	v=									2.86	2.13
	h=									1525.2	1522.4
	s=									1.7954	1.7623
1100 (**)	v=									3.06	2.29
	h=									1578.1	1575.8
	s=									1.8305	1.7977
1200 (**)	v=									3.27	2.45
	h=									1631.7	1629.6
	s=									1.8638	1.8311

** supercritical

APPENDIX
IV

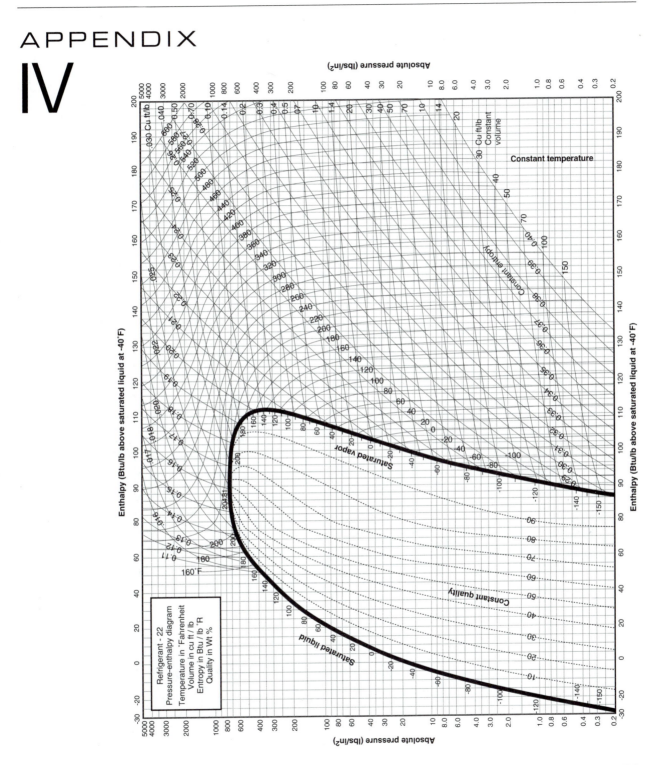

Refrigerant - 22
Pressure-enthalpy diagram
Temperature in °Fahrenheit
Volume in cu ft / lb
Entropy in Btu / lb °R
Quality in Wt %

APPENDIX

V

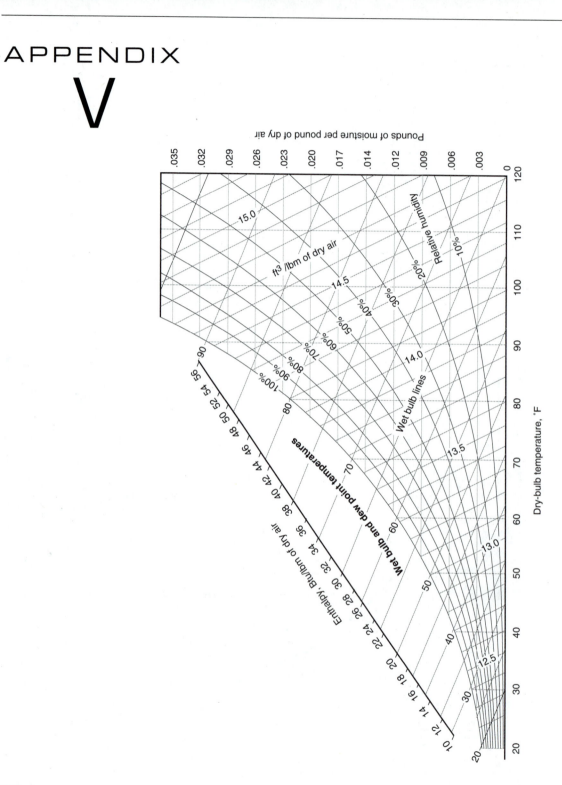

APPENDIX

VI

Conversion Factors

Length
1 m = 3.2808 ft = 39.37 in
1 cm = 0.0328 ft = 0.394 in
1 ft = 12 in = 0.3048 m
1 mile = 5280 ft = 1609.36 m = 1.609 km

Area
$1\ m^2 = 10^4\ cm^2 = 10.76\ ft^2 = 1550\ in^2$
$1\ ft^2 = 144\ in^2 = 0.0929\ m^2 = 929.05\ cm^2$

Volume
$1\ m^3 = 35.313\ ft^3 = 6.1023 \times 10^4\ in^3 = 1000\ L = 264.171\ gal$
$1\ ft^3 = 1728\ in^3 = 28.3168\ L = 0.02832\ m^3 = 7.4805\ gal$

Mass
1 kg = 1000 g = 2.2046 lbm = 0.0685 slug

Force
$1\ N = 0.225\ lb_f$
$1\ lb_f = 4.448\ N$

Energy
1 kJ = 1000 J = 0.9479 Btu = 238.9 cal
1 Btu = 1055.0 J = 1.055 kJ = 778.16 ft • lbf = 252 cal

Power
$1\ W = 1\ J/s = 1\ kg \bullet m^2/s^3 = 3.412\ Btu/hr = 1.3405 \times 10^{-3}\ hp$
1 kW = 1000 W = 3412 Btu/hr = 737.3 ft • lbf/s = 1.3405 hp

Pressure
$1\ atm = 14.696\ lb_f/in^2 = 101.325\ kPa$

Miscellaneous

Specific Heat Units   1 Btu/(lbm • °F) = 1 Btu/(lbm • R), i.e., no °F to °R conversion is needed here

1 kJ/(kg • K) = 185.8 ft • lbf/(lbm • R)

Temperature

$$T(°F) = \frac{9}{5} T(°C) + 32 = T(R) - 459.67$$

$$T(°C) = \frac{5}{9} [T(°F) - 32] = T(K) - 273.15$$

Density   1 lbm/ft^3 = 16.0187 kg/m^3

Problem Solutions

Chapter 2

3. 40 lb/ft^3, $.025 \text{ ft}^3/\text{lb}$
11. 226, 125 ft-lb
13. $E_{out} = 13,380,117$ Btu

Chapter 3

5. a. $P_2 = 1666.6$ psia
9. a. vents 13%
13. $V_2 = 8.7 \text{ ft}^3$

Chapter 4

1. $v = 13.35 \text{ ft}^3/\text{lb}$
3. $v = .08 \text{ ft}^3/\text{lb}_m = .24$ lb
5. $\mu = 18$
7. $\mu = 25$
9. $T_2 = 698°R$
11. $P_2 = 11.98$ psia
13. $T_2 = 390.4°R$

Chapter 5

11. 970°F
15. $- 132$ Btu
17. a. 13,551,360 Btu
25. 168 Btu

Chapter 6

5. 40.8°F
7. 10,400 Btu/lb
9. 247.8°F
11. 41 ft^3

Chapter 7

3. $136,800 \text{ ft}^3$
5. $T_2 = 2972°F$, $P_2 = 826$ psig
   $WORK = 455$ ft-lb, $Q = 2$ Btu
9. $T_2 = 98°F$, $Q = 39,883,242$
15. $m = .70$ lb

Chapter 8

3. $V_2 = .33 \text{ ft}^3$
7. $P_2 = 490$ psia
   $WORK = -102,233$ ft-lb
9. 1848 minutes
11. a. $P_2 = 24.5$ psig
    b. $WORK = -27,594$ ft-lb

Chapter 9

1. a. 42.5 Btu
   b. –42.5
   c. 9%
   d. 13%
3. $387 - 6$ ft
5. $P_2 = 20.9$ psia, $WORK = 7.6$ Btu
7. $P_2 = 81$ psia, $WORK = -77$ Btu
9. a. 87,500
   b. 12 ft^3
   c. 258 Btu

Chapter 10

3. 147 psia
5. 7%, change in IE = 62 Btu
9. c
11. $n = 1.018$

Chapter 11

1. 10 psia
3. $T = 70°F$, $P = 25$ psia, zero
5. $T = 70°F$ if free expansion

Chapter 12

3. $Hp = 131$, $\eta = .60$, $MEP = 17,258$
7. a. 2363°F
   b. $Q = 2.32$ Btu

Chapter 13

1. $v_1 = 7.8$, $v_2 = 15.1$
   $WORK = 32{,}533$ ft-lb, $WD = 45{,}872$
5. $Net\ WORK = -2.94$ Btu, $Q_{2-3} = -19$ Btu/lb
   $COP = 6.6$

Chapter 14

3. $\eta = .76$, $WORK = 59{,}554$ ft-lb
5. $\eta = .197$
7. $\eta = .84$

Chapter 16

1. a. $T_2 = 975.9°R$, $WORK = -105$ Btu
   b. $T_2 = 960°R$, $WORK = -104$ Btu
3. $WORK = -206$ Btu
9. a. $T_1 = 440°R$, $P_1 = 1000$, $h_1 = 105$, $P_{r1} = .677$
   $T_2 = 965$, $P_2 = 16{,}000$, $h_2 = 232$, $P_{r2} = 10.84$
   $T_3 = 1085$, $P_3 = 16{,}000$, $h_3 = 262$, $P_{r3} = 16.5$
   $T_4 = 502$, $P_4 = 1000$, $h_4 = 119.8$, $P_{r4} = 1.03$
   b. $Wd_{12} = -127$, $wd_{23} = 0$, $wd_{34} = 142$, $wd_{41} = 0$
   $q_{12} = 0$, $q_{23} = 30$, $q_{34} = 0$, $q_{41} = -14.7$
   c. $Net\ work = 15$, $\eta = .5$

Chapter 17

1. a. $T_2 = 586$
   b. $T_2 = 611$
7. a. $Net\ work = -.36$ Btu
   b. $q_{absorbed} = -1.41$ Btu
   c. $COP = 3.91$
   d. $14{,}184$ lb/hr

Chapter 18

1. a. gas
   c. vapor
   e. liquid
3. a. 5%
   c. 10%
   e. 100%

Chapter 19

1. 70% liquid
3. a. sub
   c. sat
   e. super
5. a. 63
   b. 190.47
   c. 13
   d. 76
7. a. 55
   b. 218
   c. 18
   d. 73
9. using 30° rule, $P = 285$ psig
11. R-22 model B
13. $COP = 1.56$

Chapter 20

1. $2{,}988{,}000$ Btu/hr
3. $Tsat = 218°F$, 132° superheat
7. $h_4 - h_3 = 1179 - 205 = 974$

Index